Prions and Prion Diseases

Current Perspectives

Edited by

Glenn C. Telling

University of Kentucky, Lexington, KY 40536, USA

Copyright © 2004
Horizon Bioscience
32 Hewitts Lane
Wymondham
Norfolk NR18 0JA
England

www.horizonbioscience.com

British Library Cataloguing-in-Publication Data

A catalogue record for this book is available from the British
Library

ISBN: 0-9545232-6-1

Printed and bound in Great Britain

Contents

Books of Related Interest

Contributors

*Adriano Aguzzi
Institute of Neuropathology
Universitätsspital Zürich
Schmelzbergstrasse 12
CH-8091 Zürich
Switzerland
adriano@pathol.unizh.ch

Gerald S. Baron
Laboratory of Persistent Viral
Diseases
NIAID, NIH
Rocky Mountain Laboratories
903 S. 4th St., Hamilton
MT 59840, USA

Rona M. Barron
Institute for Animal Health
Neuropathogenesis Unit
Edinburgh
UK

*Jeremy P. Brockes
Department of Biochemistry and
Molecular Biology
University College London
Gower Street
London WC1E 6BT
United Kingdom
j.brockes@ucl.ac.uk

*David R. Brown
Department of Biology and
Biochemistry
University of Bath
BA2 7AY
U.K.
bssdrb@bath.ac.uk

*Byron Caughey
Laboratory of Persistent Viral
Diseases
NIAID, NIH
Rocky Mountain Laboratories
903 S. 4th St., Hamilton
MT 59840, USA
bcaughey@niaid.nih.gov

*Yury O. Chernoff
School of Biology
and Inst Bioeng and Bioscience
Georgia Institute of Technology
315 Ferst Drive, Atlanta
Georgia 30332-0363
USA
yc22@prism.gatech.edu

Nnennaya Kanu
Dept Biochem and Molecular Biol
University College London
Gower Street
London WC1E 6BT
United Kingdom

*Jean Catherine Manson
Institute for Animal Health
Neuropathogenesis Unit
King's Buildings
West Mains Rd
Edinburgh, EH9 3JF
United Kingdom
jean.manson @bbsrc.ac.uk

Karah Nazor
Program in Gerontology
461 Health Sci Research Building
University of Kentucky
800 Rose Street
Lexington, KY 40536
USA

*Indicates Corresponding Author

Judith R. Rees
Department of Community and
Family Medicine
Dartmouth Medical School
Hanover
New Hampshire 03755
USA

Jay R. Silveira
Laboratory of Persistent Viral
Diseases
NIAID, NIH
Rocky Mountain Laboratories
903 S. 4th St., Hamilton,
MT 59840, ,USA

Jonathan O. Speare
Laboratory of Persistent Viral
Diseases
NIAID, NIH
Rocky Mountain Laboratories
903 S. 4th St., Hamilton
MT 59840, USA

*Surachai Supattapone
Departments of Biochemistry and
Medicine
7200 Vail Building
Dartmouth Medical School
Hanover
New Hampshire 03755
USA
supattapone@dartmouth.edu

*Glenn C. Telling
Sanders Brown Center on Aging
Dept Microbiol, Immunol and Mol
Genetics
332 Health Sci Research Building
University of Kentucky
800 Rose Street
Lexington, KY 40536
USA
gtell2@ uky.edu

Andrew R. Thompsett
Department of Biology and
Biochemistry
University of Bath
BA2 7AY
U.K.
bssart@bath.ac.uk

*David Westaway
Centre for Research in
Neurodegenerative Diseases
University of Toronto
Tanz Neuroscience Building
6 Queen's Park Crescent West
Toronto, Ontario
M5S 3H2
Canada
david.westaway@utoronto.ca

*R. Anthony Williamson
Dept Immunology
The Scripps Research Institute
10550 North Torrey Pines Road
La Jolla
CA 92037
USA
anthony@scripps.edu

Foreword

The unfolding of the bovine spongiform encephalopathy (BSE) epidemic in the UK and Europe and the subsequent discovery that a variant of Creutzfeldt Jakob disease (vCJD) in young adults and teenagers is caused by exposure to BSE, captured public attention and led to concerns that prions from other sources, in particular deer and elk with Chronic Wasting disease (CWD), might pose a similar risk to humans. While the economic and public health issues surrounding these diseases stimulated an enormous amount of new research in recent years, these and similar transmissible neurodegenerative diseases of the central nervous system (CNS) have been the subject of keen speculation and intense research for many years because of their extraordinary biology and the unique properties of the infectious agent.

Because they are transmissible and there are multiple strains of the agent, it was once widely believed that transmissible spongiform encephalopathies (TSE's) were caused by viruses or unconventional infectious agents containing a nucleic acid genome. Indeed, TSE research once focussed almost exclusively on controversies relating to the nature of the infectious agent. The prevailing viewpoint now attributes these diseases to subcellular pathogens called 'prions' which are defined as small proteinaceous infectious particles that lack nucleic acids. While the precise molecular structure of the infectious agent still eludes definitive identification, considerable evidence argues that prions are composed largely, if not entirely, of the scrapie isoform of the prion protein (PrP), referred to as PrP^{Sc} and that during the disease process, PrP^{Sc} acts as a template for conversion by imposing its conformation on the normally benign host-encoded version of the prion protein referred to as PrP^{C}.

This volume provides a comprehensive overview of the pathogenesis, molecular biology, biochemistry, cellular biology, animal models and immunology of prions and follows and expands upon a previous volume published in 1999 entitled "Prions: Molecular and Cellular Biology", edited by David Harris. Research into prions and prion diseases has progressed at an impressive pace in recent years. In assembling this volume I have solicited contributions not only to summarize recent progress in the field but also to focus on the remaining unresolved issues in prion biology that will shape future developments.

Perhaps the most fundamental unanswered question concerns the basic molecular mechanisms of prion replication involving the conformational conversion of PrP^{C} into PrP^{Sc} and the role that cellular factors play in this process. Also, while current models suggest that different PrP^{Sc} conformational states and/ or PrP glycosylation harbor strain-specific information, the exact mechanism by which prion strains propagate stable phenotypic traits upon serial passage remains to be determined. *In vitro* systems, including cell-free conversion assays and chronically infected cell cultures, and *in vivo* approaches including transgenesis in mice, have been the cornerstones of such studies and contributions from several key investigators summarize the major recent developments using each approach.

Since the time between inoculation and disease in animal models is extremely long, the development of *in vitro* cell-free assays in which PrP conversion can be monitored has been an important recent advance. Byron Caughey and co-workers discuss a variety of experiments which demonstrate that, while it has not yet been possible to detect newly produced infectivity *in vitro*, PrP species with properties similar to disease-associated forms do have limited capacities to propagate themselves under such circumstances. Cell-free conversion systems have provided important information relating to the molecular basis of TSE species barriers and strain variation and, because of their specificity and relatively high throughput, will likely be successfully exploited as screens for anti-TSE drugs and diagnostic tests in the future. Cell biological approaches afford the potential to analyze the biosynthesis, posttranslational processing, cellular localization, and trafficking of PrP in prion infected cells. In practice, however, the details of PrP^{Sc} formation in the host cell are not well understood and the ability to study infectious prions in cell culture has been limited by the ability to infect most cell types by *in vitro* challenge. Consequently, unlike most animal viruses which can be readily propagated and titrated in cultured cells, there are only a handful of rodent cell lines in which experimentally-adapted scrapie prions can be studied and there are currently no means by which human CJD prions or prions from other species (BSE, CWD or scrapie) can be propagated in cell culture. These difficulties notwithstanding, various cell culture models of prion diseases have been successfully used to study the basic molecular cell biology of PrP and its various mutants and conformers. Jeremy Brockes and Nnennaya Kanu elegantly summarize such mechanistic studies relating to prion formation in cell culture.

Because prion infectivity can only be assayed in experimental animals, studies involving the manipulation of PrP genes by transgenesis in mice have provided crucial insights into the mechanism of prion propagation and the pathogenesis of prion diseases. My colleague Karah Nazor and I summarize studies of prion pathogenesis from a variety of transgenic approaches and describe how such models have furthered our understanding of the molecular basis of prion species barriers, prion strains and inherited human prion diseases. In their chapter on PrP deletion mutants, Surchai Supattapone and Judith Rees summarize structure-function deletion mutagenesis studies of the PrP gene, performed in large part in transgenic mice, which have had an enormous impact on elucidating the mechanism of prion replication and disease pathogenesis. These studies complement structural investigations of PrP and have identified distinct domains with roles in PrP folding and conformational conversion as well as PrP^{Sc}-induced neurotoxicity. Rona Barron and Jean Manson describe recent advances using elegant gene targeting approaches in which specific mutations have been introduced into the endogenous murine PrP gene. These studies have definitively shown that specific amino acids in PrP control scrapie incubation times in mice, and that the P101L mutation alters the susceptibility of the mice to several prion strains from different species.

The gene replacement studies described by Baron and Manson are closely related to experiments designed to knock out PrP function. While $Prnp^{0/0}$ knockout mice were fundamentally important in substantiating the protein only hypothesis of prion replication, initial studies of $Prnp^{0/0}$ knockout mice did not reveal any major disruptions to either development or behavior that could be ascribed to normal PrP^C function. While this issue has proven challenging, in recent years it has become clear that PrP binds copper raising the possibility that PrP^C functions in copper homeostasis, especially in the nervous system. David Brown and his colleague Andrew Thompsett summarize the various approaches that have characterized PrP as a copper binding protein as well as evidence that PrP displays antioxidant activity. While initial $Prnp^{0/0}$ knockout mice were less useful than anticipated for discerning the exact function of PrP^C, studies of subsequent ataxic $Prnp^{0/0}$ knockout mice led to the surprising discovery that a PrP-related protein, doppel (Dpl), produces neurodegeneration. These and other studies which are detailed in David Westaway's chapter on $Prnd$ and the Doppel protein, suggest cross-talk between Dpl and PrP^C in certain physiological and pathogenic situations.

Since prion infection fails to induce a humoral or cellular immune response, the role of the immune system in disease pathogenesis remained largely unexplored until relatively recently. Two contributions in this volume highlight immunological aspects of prion disease pathogenesis. Although the pathological consequences of prion infections occur in the CNS and experimental transmission of these diseases is most efficiently accomplished by intracerebral inoculation, most natural infections occur as a result of exposure to infectivity in the periphery. The exact mechanisms and sequence of events resulting in infection of the CNS following prion neuroinvasion involving various components of the immune and peripheral nervous systems is comprehensively reviewed by Adriano Aguzzi. Various *in vitro* and *in vivo* studies have recently demonstrated that antibody-mediated immunotherapy might be a viable approach to prevent peripheral prion pathogenesis. These and other immunological advances in prion diseases are described in detail by Anthony Williamson. Antibodies specifically binding to certain regions of PrP have not only emerged as a class of potent inhibitors of prion replication, but have also provided important mechanistic information about PrP^C-PrP^{Sc} interactions.

Finally, the discovery in the mid-90's that certain phenotypic traits in yeast propagated by a mechanism similar to mammalian prions did much to solidify the prion hypothesis and suggested that such epigenetic forms of information transfer were more widespread in nature than was once appreciated. Yeast prions provide a model for understanding both the molecular mechanisms of mammalian amyloidoses and the general principles of protein-based inheritance and, because yeast is such a tractable experimental system that readily lends itself to genetic, cell biological and biochemical analysis, progress on the study of yeast prions has been extremely rapid. The mechanism of yeast prion formation involving cellular stress-defense systems, the cytoskeleton, and functional partners of the specific prion-forming proteins as well as the mechanism of prion curing by guanidine hydrochloride are exhaustively summarized by Yury Chernoff.

In closing, I would like to extend my warm thanks to all the contributors who agreed to contribute to this volume despite heavy demands on their time. The result is a comprehensive overview of recent developments and future challenges by researchers at the forefront of one of the most exciting and challenging areas of biology.

Glenn Telling
February 2004

From: Prions and Prion Diseases: Current Perspectives
Edited by: Glenn C. Telling

Chapter 1

A Functional Role for a Copper Binding Prion Protein

Andrew R. Thompsett
and David R. Brown

Abstract

It is now widely accepted that the normal cellular prion protein (PrP^C) binds copper. While the extent, affinities and sites of binding are still disputed evidence suggests that PrP binds four coppers in the highly conserved histidine rich repeat domain and possibly at a fifth site outside the repeat region. Here we review the characteristics of the cupro-prion protein in relation to the normal function of the PrP and discuss a role for PrP in the relief of neuronal oxidative stress. Various mechanisms including copper signaling, metal ion homeostasis, and roles for PrP as a copper chaperone and enzymatic superoxide dismutase are considered to explain how PrP may perform this function. Finally, loss of PrP function is compared to the disease state and related to other neurological disorders in which copper is implicated.

1. Introduction

The transition metal copper is an essential element for life. Many enzymes and copper containing proteins are vital to cellular function (Fenton 1995). As a result there has been much interest since the discovery that PrP binds copper. Expression of the normal cellular isoform of PrP, referred to as PrP^C, is essential for the development of prion diseases (Büeler *et al.* 1992; 1993). The conversion of this cell surface glycoprotein to the diseased form, referred to as PrP^{Sc}, is closely linked to pathogenesis and neuronal death (Prusiner, 1982). For a long time the biological function of the normal prion protein was unknown, partly because research tended to focus on the conversion of PrP^C to PrP^{Sc}. Along with the discovery of copper binding have come suggestions that PrP may play a key role in copper homeostasis, especially in the nervous system. Perhaps more importantly, there is evidence that PrP displays considerable antioxidant activity.

The exact binding, function and biological implications of cupro-prion proteins are still, however, a matter for intense debate. In fact perhaps the only undisputed fact is that mammalian PrP can bind copper. Current awareness and concern over prion diseases is still high following the bovine spongiform encephalophathy (BSE) epidemic in the UK, especially in light of its perceived future impact on human health through variant Creutzfeldt-Jakob disease (vCJD) (Highfield, 2002). Furthermore, the finding that a key molecular player in the disease, the prion protein, has a cofactor, copper, which itself is implicated in other neurodegenerative diseases adds to the intrigue. In this chapter the evidence for a role for copper in PrP^C function is examined and related to prion disease and the putative pathogenic, PrP^{Sc} form of the protein.

2. The Copper Paradox and Metal Ion Homeostasis

2.1. Life and Death

Copper is both an essential element and toxic to life. Both of these characteristics are intimately linked to the presence of two biologically available oxidation states for copper (Fenton, 1995). Dual oxidation states make copper both an important component of catalytic centers or electron transporters, but also a mediator of oxidative stress and cellular damage. The risk from copper comes largely from Fenton like chemistry which was first applied to iron and involves a complex interplay of redox active metal ions and oxygen species. The outcome is reactive oxygen species such as superoxide (O_2^-) and hydroxyl radicals (OH·) which can cause severe damage to lipids, proteins and therefore cells (Halliwell, 1992).

This chapter will begin with a brief summary of copper homeostasis and related diseases to provide a context for considering the cupro-prion protein. The copper paradox of life and death has lead nature to provide cells with an extremely tightly regulated transport system to deal with copper homeostasis. Mechanisms exist in prokaryotes and eukaryotes to ensure that potentially dangerous free copper is sequestered and/or removed while providing routes to supply copper to essential enzymes (Arnesano *et al.*, 2002a). In fact the amount of free copper in yeast has been estimated to be less than one ion per cell, while the total cellular copper content is 3.9×10^5 atoms (Rae *et al.*, 1999). Clearly nature prefers copper to be accounted for and carefully distributed rather than allowed to act as a free agent in the cellular environment. Even in the plasma, copper is associated with low molecular weight carriers such as histidine and carrier proteins such as albumin and ceruloplasmin.

2.2. Homeostasis

Multiple mechanisms for copper homeostasis have recently been elucidated (Rosenzweig, 2001; Puig *et al.*, 2002; Arnesano *et al.*, 2002a; 2002b). In yeast, three specific copper delivery pathways are under investigation (Rosenzweig, 2001; Puig *et al.*, 2002). The uptake proteins Ctr1 and 3 pass copper (I) to chaperones such as Atx1, Cox17 and CCS. A class of proteins designated metallochaperones specifically delivers copper to their final destinations, Atx1 to Fet3 and 5 via Ccc2, Cox17 to Cytochrome *c* oxidase and CCS to the enzyme Cu/Zn Superoxide dismutase (SOD1 or Cu/Zn-SOD). Other proteins such as Ace1, Mac1 and Aft1 are the copper dependant transcription factors of these systems while metallothioneins are thought to be scavengers and sequesters of free copper. Metallothioneins are small cystine rich proteins, which can bind up to 12 copper ions, providing a large cellular sink. It should also be noted that these studies are not exhaustive and that other possible pathways are probably still awaiting discovery. In higher eukaryotes specific cell types may indeed have there own specially adapted homeostatic pathways.

2.3. Copper and Disease

Breakdowns in copper homeostasis result in various human diseases such as Menkes syndrome and Wilson disease (Strausak *et al.*, 2001). The so-called Menkes and Wilson proteins associated with these diseases are ATPases sharing 55% amino acid identity with each other. In turn they are human homologues of the yeast Ccc2 protein that interacts with the Atx1 metallochaperone. In the case of Menkes syndrome there is a systemic copper deficiency, which results in characteristic kinky hair, altered brain development and neurodegeneration. Up to 200 mutations in the ATP7A gene have been documented in Menkes syndrome. These disruptions of the Menkes protein prevent copper export from the intestines leading to copper build up in the intestines and kidneys but deficiencies in the liver and brain. On the other hand, Wilson disease patients exhibit an accumulation of liver and brain copper. Damage to the brain in the case of Wilson disease takes the form of spongiform degeneration, resembling the neuropathology observed in most prion diseases (Tanzi *et al.*, 1993).

Amyotrophic lateral sclerosis (ALS) is a late onset disease of motor neurons. Approximately 10% of cases of ALS are familial (FALS), in about 20% of these cases mutations have been mapped to the gene for SOD1. This copper and zinc containing protein catalyses the diproportionation of superoxide O_2^- to oxygen and hydrogen peroxide. The enzyme is extremely efficient; operating closely to the diffusion control limit and as such is a major cellular defense against oxidative stress. SOD1 gene mutations that produce the most severe symptoms are not usually those that result in reduced enzymatic function; therefore it is appears that there is an unwanted pathogenic gain of function related to the change in the protein. The results of the failure of copper containing enzymes are seen in familial amyotrophic lateral sclerosis (FALS) (Multhaup *et al.*, 2002).

More recently copper has been associated with other neurological diseases such as Parkinson's and Alzheimer's diseases (PD and AD) (Multhaup *et al.*, 2002). These diseases are characterized by the build up of insoluble protein aggregates and therefore are often compared to the prion diseases. Exposure to heavy metals is considered a risk factor in PD, due in particular to the associated risk of oxidative stress. Additionally reduced levels of the antioxidant glutathione are observed in PD. Defects have also been documented in the copper enzymes cytochrome c oxidase and ceruloplasmin. The major component of the insoluble Lewy bodies of PD is α-synuclein. This protein has an acidic C-terminus which, in the presence of copper, be can induced to oligomerize. Whether this is an *in vitro* artifact or important process of disease pathogenesis is not yet clear (Paik *et al.*, 1999). In addition to the amyloid-β (Aβ) component, the plaques in AD have a non-amyloid component consisting of α-synuclein. Copper, iron and zinc modulate the precipitation and redox activity of Aβ (Bush *et al.*, 2002) and are also enriched in the core of AD plaques (Lovell *et al.*, 1998). It has been considered that copper ion binding to Aβ is a cause of oxidative stress in AD. Links between AD and copper intensified with the discovery of copper binding potential of the amyloid precursor protein (APP) (Hesse *et al.*, 1994). The physiologically relevant copper binding capability of the extracellular domain of APP has lead it to be considered as a neuronal copper transporter. The copper, however, is not redox inactive; APP has been shown to reduce copper (II) to copper (I) in a cell free system, possibly leading to the formation of an intramolecular disulfide bridge (Multhaup *et al.*, 1996). It has been suggested that copper (I) may form hydroxyl radicals from hydrogen peroxide which in turn fragment APP, however reduction to copper(I) maybe important for copper uptake. Finally in cell culture, copper has been shown to influence APP processing leading to less Aβ secretion (Borchardt *et al.*, 1999).

3. Copper and the Prion Protein

3.1. A Historical Perspective

More than 30 years ago mice were treated with a copper chelator, cuprizone (bis-cyclohexanone oxaldihydrazone). Charging the chelator with copper negated the effects of depleting brain copper and modifying copper enzyme activity, showing the mode of action to be through copper chelation (Pattison *et al.*, 1971a; 1971b). The resultant histopathology of the treated mice included spongiform encephalopathy and gliosis closely resembling mouse scrapie. Unlike scrapie, the resultant encephalopathy was not transmissible and showed no proteinaceous fibrils (Pattison *et al.*, 1973). The expression of PrP^C and its conversion to PrP^{Sc} is linked to neurodegeneration in prion diseases therefore cuprizone treatment cannot be said to have caused prion disease in these mice. However, it is notable that these scrapie resembling pathologies can be both drug induced (cuprizone)

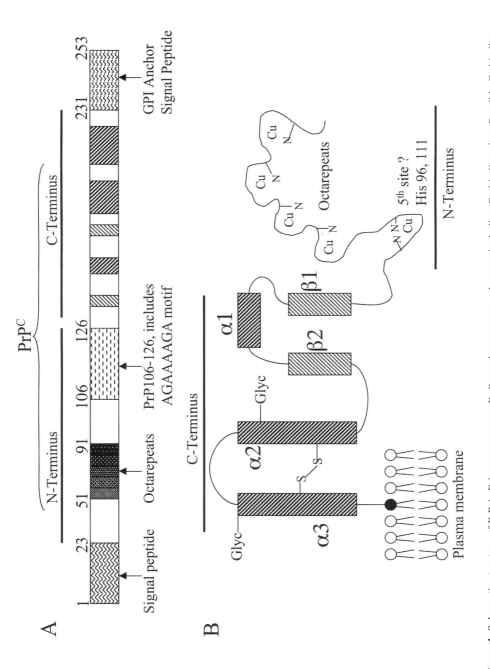

Figure 1. Schematic structure of PrP A, Primary structure; B, Secondary structure elements including Cu-binding sites. Possible Cu binding histidine residues shown as N, sites of glycosylation as Glyc.

or genetically predisposed (Wilson disease) as a result of altered copper homeostasis.

Deer and elk in certain regions of North America are susceptible to Chronic wasting disease (CWD). Until recently CWD was considered to be a disease related to copper deficiency, but is now widely recognized to be a prion disease (Bahamanyar *et al.*, 1985). The spongiform pathology observed is, however, very similar to that induced by copper deficiency. Coincidentally, CWD occurs in areas with low soil copper content (Purdey, 2000). PrP has however been detected in brain plaques of infected animals to confirm the prion disease diagnosis (Guiroy *et al.*, 1991). The similarities between copper deficiency and prion disease are once again striking.

Immobilized metal ion affinity chromatography (IMAC) using copper (II) has been used as a means of purifying native prion protein. In fact the method was able to separate full length PrP^C from the N-terminal truncated form, perhaps giving the first clues as to metal binding sites of PrP^C (Pan, 1992). However, the first direct evidence that the prion protein might bind copper came in the mid 1990s (Hornshaw *et al.*, 1995a; 1995b). In these studies, it was found that peptides based on unusual repeating segments of the prion protein, known as the repeat region (commonly octarepeat in mammals or hexarepeat in avian species), bound copper with relatively high affinity. Especially notable is that peptides based on the fairly diverse human and chicken prion protein sequences both showed this copper binding ability. In both species this region is rich in histidine, a common ligand in cupro-proteins.

3.2. Prion Protein Structure

As with many subsequent studies the first study was based on peptide data as opposed to considering the protein as a whole. To consider the potential metal binding of prion proteins it is worth considering both the PrP gene (referred to as *Prnp* in mice) and PrP structure (Figure 1). The major native form of the protein is a cell surface glycoprotein, but there is also evidence for a trans membrane form (Hegde *et al.*, 1998) as well as the protease resistant PrP^{Sc} form. During protein maturation the N-terminal signal sequence is processed for entry into the endoplasmic reticulum, as is a C-terminal glycosylphosphatidyl-inositol (GPI) signal peptide to attach the protein to the cell membrane by a GPI anchor (Borchelt *et al.*, 1993). The N-terminal domain of the protein shows amino acid insertions and deletions between species, whereas the C-terminal domain contains interspecies amino acid residue substitutions (Wopfner *et al.*, 1999).

The solution structure of the C-terminal domain has been determined for mouse, hamster and human PrPs (Riek *et al.*, 1996; Donne *et al.*, 1997; Zahn *et al.*, 2000). The C-terminal domain is a globular monomer comprised of three α-helices and two β-strands and appears to be stable over a range of temperature and pH (Wildegger *et al.*, 1999). The regions of secondary structure are largely, but not completely conserved, excepting the short first β-strand which is entirely conserved between species (Wopfner *et al.*, 1999). An intramolecular disulfide bond bridges the second and third helices and two asparagine residues provide potential glycosylation sites. Native PrP^C exists with zero, one or both sites glycosylated. An X-ray crystal structure exists for a fragment of recombinant human PrP (Knaus *et al.*, 2001). During the crystallization process the N-terminus of the protein (to Gly119) was cleaved and so its structure is not available. However, the C-terminal domain forms a homodimer where the intramolecular disulfide bridge is broken to form two intermolecular bridges and the third helices are swapped.

In contrast, the N-terminal domain has been found to be largely unstructured in solution (Hornemann *et al.*, 1997; Zahn *et al.*, 2000). It is within this unstructured domain that a hydrophobic core containing the highly conserved palindromic "AGAAAAGA" signature sequence of prion proteins is found. Considering the exceedingly large homology between species this sequence is probably essential for the function of the normal protein (Schätzl *et al.*, 1995). Furthermore the palindromic sequence follows a cleavage site, probably involved in either the maturation or normal metabolic clearance of PrP. The resultant 18kDa peptide has been designated PrP fragment C1 (Pan *et al.*, 1992; Chen *et al.*, 1995). Peptides based on the highly hydrophobic core and its surrounding residues (residues 106 to 126 of the murine protein) are solely able to initiate neuronal death in PrP^C expressing cells. The PrP (106-126) peptide increases neuron sensitivity to oxidative stress and copper toxicity (Forloni *et al.*, 1993; Brown *et al.*, 1994; 1996; 1997a). In fact the peptide PrP (106-126) shows many characteristics of PrP^{Sc} such as high β-sheet content, protease resistance and a tendency to fibril formation (Selvaggini *et al.*, 1993). Additionally it is in this domain, just prior to the hydrophobic core, that many species have the repeat region (see Figure 2) on which Hornshaw *et al.* (1995a; 1995b) based their copper binding peptide. In humans, as in many mammals the repeat sequence takes the form PHGGGWGQ, which is repeated between four and six times the first being a histidine-free nonamer PQGGGGWGQ. Among mammals the repeats are largely conserved, the most diverse mammal being a marsupial with four repeats of 9 or 10 residues (Wopfner *et al.*, 1999). The repeats are also found in the sequences of avian species so far reported although they take a slightly different form, having an even higher proline and lower glycine content based around an XP(S/G)YPX motif. The histidine content, as for mammals is high, with between 4 and 6 distributed through 6 to 9 repeats (Wopfner *et al.*, 1999). The conservation of repeat regions rich in histidine and proline and especially the hydrophobic core suggest that despite the apparently unstructured nature the N-terminus of the protein is important for function.

The octarepeat regions also figure in inherited human prion disease. Insertions of 2, 4-9 octarepeats have been reported in cases of familial Creutzfeldt-Jakob disease (Owen *et al.*, 1990, 1992; Goldfarb *et al.*, 1991; 1993; Campbell *et al.*, 1996). These familial cases of CJD are usually characterized by early onset and slow disease progression. Mutant hamster PrP with 9 or 11 copies (4 or 6 inserts) of the repeat motif expressed in mouse neuroblastoma cells had abnormal properties of aggregation and increased protease resistance. However, in a cell-free system these molecules were at least 1000-fold less protease-resistant than PrP^{Sc} from infected brain tissue, and showed no increased ability to form PrP^{Sc} (Priola *et al.*, 1998). Approximately 0.5% of the population has a deletion of one octarepeat, which appears to have no correlation to disease (Palmer *et al.*, 1993). One individual has been identified with four inserted repeats but showed no neurological disorder (Goldfarb *et al.*, 1991).

3.3. Attempts to Define the Binding Sites

3.3.1. Copper Binding to the Octarepeats

Following the initial reports of copper binding to peptides based on the mammalian and avian octameric repeats (Hornshaw *et al.*, 1995a; 1995b), further studies have confirmed these findings both using synthetic peptides and recombinant protein. However, the variety of protein fragments, experimental conditions and techniques used can be bewildering and make comparisons of data difficult.

A

```
Human     PQGGGWGQ PHGGGWGQ PHGGGWGQ PHGGGWGQ -------- PHGGG-WGQ
Sheep     PQGGGWGQ PHGGGWGQ PHGGGWGQ PHGGGWGQ -------- PHGGGGWGQ
Pig       PQGGGWGQ PHGGGWGQ PHGGGWGQ PHGGGWGQ -------- PHGGGGWGQ
Cow       PQGGGWGQ PHGGGWGQ PHGGGWGQ PHGGGWGQ PHGGGWGQ PHGGGGWGQ
Giraffe   PQGGGWGQ PHGGGWGQ PHGGGWGQ PHGGGWGQ PHGGGWGQ PHGGGGWGQ
Rabbit    PQGGG-WGQ PHGGGWGQ PHGGGWGQ PHGGGWGQ -------- PHGGG-WGQ
Mouse     PQGGT-WGQ PHGGGWGQ PHGGGWGQ PHGGSWGQ -------- PHGGG-WGQ
Dolphin   PQGGGWGQ PHGGGWGQ PHGGGWGQ PHGGGWGQ -------- PHGGGGWGQ
```

B

```
Human     PQGG--GGWGQPHGGG--WGQPHGGG--WGQPHGGG-WGQ
Possum    PGGNRYPGWGHPQGGGTNWGQPHPGGSNWGQPHGGSNWGQ
```

C

```
Chicken     QPGYPH NPGYPH NPGYPH ------ NPGYPH NPGYPQ NPGYPH NPGYPG
Prion bird  QPGYPQ NPGYPH NPGYPH ------ NPGYPH NPGYPQ NPG--  ------
Duck        QPGYPQ NPGYPH NPGYPH NPGYPH ------ NPGYPH NPG--  ------
Turtle      NPGYPQ NPSYPH NPAYPP NPAYPP NPGYPH NPSYPR NPSYPQ NPGYPG
```

D

```
Human       -----GGGT--HSQ--WN-KPSKP-KTNMKHMAGAAAAGA
Cow         -----GGT--HGQ--WN-KPSKP-KTNMKHVAGAAAAGA
Giraffe     -----GGT--HGQ--WN-KPSKP-KTNMKHMAGAAAAGA
Sheep       -----GGS--HSQ--WN-KPSKP-KTNMKHVAGAAAAGA
Pig         -----GGGS--HGQ--WN-KPSKP-KTNMKHVAGAAAAGA
Mouse       -----GGGS--HNQ--WN-KPSKP-KTNLKHVAGAAAAGA
Rabbit      -----GGT--HNQ--WG-KPSKP-KTSMKHVAGAAAAGA
Dolphin     -----GGGT--HNQ--W--KPSKP-KTNMKHVAGAAAAGA
Possum      -----G--Y-N-K--W--KPDKP-KTNLKHVAGAAAAGA
Chicken     GYNPSSSGGS-YHNQKPW--KPPK---TNFKHVAGAAAAGA
Prion bird  GYNPSSSGGT-YHNQKPW--KPPKS-KTNFKHVAGAAAAGA
Duck        -YNPSSSGG-NYHHQKPW--KPPK---TNFKHVAGAAAAGA
Turtle      HYNPAGGTNFKNQKPW--KPDKP-KTNMKAMAGAAAAGA
Xenopus     PYNP--SG--YNKQ--W--KPPKS-KTNMKSVAIGAAAGA
Takifugu    ---GGGGA--GQAYRPVQSSNP--SAGK-VAGAAAAGA
```

Figure 2. Conserved Domains in the PrP N-Terminus Sequence homology alignments of: A, mammalian octarepeats; B, marsupial and human; C, avian and reptile; D, the possible 5th Cu binding site and palendromic sequence (in italics). Using the Clustal program, http://www.ebi.ac.uk/clustalw/index.html. Possible Cu binding histidine residues are shown in bold.

The first peptide studies of the human and chicken used circular dichroism (CD) to follow structural changes induced by copper (II). Mass spectrometry also confirmed selectivity for copper above other divalent metal ions. Fluorescence spectroscopy of these peptides gave surprisingly low affinity constants, in the μM range for copper binding perhaps suggesting an *in vitro* artifact rather than physiological relevance (Hornshaw *et al.*, 1995a; 1995b).

The next study to address copper binding made use of a recombinant fragment of the human PrP N-terminal domain between residues 23 and 98 (PrP23-98). An equilibrium dialysis of PrP23-98 with $Cu^{2+}(glycine)_2$ suggested a saturation stoichiometry of 5.6 coppers per peptide (Brown *et al.*, 1997b). The PrP23-98 fragment contains five histidine residues (4 within the octarepeats) and so it would appear that each is associated with a copper ion. Additionally, copper addition was cooperative with a Hill coefficient of 3.4. The first copper binds with low affinity, but further binding occurs with a higher K_d. Further experiments with both recombinant PrP and PrP^C confirmed the copper binding nature of the mouse prion protein and several additional studies have confirmed these findings with other methods. Comparison of recombinant mouse prion protein containing a $(His)_6$ tag to aid purification, which incidentally also acted as an internal control for copper binding, showed that a mutant lacking the octarepeat region bound 4 copper atoms less than that with the wild type sequence as shown by total X-ray fluorescence spectroscopy (Brown *et al.*, 1999a).

A further battery of spectroscopic techniques was brought to bear on the octarepeats in a combination of ^1H NMR, ultra violet (UV) and visible CD and electron spin resonance (ESR) (Viles *et al.*, 1999). This extensive study of octarepeat containing peptides again suggested a relatively low initial affinity ($K_d = 6μM$), confirmed cooperativity of copper binding and, for the first time, directly implicated the histidines. The pH sensitivity of binding was also noted and suggested a reason why previous studies of full length recombinant PrP gave a lower (ca 2Cu) binding capacity. Raman spectroscopy has lead to similar conclusions concerning stoichiometry and pH but suggested a different coordination environment (Miura *et al.*, 1999).

Further mass spectrometry of octarepeat peptides (Whital *et al.*, 2000) demonstrated binding of 4 copper ions to the complete repeat domain in the μM range. This has been confirmed by mass spectrometry of full length and fragmented recombinant protein

(Kramer *et al.*, 2001), while fluorescence spectroscopy of these polypeptides also indicates cooperative binding of 4 copper ions with μM affinity. An elegant method of "footprinting" histidine dependent copper binding using mass spectrometry pinpointed the four histidine residues of the octarepeats as each directly coordinating a copper ion (Qin *et al.*, 2002).

An alternative view suggests that the octarepeat region contains only one high affinity copper-binding site (Jackson *et al.*, 2001). This is proposed to have *f*M affinity that is masked by a second lower affinity site, excluded in this study by competition with glycine. Such a high affinity site would certainly seem a more attractive proposition from the point of view of physiological relevance. The authors propose this site to consist of 4 of the 5 histidines in their peptide. Presumably these come from the four octarepeats, as in the same article they suggest the fifth histidine is in a second high affinity site out of the octarepeat region.

In conclusion, there is consensus that the octarepeats of PrP bind copper and it is argued that even a μM affinity would be sufficient for an extracellular protein (Kramer *et al.*, 2001; Lehmann, 2002). That the histidines are strongly involved and hence make copper binding pH sensitive is also generally agreed. Additionally a common feature of many of these studies is the finding that the metal binding of the octarepeats is specific to copper over other metals by several orders of magnitude.

3.3.2. The Putative Binding Site

The fact that the mammalian octarepeat or avian hexarepeat sequence motif is novel is not in dispute. However, to say that it constitutes a novel metal binding site may be premature. Both type 1 and type 2 copper sites for example are often comprised of ligands distant in the primary sequence but brought into close proximity by the tertiary structure. Under these circumstances sequence motifs for copper binding are difficult to assess (Fenton, 1995; Arnesano *et al.*, 2002b).

Several of the studies above, however, have made suggestions as to the possible coordination environment. Perhaps the most conventional, although disputed because it suggests a unitary copper stoichiometry, is a tetragonal distribution of the 4 histidine imidazole groups about a single copper, reminiscent of a type 2 copper center

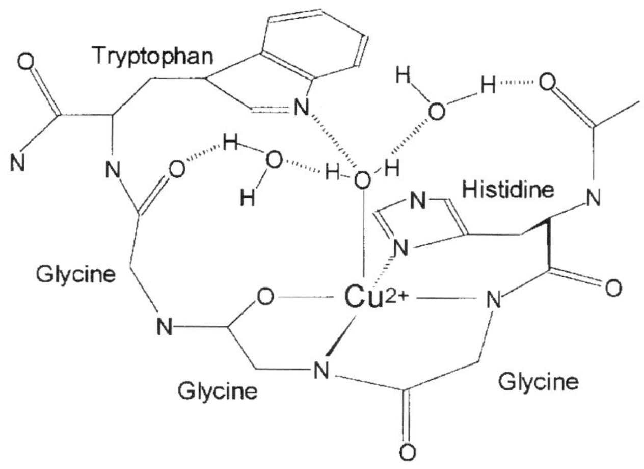

Figure 3. The proposed octarepeat His-Gly-Gly-Gly-Trp copper binding site, including H-bonded water molecules as suggested by X-ray crystallography (Burns *et al.*, 2002).

(Jackson *et al.*, 2001). Cooperativity of additional copper binding has suggested to others that the binding sites specifically interact. This has lead to a suggestion of an unusual planar ring structure consisting of bridging imidazoles. There is precedence for such a structure in the Cu/Zn SOD dimer. Two imidazole and the histidine amide nitrogen would therefore coordinate each copper with water filling the vacant space in a square planar geometry. The intervening peptides between histidine ligands could form specific but irregular loop structures (Viles *et al.*, 1999). Meanwhile Raman studies indicate that each octarepeat independently binds a single copper through the imidazole group and the amide nitrogens of the second and third glycines. In this case HGGG is all that is required for metal binding (Miura *et al.*, 1999). In a similar vein CD and ESR have suggested the HGGG motif provides imidazole and three amide nitrogens as ligands (Bonomo *et al.*, 2000).

An initial study of octarepeat based peptides studied with visible CD, ESR and electron spin-echo envelope modulation (ESEEM) revealed spectra of the motif HGGGW-Cu to be identical to those obtained from the full octarepeat sequence (Aronoff-Spencer *et al.*, 2000). More recently the copper (II) adduct of the HGGGW peptide has been crystallized and the structure solved. This is claimed to be the first high-resolution structure of the octarepeat binding site (Burns *et al.*, 2002). Copper is coordinated through the imidazole ring of the histidine, the backbone amide nitrogen of the next two glycines as well as the second glycine's carbonyl oxygen (See Figure 3). These ligands sit in the equatorial plane with the copper ion raised slightly above. A water molecule fills a fifth axial coordination site. There is the suggestion that the tryptophan, itself raised above the equatorial plane, forms a hydrogen bond with this water through the indole nitrogen. It should be noted that the glycine amide copper bond is very pH sensitive, more so than the histidine link and so may have implications for physiological binding of copper. Furthermore the third glycine has no direct interaction with copper or the axial water molecule and this position is the only non-conserved amino acid of the octarepeat amongst mammals. The authors also envisage the complete octarepeat domain ordering, based on detected hydrogen bonding. This appears to resemble a "string of beads" as opposed to a more compact structure.

While it should be remembered that this structure is the result of peptide studies rather than wild type protein it is probably the closest model currently available to the metal binding site contained within the octarepeats. As such it defines a unique copper binding site differing greatly from the well-established modes of copper coordination known to bioinorganic chemists.

3.3.3. The Fifth Site

Several studies have pointed to a fifth copper site, located outside of the octarepeats. There have been suggestions that this site of coordination could be on exposed histidine residues in the structured C-terminal domain (Cereghetti *et al.*, 2001). Other studies however have pointed to the involvement of one or both of two histidine residues situated between the octarepeats and the palindromic sequence (Jackson *et al.*, 2001). The first of these histidines is situated in what has been described as a pseudo repeat. The data implicating these residues in copper binding seems to be the strongest. One study has put the affinity in the *f*M range and the same group have suggested a structure based on spectroscopic and X-ray absorption fine structure (XAFS) data of a C-terminal recombinant fragment which excludes the octarepeats (Hasnain *et al.*, 2001). Both of the histidine residues at positions 96 and 111 are suggested to be in the coordination sphere along with a sulfur and oxygen, potentially of methionine 109 and glutamine 98 respectively. Two further oxygen donating ligands could be water or from a further amino acid ligand. Such a coordination is broadly in line with the common type 2 copper sites where copper(II) is coordinated with predominantly nitrogen (often histidine) or oxygen ligands and sulfur involvement is through methionine rather than cystine.

3.4. Physiological Relevance of Copper Binding

The body of evidence suggesting that copper can bind to PrP^c is not in doubt. Evidence that this is both within the octarepeats and external to them is growing. A putative copper moiety of the octarepeats has also been crystallized, showing that such a binding site can exist within each octarepeat. While these data suggest that the 5:1 copper to protein stoichiometry is correct, the question of physiological relevance of this interaction remains. The conservation of repeat regions between many species suggests an advantage, if not a necessity to their presence. The histidine rich nature also implies a metal binding site, while the demonstrated specificity of this binding site for copper strengthens this argument. The disparity between binding affinities is, however, still a matter of controversy. Some groups suggest high affinity while others maintain that a lower affinity would still enable the protein to function as a metalloprotein. Perhaps this discrepancy is most easily explained as a consequence of attempting to bind copper to non wild type protein under non-physiological conditions. To support this argument an expression study has shown that PrP^C expression allows cellular uptake of copper in the nM range, such an affinity for copper would certainly place it in the range of other cupro-proteins (Brown *et al.*, 1999b).

Perhaps the most convincing evidence that PrP^C is a cupro-protein is that the native protein isolated from mouse brain can be purified with 3 bound copper atoms. In these studies there was no evidence of any other divalent metal ion coordination to the isolated protein. Furthermore, PrP from cultured cells may be isolated with up to 4 copper atoms per molecule, but even cell culture in a low copper media yields a minimum of one bound copper (Brown *et al.*, 2001a). Failure to observe a 5 copper form from cell culture may suggest that a fifth site is not physiological, although it could be argued that that the method of purification results in a lowering of pH to between 5 and 6 which could result in copper loss. In fact the backbone Gly-Cu linkage shown in the octarepeat crystal structure is suggested to be unstable below a pH of approximately 6.5. The three atoms of copper maybe removed from the native protein by denaturation in urea and refolding in water; however, subsequent refolding in the presence of copper coordinates five coppers. The native protein (3Cu) is inert to further additions by aqueous copper; however, incubation with a histidine-copper complex again provides a 5 to 1 stoichiometry.

Since PrP^C is a cupro-protein, it is reasonable to assume that it will function optimally in the metal bound state. Many early studies of PrP tended to neglect the native function of the protein instead focusing on the conversion to the abnormal PrP^{Sc}.

3.5. Properties of Cupro-prion Proteins

3.5.1. Physical and Structural

There has been a struggle to define the normal function of PrP *in vitro*. Again most studies rely on peptides and fragments of recombinant protein. Even with full length recombinant PrP purified from bacterial inclusion bodies there is a distinct difference in properties dependant on the refolding protocol used. Addition of copper during the removal of urea results in an enzymatic activity to be discussed shortly, absent from protein refolded in water and then copper treated (Brown *et al.*, 1999a). Copper added during refolding has been reported to stabilize the molecule making it more soluble while addition of copper salts to apoprotein can lead to aggregation (Daniels *et al.*, 2002; Viles *et al.*, 1997). The copper refolded recombinant PrP has properties comparable to native protein (isolated containing 3 copper atoms) missing from refolded recombinant protein to which copper has been added (Brown *et al.*, 1999a; 2001a). Perhaps this highlights the problems of the addition of prosthetic metal under non-physiological conditions. The absence of specific metal chaperones that deliver their ligand specifically or the failure to incorporate copper at the correct point in the refolding procedure can easily lead to redundant conformations.

Overall there appears to be a conformational change to the protein on copper coordination. Unsurprisingly this has been largely mapped to the N-terminal domain and probably represents an increase in tertiary structure (Viles *et al.*, 1999; Miura *et al.*, 1999; Bonomo *et al.*, 2000). Furthermore an aggregation independent increase in resistance to protease activity is observed again indicative of increased N-terminal structure.

An important consideration of metalloprotein function involves the oxidation state of the metal and the redox behavior of the protein. For example the Ctr cellular copper uptake proteins import copper in the reduced Cu^+ form (Puig *et al.*, 2002). The nature of the copper-binding site from structural studies would seem to suggest copper is present as Cu^{2+}. Cystines often feature in copper (I) binding proteins (Fenton, 1995), while methionine rich motifs may also make up copper (I) motifs (Arnesano *et al.*, 2002). Copper is however found in Cu/Zn SOD among others, with histidine ligands in both oxidized and reduced states and so copper (I) coordination cannot be discounted. The peptide studies that have used EPR, visible spectroscopy and visible CD have indicated that copper added as copper(II) remains as such. Two studies of copper redox states in prion proteins both using the chelator bathocuproine disulfonate have argued for copper (II) in one instance and reduced copper in the other (Ruiz *et al.*, 2000; Shiraishi *et al.*, 2000).

Oxidation of recombinant PrP with bound copper has been observed by ascorbate and dopamine resulting in carbonyl formation (Requena *et al.*, 2001; Shiraishi *et al.*, 2002). In one of these studies, oxidation was shown to induce aggregation and precipitation of full-length recombinant PrP (Requena *et al.*, 2001). Meanwhile full-length recombinant chicken and mouse PrPs show specific oxidation of methionine residues presumably due to the presence of copper, indicating a redox activity (Wong *et al.*, 1999).

An ability to bind copper may indicate a role for PrP^C in copper homeostatic pathways. However, redox activity could suggest not only a role for copper uptake but also enzymatic function.

3.5.2. Superoxide Dismutase Activity of Recombinant PrP^C

The enzyme Cu/Zn-SOD catalyses the disproportionation of superoxide (O_2^-) to hydrogen peroxide and oxygen.

$$Cu(II)SOD + O_2^- \rightarrow Cu(I)SOD + O_2$$

$$Cu(I)SOD + O_2^- + 2H^+ \rightarrow Cu(II)SOD + H_2O_2$$

It is significant therefore that the recombinant forms of both chicken and mouse PrP have been found to display superoxide dismutase activity. The activity of recombinant PrP has been found to be approximately 10 % of that of Cu/Zn-SOD based on protein concentration. The fact that Cu/Zn-SOD is an extremely potent enzyme means that even this lower SOD like activity is significant. The SOD activity could be attributed to copper contained in the octarepeat regions as both apoprotein and a deletion mutant lacking octarepeats failed to display the activity. Additionally the octarepeat deletion mutant contained a $(His)_6$ tag to aid purification which bound 6 copper atoms so acting as an internal control for non-specific copper SOD activity. This SOD activity is only observed when the protein is refolded in the presence of copper; addition of copper to refolded apoprotein is insufficient (Brown *et al.*, 1999a). Rather than being an artifact of an unusual *in vitro* folding procedure, native PrP^C extracted from mouse brain was shown to exhibit a similar activity (Brown *et al.*, 2001a). PrP extracted from cultured mouse cerebellar cells showed an SOD activity in proportion to copper content. The lowest copper to protein ratio isolated was 1:1 and this failed to be enzymatically active; however, activity increased with each subsequent increase in copper up to a 4:1 ratio. The activity was independent of added copper and of an excess of EDTA, the latter implying tight binding site. The prion peptide fragment PrP106-126, itself considered a mimic of PrP^{Sc} and toxic to neurons was able to inhibit SOD activity of the native protein (Brown *et al.*, 2001a). As previously noted, the PrP (106-126) peptide increases neuron sensitivity to oxidative stress and copper toxicity, inhibition of PrP^C SOD activity could clearly account for such behavior.

The above mentioned oxidation of methionines observed on recombinant PrPs occurred under similar conditions to the SOD activity, being absent from apoprotein, although copper added to apoprotein shows 10% of the oxidation compared to copper refolded protein. The histidine residues present in the octarepeats and also in the added $(His)_6$ tag are unaffected by oxidation (Wong *et al.*, 1999). Such results imply that the specific oxidation of methionines is related to SOD activity and as such are part of the reactive center. Such a view is consistent with the observation that the octarepeats alone are not sufficient for SOD activity, but that the complete protein is required. This is in itself supported by the observation that reduction of the intramolecular disulfide bond (but still containing Cu^{2+}) of PrP is sufficient to ablate SOD activity (Wong *et al.*, 2000a). Thus the complete PrP is required, correctly loaded with at least two copper atoms and a structured C-terminus.

4. Cellular Copper and PrP

4.1. Normal and Knock Out Implications for Prion Function

The PrP^C glycoprotein is expressed on the surface of many different cell types. While expression in neurons is elevated compared to other cell types and it is reasonable to assume that PrP expression is of increased importance to neurons, PrP should not be considered to have a neuron only function (Brown, 2001b). PrP has been shown to be highly concentrated in synapses, both in the central nervous system and the endplates (Salès *et al.*, 1998; Fournier *et al.*, 2000). PrP^C has been isolated along with PrP^{Sc} in insoluble rafts, suggested to be cholesterol rich microdomains of the synaptic membrane (Naslavsky *et al.*, 1997). PrP^C is expressed both pre- and post-synaptically (Salès *et al.*, 1998; Fournier *et al.*, 2000; Herms *et al.*, 1999). There is evidence of specific transport of glycoforms perhaps indicating that there could be a preferred presynaptic form (Rodolfo *et al.*, 1999).

Initial studies of *Prnp* knockout mice did not highlight any major disruptions to either development or behavior. These knockout mice do not develop prion disease through lack of PrP^C function. Despite these findings it would appear unlikely that such a protein, conserved among higher species and with characteristic expression profiles should be redundant. Therefore it is likely that the function of PrP is duplicated by other mechanisms (Brown, 2001b; Lehmann, 2002).

Further studies revealed subtle distinctions between wild type and *Prnp* knockout animals. Behavior has been affected by changes in circadian rhythms (Tobler *et al.*, 1996) and there appears to be some electrophysiological changes in the nervous system (Collinge *et al.*, 1994). These latter changes suggest a link between PrP expression and neurotransmission (Brown, 2001b). PrP expression in the synapses peaks during synaptogenesis (Salès *et al.*, 2002) it is therefore interesting that PrP knockout mice also show structural changes in synapse formation (Colling *et al.* 1997).

Sensitivity to oxidative stress has been seen in *Prnp* knockout mice (Herms *et al.*, 1999; Brown *et al.*, 1997b), resulting in increased oxidation of lipid and protein. Decreased SOD activity and changes in catalase (H_2O_2 disproportionation) and ornithine decarboxylase (stress adaptation) activities were also observed (Klamt *et al.*, 2001). Additional effects in neuronal cell cultures of knockout mice involve glutathione metabolism resulting from reduced glutathione reductase activity (White *et al.*, 1999), while more recently melatonin and the extracellular SOD3 have also been shown to be reduced (Brown *et al.*, 2002). In general, cultured cells lacking PrP tend to be less viable under culture conditions and are also more sensitive to oxidative damage and toxicity (Brown *et al.*, 1997a; 1998a; Kuwahara *et al.*, 1999; White *et al.*, 1999). Further phenotypic changes include glutamate uptake by astrocytes (Brown *et al.*, 1999b) and microglia sensitivity to activation (Brown *et al.*, 1998b).

4.2. Copper Uptake

A further finding to come from *Prnp* knockout studies is that the synaptosomal fractions of wild type mouse brain contain more copper than their knockout counterparts (Brown *et al.*, 1997b). This is significant considering the high levels of synaptic PrP^C expression. Localized copper at high concentrations is released in vesicles (Demeter *et al.* 1979) from the synapses in high concentration, suggested to be as much as 250 µM (Kardos *et al.*, 1989) upon polarization. Although well established, the reason for this release is unknown (Colburn *et al.*, 1965). The released copper is rapidly taken up by neurons and redistributed within 30 minutes (Barnea *et al.*, 1989). Interestingly PrP has a fast turnover with a half-life of less than one hour in the presence of copper (Pauly *et al.*, 1998). The presence of copper at the synapse appears important in the modulation of several processes such as GABA and glutamate uptake (Gabrielsson *et al.*, 1986; Brown *et al.*, 1999b). Copper can also inhibit transmission at NMDA receptors (Vlachová *et al.*, 1996). This correlates with synaptic defects of *Prnp* knockout mice involving NMDA receptors and GABA type inhibitor currents (Collinge *et al.*, 1994).

To examine the copper-PrP^C-synapse relationship more fully, wild type, knockout and over expressing PrP^C cells were loaded with radioactive ^{67}Cu copper and allowed to stabilize. The cells were treated with a depolarizing agent vetridine, an agent that causes the cells to mimic synaptic release. Levels of copper release were directly proportional to the expression of PrP^C, with the knockout cells releasing the least copper and PrP overexpressing cells the most (Brown, 1999c). A protection against the effects of exogenous copper to inhibitory currents has been observed in wild type Purkinje cells compared to *Prnp* knockouts (Brown *et al.* 1997b).

The difference between the copper content of the synaptosomal fractions of wild type and knockout strains are considerable and are mirrored when cultured cell lines are compared. Cerebellar cells from wild type mice expressing PrP grown in low copper culture media retain more copper than cells from knockout mice. However, treatment of the cultured cells with phosphatidylinositol-specific phospholipase C (PIPLC), results in equivalent copper uptake between wild type and knockout cells (Brown *et al.* 1997b). Similar experiments to study copper transport by PrP used the ^{67}Cu isotope to monitor copper uptake *in vivo* (Brown, 1999c). No differences were observed when aqueous copper was added, however when chelated to histidine, copper uptake was found to be proportional to the level of PrP expression. Such a dependence on chelated as opposed to free copper is reminiscent of copper coordination to native PrP^C (*vide supra*). Increased PrP expression increased both cytosolic and membrane associated copper. The inference from such results is that the number of copper binding sites within cerebellar cells increases in relation to PrP expression. A study of the Cu/Zn-SOD enzyme from these cells showed that ^{67}Cu was incorporated into this protein in proportion to the level of PrP expression (Brown *et al.*, 1998c).

Endocytosis of PrP occurs via clathrin coated pits to an early endocytic compartment and then is returned to the plasma membrane (Shyng *et al.*, 1994; 1995). Copper and zinc have been shown to stimulate reversible endocytosis and as a result could enter the cell. N-terminal deletions of the PrP failed to show the rapid internalization of PrP as observed in the case of the wild type. Furthermore specific mutations in or deletions of the octameric repeats that remove the copper binding histidines result in poor endocytosis, indicating that this is a direct copper binding effect (Pauly *et al.*, 1998). A recent paper has challenged this interpretation with the observation that although truncation of the N-terminus inhibited internalization of mouse PrP, wild type function was restored when the *Xenopus* N-terminus was fused to the shortened protein. Significantly, *Xenopus* PrP lacks obvious copper binding residues in the N-terminus (Nunziante *et al.*, 2003).

4.3. Copper and Oxidative Stress

As outlined above, a decreased resistance to conditions of oxidative stress appears to be a phenotype of *Prnp* knockout mice. Furthermore, Cu-PrP^C shows superoxide dismutase activity *in vitro* (Brown *et al.*, 1997b). A finding which has been disputed (Waggoner *et al.* 2000), but which has been repeated by several groups is that *Prnp* knockout mice not only have lower brain copper levels but also have approximately 50% Cu/Zn-SOD activity of the wild type mice (Brown *et al.*, 1997b; White *et al.*, 1999; Wong *et al.*, 2000b). Expression of the SOD protein appears unaffected in *Prnp* knockout mice and so the shortfall in activity is related to incomplete copper loading of the enzyme. Related experiments with PC12 cells correlated high levels of PrP^C expression to increasing resistance to both oxidative stress and copper toxicity (Brown *et al.*, 1997c). It appears that in general increased expression of PrP leads to protection from oxidative stress. Conversely PrP expression in PC12 cells increases when they are cultured under conditions of oxidative stress (Brown *et al.*, 1997c). Peptides based on the copper binding octameric repeat region of the PrP are able to protect cell cultures against copper and superoxide toxicity, presumably through chelation of copper. The protective effect was greatest in the case of cultured neurons from *Prnp* knockout mice (Brown *et al.*, 1998a). Copper and oxidative stress are linked by Fenton type chemistry, whereby copper can produce many reactive oxygen species; this chemistry is not however linked to copper alone. Interestingly, in the presence of copper and hydrogen peroxide a site specific cleavage of PrP^C occurs within the octarepeat region that appears to be a specific response to oxidative stress (McMahon *et al.*, 2001)

5. The Role of the Metalloprotein

Analysis of the primary structure gives no true clues to the role of the normal PrP through sequence homology to proteins of known function. The cell surface location indicated by the presence of a GPI anchor may indicate a role in cell-cell signaling, adhesion or cellular defense (Sendo *et al.*, 1998). The three dimensional structure of a protein can give clues to function and is often more conserved than primary sequence between related proteins. The solution structure of the C-terminal domain fails to clarify any cellular role. However, the presence of related sequences in so many vertebrates suggests that PrP is not redundant. Furthermore the properties of the copper associated protein outlined above certainly give clues as to its function (See Figure 4).

Three classes of membrane bound copper proteins are known in eukaryotes. These consist of oxidases such as the mitochondrial cytochrome *c* oxidase (Cox), copper translocators such as the Ctr proteins and finally extracellular SOD (SOD3) (Linder, 1991). It is possible therefore that GPI anchored PrP^C may have an activity such as these, especially the last two considering the mitochondrial location of Cox.

The mechanism of PrP copper loading is not clear and could occur internally, extracellularly or both. This could suggest that at the very least PrP acts as a chelator of copper, acting to sequester copper. Such a role could prevent copper from taking part in Fenton like chemistry and producing reactive oxygen species. As such it would be similar in function to proteins such as cellular metallothioneins or extracellular ceruloplasmin and albumin. Chelated copper could be internalized, or, by processing of the PrP N-terminus, held in a redox inactive state in the plasma for excretion. Increased levels of PrP expression are therefore found where copper levels are most likely to cause harm.

Other theories take the chelator properties a step further. The proposed copper dependent endocytosis has been used as a strong argument for a role in copper uptake (Pauly *et al.*, 1998). This is indeed supported by evidence of cellular copper content being linked to the level of PrP expression (Brown, 1999c). It has been suggested that changes induced by copper binding in the N-terminal can start a signaling pathway, perhaps by the intermolecular interactions of the conserved, but non metal binding glutamines of the octarepeats (Burns *et al.*, 2002). A plausible theory is that endocytosed Cu-PrP is then demetallated by a pH drop in the vesicle, followed by copper(II) reduction and redistributed by copper chaperones (Miura *et al.*, 1999). Such a mechanism could either be simply a copper sensor/buffer (*cf* calmodulin and calcium) or perhaps signal transduction pathway in response to conditions of oxidative stress or finally a more specific copper chaperone. Evidence for a role as a copper chaperone is the incorporation of radioactive copper into Cu/Zn-SOD through PrP^C copper uptake. In such a chaperone role, PrP could pass copper to the Cu/Zn-SOD chaperone CCS, another unidentified chaperone or directly to Cu/Zn-SOD. It has been shown that in conditions rich in

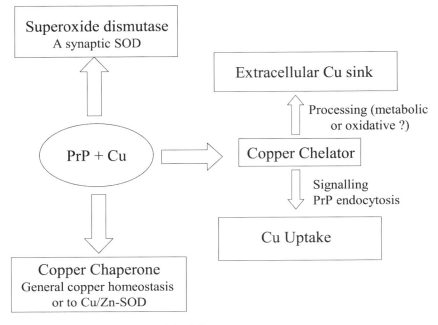

Figure 4. Potential functions of Cu-PrP

copper Cu/Zn-SOD does not require expression of CCS to obtain its copper (Rae *et al.*, 1999). CCS has also been shown to display an interaction involving its copper-binding domain with the X11α neuronal facilitator protein through the C-terminal PDZ2 domain. This alternative interaction of CCS can regulate Cu/Zn-SOD activity (McLoughlin *et al.*, 2001). In fact it is becoming clear that copper homeostasis in the nervous system and perhaps other cell types is not as simple as the general mechanisms outlined in section 2.2. The amyloid precursor protein APP can interact with a domain of the X11α neuronal facilitator protein, it is not known if this is copper related, but provides an indirect link between APP and Cu/Zn-SOD (Mueller *et al.*, 2000). In addition, several of the EF hand calcium binding proteins specific to the CNS have been shown to have copper binding activity and have been suggested to have secondary roles as copper chaperones (Shiraishi *et al.*, 1998; Schafer *et al.*, 2000; Groves *et al.*, 2001; Landriscina *et al.*, 2001). Under these circumstances and the evidence that PrP is involved in copper delivery, a chaperone function of PrP cannot be easily discounted.

Finally, and probably the most striking feature of PrP, is the ability to act as a superoxide dismutase in its copper bound (native) state.

This feature is shown clearly *in vitro* and although the physiological relevance has been disputed the majority of evidence points to a relationship between copper, PrP and SOD activity. Furthermore, it has been suggested that although this enzymatic function is always present in native PrP it is most important in the synapses where it is highly expressed. Operating as a synaptic SOD, Cu-PrP could have a role preventing disruption of neurotransmission by superoxide, a known phenomenon. Such a role would be of great importance in a situation where copper fluxes are high and therefore the risk of reactive oxygen species formation increased. In support of this argument it is interesting to note that of the three mammalian SOD enzymes, the third SOD3 or extracellular SOD is low in the brain (Ookawara *et al.*, 1998). Conversely cytosolic Cu/Zn-SOD (SOD1) and manganese SOD (SOD2) of the mitochondria are well represented in the nervous system. Extracellular SOD is also a copper binding protein (4 atoms) and has 3 isoforms. It can either be surface bound or released (Marklund, 1982). There is therefore a strong argument that PrP^C expression is increased in areas of low SOD cover, such as the synapses to protect these environments from oxidative stress. The quick turn over of PrP^C may also represent a rapid response to such stresses.

6. Copper and Prion Disease

Conversion of normal extracellular PrP^C to the abnormal PrP^{Sc} is characteristic of prion disease. For PrP^{Sc} to be toxic, cells must express their own PrP^C; additionally *Prnp* knockouts do not have a disease phenotype. This isoform conversion is key to disease progression.

6.1. Metals and Protein Misfolding

As mentioned above, aggregation and oxidation has been observed with copper loaded PrPs in the presence of oxidative stress. Further studies have shown a tendency for recombinant PrP to precipitate in the presence of copper salts. Indeed, aged proteins exposed to high copper concentrations have lead to protease resistance, a characteristic of the PrP^{Sc} form of the protein (Qin *et al.*, 2000). A further study however suggested that despite being protease resistant, copper treated recombinant PrP did not have the same conformation as PrP^{Sc} (Quaglio *et al.*, 2001).

Studies looking at the infectivity and conformation of PrP^{Sc} have linked these characteristics to metal ion content. A brain extract high in PrP^{Sc} was treated to be less infective (not non-infective). Subsequent incubation of the extract with copper restored complete infectivity (McKenzie *et al.*, 1998). As pointed out earlier, through Fenton like chemistry copper can be an agent of oxidative stress. A more recent paper demonstrated two phenotypically distinct strains of CJD had distinct forms of PrP^{Sc}. The distinction was suggested to be the metal ion (copper/zinc) occupancy of the PrP^{Sc}. By adjusting the metal ion contents the authors were able to interconvert the two PrP^{Sc} forms. Although interesting, such an experiment does not prove the initial formation of PrP^{Sc} to be due to the metal ion occupancy (Wadsworth *et al.*, 1999).

6.2. Metal Imbalance and Oxidative Stress

The normal PrP appears from the evidence outlined above to provide resistance to oxidative stress; it is therefore interesting to note that oxidative stress is apparent in prion disease. The peptide PrP106-126 is a much-studied mimic of PrP^{Sc} *vide supra*. The peptides toxicity to neurons in culture appears to result from the release of toxic radicals from microglia, while at the same time reducing resistance to neurons. The neurons are, however, only sensitive to PrP106-126 if they express PrP^{C} (Brown *et al.*, 1996; 1997a). From these results it appears that a major factor in neuron death is an increased sensitivity to disease induced oxidative stress. In support of this statement, PrP106-126 treated neurons are less resistant to copper toxicity and oxidative stress showing less SOD activity (Brown *et al.*, 1997a). It has been suggested that PrP106-126 binding to PrP^{C} can disrupt cellular uptake of copper, leading to low copper incorporation into Cu/Zn-SOD in addition to inhibiting PrP^{C} own SOD activity. On the other hand, chelation of copper from cultures treated with PrP106-126 prevents cell death, implying that copper is necessary for peptide toxicity (Brown, 2000a).

Two alleles of the mouse PrP gene that control prion incubation times, designated *Prnp-a* or -*b*, differ by only two amino acid residues, the *b* allele generally producing long prion incubation times. Studies of recombinant PrP-A and B suggest that PrP-B has higher SOD activity and that it is also more labile. It is possible that one or both of these factors influence resistance to disease (Brown *et al.*, 2000b).

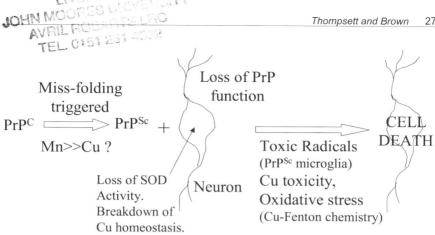

Figure 5. Loss of Cu-PrP function and neuronal death

An area of increasing interest in prion disease is perturbations in the concentrations of transition metals in diseased brains. These studies have indicated quite startling findings about metal levels in prion disease and are making researchers consider environmental factors in the incidence of sporadic disease. Levels of brain copper in patients with CJD are significantly decreased (Wong *et al.*, 2001a). This finding is true also in mice infected with scrapie (Wong *et al.*, 2001b). The mice also showed an increase in liver copper, the organ through which copper is excreted. The brain copper concentration decreases prior to the on set of symptoms, this is true also of an increase in manganese levels, both in brain and blood. Analysis of PrP from the infected brain tissue showed a decrease in bound copper associated with the PrP^{Sc} isoform, this was coupled with a significant loss of SOD activity. Further effects of this apparent replacement of copper with manganese in the brain were loss of Cu/Zn-SOD activity and increasing levels of nitric oxide and superoxide. The metal imbalance therefore appears to have dramatic consequences to the response to oxidative stress (Figure 5).

Further work with recombinant PrP has indicated that manganese can compete with copper to bind PrP^{C}. The Mn-PrP has a lower SOD activity than Cu-PrP, but the activity is not ablated. However, on aging all activity is lost, furthermore the protein becomes protease resistant and the CD spectrum shows an increase of β-sheet content. Such behavior is reminiscent of PrP^{Sc}. Such results could suggest that a metal imbalance can lead to misincorporation of metal into the native protein leading to formation of misfolded protein, loss of stress protection and breakdowns in metal ion homeostasis exacerbating the situation (Brown *et al.*, 2000c). It has been noted that hot spots of prion disease are associated with soil high in manganese and low in copper (Purdey, 2000).

6.3. Relationships to Other Neurodegenerative Diseases

The similarities of the brain pathology of Wilson's disease and some prion diseases have already been noted. It is of interest that these symptoms are as a result of disrupted copper homeostasis in Wilson's disease, although there is no protein deposition as in prion diseases. It does, however, strengthen the argument that altered copper homeostasis is a feature of prion disease.

The Wilson's disease gene ATP7B in addition to coding the Wilson disease protein expresses by a novel splice variant PINA. PINA is a copper binding protein expressed solely in the pineal gland and whose expression coincides with production of the antioxidant and soporific compound melatonin. The pineal gland has one of the most studied circadian rhythms, whereby melatonin is only produced at night (reviewed by Borjigin *et al.*, 2002). It is known that *Prnp* knockout mice have disrupted circadian rhythms (Tobler *et al.*, 1996) and it has been noted that the prion disease familial fatal insomnia (FFI), shows altered sleep patterns. *Prnp* knockout mice also produce low levels of melatonin (Brown *et al.*, 2002). It is tempting to speculate that altered PrP expression may disrupt copper delivery to PINA, which in turn affects melatonin synthesis. Such disruption would have repercussions both for sleep patterns and responses to oxidative stress.

Recently a link has been reported between ALS and PrP expression. ALS is thought to be the result of a gain of function of an aberrant isoform of Cu/Zn-SOD. In a mutant mouse model of ALS where SOD is over expressed, expression of PrP^C is down regulated (Dupuis *et al.*, 2002). Such behavior fits either the role of a metal chaperone shutting down the supply of copper to a malfunctioning enzyme or the regulation of an enzyme of similar function.

Perhaps the most intriguing comparisons can be made with Alzheimer's disease (AD), the formation of insoluble plaques being an obvious comparison. The presence of copper within these plaques and its ability to induce aggregation of amyloidβ (Aβ) in turn increases interest in the metal content of PrP^{Sc} fibrils. Initial findings seem to suggest that PrP^{Sc} is not a huge sink of precipitated or redox active copper, moreover perhaps that copper is lost on PrP^{C} to PrP^{Sc} conversion (Wong *et al.*, 2001b). However, copper and zinc binding has been shown to have a positive affect on the aggregation of the PrP106-126 peptide (Jobling *et al.*, 2001). Also of interest is the similarity, at a superficial level, of the APP protein and the PrP. Both are copper binding, probably show some redox behavior and may have an enzymatic function. Furthermore an event, which may include aberrant metal binding, can turn them both into a disease related isoform.

7. Conclusion

Prion diseases continue to be of interest to the public at large. Studying the role of the native protein can bring significant contributions to the knowledge of prion disease. In this chapter we hope to have convinced the reader that this can only be done by considering the PrP as a true cupro-protein. The copper form of the protein can show why it is so important in combating oxidative stress and gives further clues as to why prion disease results in neuronal death.

References

Arnesano, F., Banci, L., Bertini, I., Ciofi-Baffoni, S., Molteni, E., Huffman, D.L., and O'Halloran, T.V. 2002a. Metallochaperones and metal-transporting ATPases: a comparative analysis of sequences and structures. Genome Res. 12: 255-71.

Arnesano, F., Banci, L., Bertini, I., and Thompsett, A.R. 2002b. Solution structure of CopC: a cupredoxin-like protein involved in copper homeostasis. Structure (Camb). 10: 1337-1347.

Aronoff-Spencer, E., Burns, C.S., Avdievich, N.I., Gerfen, G.J., Peisach, J., Antholine, W.E., Ball, H.L., Cohen, F.E., Pruisner, S.B., and Millhauser G.L. 2000. Identification of the Cu^{2+} binding sites in the N-terminal domain of the prion protein by EPR and CD spectroscopy. Biochemistry. 39: 13760-13771.

Bahamanyar, S., Williams, E.S., Johnson, F.B., Young, S., and Gajdusek, D.C. 1985. Amyloid plaques in spongiform encephalopathy of mule deer. J. Comp. Path. 85: 1-5.

Barnea, A., Hartter, D.E., and Cho, G. 1989 High affinity uptake of Cu^{67} into veratridine releasable pool in brain tissue. Am. J. Physiol. 257: C315-C322.

Bonomo, R.P., Impellizzeri, G., Pappalardo, G., Rizzarelli, E., and Tabbi, G. 2000. Copper(II) binding modes in the prion octapeptide PHGGGWGQ: a spectroscopic and voltammetric study. Chem-Eur. J. 6: 4195-4202.

Borchardt, T., Camakaris, J., Cappai, R., Masters, C.L., Beyreuther, K., and Multhaup, G. 1999. Copper inhibits beta-amyloid production and stimulates the non-amyloidogenic pathway of amyloid-precursor-protein secretion. Biochem J. 344 Pt 2: 461-467.

Borchelt, D.R., Rogers, M., Stahl, N., Telling, G., and Prusiner, S.B. 1993. Release of the cellular prion protein from cultured cells after loss of its glycoinositol phospholipid anchor. Glycobiology. 3: 319-329.

Borjigin, J., Sun, X., and Wang, M.W. 2002. The role of PINA in copper transport, circadian rhythms, and Wilson's disease. In Handbook of Copper and Pharmacology and Toxicology. E.J. Massaro, ed. Humana Press, New Jersey.

Brown, D.R., Herms, J., and Kretzschmar, H.A. 1994. Mouse cortical cells lacking cellular PrP survive in culture with a neurotoxic PrP fragment. Neuroreport. 5: 2057-2060.

Brown, D.R., Schmidt, B., and Kretzschmar, H.A. 1996. Role of microglia and host prion protein in neurotoxicity of a prion protein fragment. Nature. 380: 345-347.

Brown, D.R., Schulz-Schaeffer, W.J., Schmidt, B., and Kretzschmar, H.A. 1997a. Prion protein-deficient cells show altered response to oxidative stress due to decreased SOD-1 activity. Exp Neurol. 146: 104-112.

Brown, D.R., Qin, K., Herms, J.W., Madlung, A., Manson, J., Strome, R., Fraser, P.E., Kruck, T., von Bohlen, A., Schulz-Schaeffer, W., Giese, A., Westaway, D., and Kretzschmar, H. 1997b. The cellular prion protein binds copper *in vivo*. Nature. 390: 684-687.

Brown D.R., Schmidt B., and Kretzschmar H.A. 1997c. Effects of oxidative stress on prion protein expression in PC12 cells. Int. J. Dev. Neurosci. 15: 961-72.

Brown, D.R., Schmidt, B., and Kretzschmar, H.A. 1998a. Effects of copper on survival of prion protein knockout neurones and glia. J. Neurochem. 70: 1686-1693.

Brown, D.R., Besinger, A., Herms, J.W., and Kretzschmar, H.A. 1998b. Microglial expression of the prion protein. Neuroreport. 9: 1425-1429.

Brown, D.R., and Besinger, A. 1998c. Prion protein expression and superoxide dismutase activity. Biochem. J. 334: 423-429.

Brown, D.R., Wong, B.S., Hafiz, F., Clive, C., Haswell, S.J., and Jones, I.M. 1999a. Normal prion protein has an activity like that of superoxide dismutase. Biochem. J. 344 Pt 1:1-5.

Brown, D.R., and Mohn, C.M. 1999b Astrocytic glutamate uptake and prion protein expression. Glia. 25: 282-292.

Brown, D.R. 1999c. Prion protein expression aids cellular uptake and veratridine-induced release of copper. J. Neurosci. Res. 58: 717-725.

Brown D.R. 2000a. PrP^{Sc}-like prion protein peptide inhibits the function of cellular prion protein. Biochem. J. 352: 511-518.

Brown, D.R. , Iordanova, I.M., Wong, B.-S., Vénien-Bryan, C., Hafiz, F., Glasssmith, L.L., Sy, M.-S. , Gambetti, P., Jones, I.M., Clive, C., and Haswell, S.J. 2000b. Functional and structural differences between the prion protein from two alleles prnp^a and prnp^b of mouse. Eur. J. Biochem. 267: 2452-2459.

Brown, D.R., Hafiz, F., Glasssmith, L.L., Wong, B.-S., Jones, I.M., Clive, C., and Haswell, S.J. 2000c. Consequences of manganese replacement of copper for prion protein function and proteinase resistance. EMBO J. 19: 1180-1186.

Brown, D.R., Clive, C., and Haswell, S.J. 2001a. Antioxidant activity related to copper binding of native prion protein. J. Neurochem.. 76:69-76.

Brown, D.R. 2001b. Prion and prejudice: normal protein and the synapse. Trends Neurosci. 24: 85-90.

Brown, D.R., Nicholas, R.St.J., Canevari, L. 2002. Lack of prion protein expression results in a neuronal phenotype sensitive to stress. J. Neurosci. Res. 67: 211-224.

Büeler, H., Fischer, M., Lang, Y., Bluethmann, H., Lipp, H.P., DeArmond, S.J., Prusiner, S.B., Aguet, M., and Weissmann, C., 1992. Normal development and behaviour of mice lacking the neuronal cell-surface PrP protein. Nature. 356: 577-582.

Büeler, H., Aguzzi, A., Sailer, A., Greiner, R.A., Autenreid, P., Aguet M., and Weissmann C., 1993. Mice devoid of PrP are resistant to scrapie. Cell. 73: 1339-1347.

Burns, C.S., Aronoff-Spencer E., Dunham, C.M., Lario, P., Avdievich, N.I., Antholine, W.E., Olmstead, M.M., Vrielink, A., Gerfen, G.J., Peisach, J., Scott, W.G., and Millhauser G.L. 2002. Molecular features of the copper binding sites in the octarepeat domain of the prion protein. Biochemistry. 41: 3991-4001.

Bush, A.I. and Tanzi, R.E. 2002. The galvanization of β-amyloid in Alzheimer's disease. Proc. Natl. Acad. Sci. U.S.A. 99: 7317-7319.

Campbell, T.A., Palmer, M.S., Will, R.G., Gibb, W.R., Luthert, P.J., and Collinge, J. 1996. A prion disease with a novel 96-base pair insertional mutation in the prion protein gene. Neurology. 46: 761-766.

Cereghetti, G.M., Schweiger, A., Glockshuber, R., and van Doorslaer, S. 2001. Electron paramagnetic resonance evidence for binding od Cu^{2+} to the C-terminal domain of the murine prion protein. Biophys. J. 81:516-525.

Chen, S.G., Teplow, D.B., Parchi, P., Teller, J.K., Gambetti, P., and Autilio-Gambetti, L. 1995. Truncated forms of the human prion protein in normal brain and in prion diseases. J. Biol. Chem. 270: 19173-19180

Colburn, R.W., and Maas, J.W. 1965. Adenosine triphosphate-metal-norepinephrine ternary complexes and catecholamine binding. Nature. 208: 37-41.

Colling, S.B., Collinge, J., and Jefferies, J.G.R. 1996. Hippocampal slices from prion protein null mice: disruption of Ca^{2+} activated K^+ currents. Neurosci. Lett. 209: 49-52.

Collinge, J., Whittington, M.A., Sidle, K.C., Smith, C.J., Palmer, M.S., Clarke, A.R., and Jefferys, J.G. 1994. Prion protein is necessary for normal synaptic function. Nature. 370: 295-297.

Daniels, M., and Brown, D.R. 2002. Purification and preparation of prion protein: synaptic superoxide dismutase. Methods Enzymol. 349: 258-267.

Demeter, I., Szokefalvi-Nagy, G., Varga, L., Keszthelyi, L., Hollos-Nagy, K., and Nagy A., 1979. Ion content of synaptic vesicles. Acta Biochim. Biophys. Acad. Sci. Hung. 14: 189-195.

Donne, D.G., Viles, J.H., Groth, D., Mehlorn, I., James, T.L., Cohen, F.E., Pruisner, S.B., Wright, P.E., and Dyson, H.J. 1997. Structure of the recombinant full-length hamster prion protein PrP(29-231). Proc. Natl. Acad. Sci. U.S.A. 94: 13452-13457.

Dupuis, L., Mbebi, C., Gonzalez de Aguilar, J.L., Rene, F., Muller, A., de Tapia, M., and Loeffler, J.P. 2002. Loss of prion protein in a transgenic model of amyotrophic lateral sclerosis. Mol. Cell Neurosci. 19:216-224.

Fenton, D.E. 1995. Biocoordination Chemistry. Oxford University Press, Oxford, UK.

Forloni, G., Angeretti, N., Chiesa, R., Monzani, E., Salmona, R. Bugiani, O., and Tagliavini, F. 1993, Neurotoxicity of a prion protein fragment. Nature. 362: 543-546.

Fournier, J.G., Escaig-Haye, F., and Grigoriev, V. 2000. Ultrastructural localization of prion proteins: physiological and pathological implications. Microsc. Res. Tech. 50: 76-88.

Gabrielsson, B., Robson, T., Norris, D., and Chung, S.H. 1986. Effects of divalent metal ions on the uptake of glutamate and GABA from synaptosomal fractions. Brian Res. 384: 218-223.

Goldfarb, L.G., Brown, P., McCombie, W.R., Goldgaber, D., Swergold, G.D., Wills, P.R., Cervenakova, L., Baron, H., Gibbs, C.J., and Gajdusek, D.C. 1991. Transmissible familial Creutzfeldt-Jakob disease associated with five, seven, and eight extra octapeptide coding repeats in the PRNP gene. Proc. Natl. Acad. Sci. U.S.A. 88: 10926-10930.

Goldfarb, L.G., Brown, P., Little, B.W., Cervenakova, L., Kenney, K., Gibbs, C.J., and Gajdusek, D.C. 1993. A new (two-repeat) octapeptide coding insert mutation in Creutzfeldt-Jakob disease. Neurology. 43: 2392-2394.

Groves, P., and Palczewska, M.2001. Cation binding properties of calretinin, an EF-hand calcium-binding protein. Acta Biochim. Pol. 48:113-119.

Guiroy, D.C., Williams, E.S., Yanagihara, R., and Gajdusek, D.C. 1991. Immunolocalization of scrapie amyloid PrP27-30 in chronic wasting disease of Rocky Mountain elk and hybrids of captive mule deer and white tailed deer. Neurosci. Lett. 126: 195-198.

Halliwell, B. 1992. Reactive oxygen species and the central nervous system. J. Neurochem. 59: 1609-1623.

Hasnain, S.S., Murphy, L.M., Strange, R.W., Grossmann, J.G., Clarke, A.R., Jackson, G.S., and Collinge, J. 2001. XAFS study of the high affinity copper binding site of human PrP(91-231) and its low resolution structure in solution. J. Mol. Biol. 311: 467-473.

Hegde, R.S., Mastrianni, J.A., Scott, M.R., Defea, K.D., Tremblay, P., and Torchia, M. 1998. A transmembrane form of the prion protein in neurodegenerative disease. Science. 279: 827-834.

Herms, J., Tings, T., Gall, S., Madlung, A., Giese, A., Siebert, H., Schurmann, P., Windl, O., Brose, N., and Kretzschmar H. 1999. Evidence of presynaptic location and function of the prion protein. J. Neurosci. 19: 8866-8875.

Highfield, R. 2002. Every Briton should have a CJD test, says scientist. Daily Telegraph, 2nd December 2002.

Hornshaw, M.P., McDermott, J.R., and Candy, J.M. 1995a. Copper binding to the N-terminal tandem repeat regions of mammalian and avian prion protein. Biochem. Biophys. Res. Comm. 207: 621-629.

Hornshaw, M.P., McDermott, J.R., Candy, J.M., and Lakey, J.H. 1995b. Copper binding to the N-terminal tandem repeat regions of mammalian and avian prion protein: structural studies using synthetic peptides. Biochem. Biophys. Res. Comm. 214: 993-999.

Hesse, L., Beher, D., Masters, C.L., and Multhaup, G. 1994. The beta A4 amyloid precursor protein binding to copper. FEBS Lett. 349: 109-116.

Hornemann, S., Korth, C., Oesch, B., Riek, R., Wider, G., Wurthrich, K, and Glockshuber, R. 1997. Recombinant full length murine prion protein mPrP(23-231): purification and spectroscopic characterization. FEBS Lett. 413, 277-281.

Jackson, G.S., Murray, I., Hosszu, L.L., Gibbs, N., Waltho, J.P., Clarke, A.R., and Collinge, J. 2001. Location and properties of metal-binding sites on the human prion protein. Proc. Natl. Acad. Sci. U.S.A. 98: 8531-8535.

Jobling, M.F., Huang, X., Stewart, L.R., Barnham, K.J., Curtain, C., Volitakis, I., Perugini, M., White, A.R., Cherny, R.A., Colin L. Masters, C.L., Barrow, C.J., Collins, S.J.,Bush, A.I., and Cappai, R. 2001 Copper and zinc binding modulates the aggregation and neurotoxic properties of the prion peptide PrP106-126. Biochemistry. 40: 8073-8084.

Kardos, J., Kovacs, I., Hajos, F., Kalman, M., and Simonyi, M. 1989. Nerve endings from rat brain tissue release copper upon depolarization. A possible role in regulating neuronal excitability. Neurosci. Lett. 103: 139-144.

Klamt, F., Dal-Pizzol, F., Conte da Frota, M.L., Walz, R., Andrades, M.E., da Silva, E.G., Brentani, R.R., Izquierdo, I., Fonseca and Moreira, J.C. 2001. Imbalance of antioxidant defense in mice lacking cellular prion protein. Free Radic. Biol. Med. 30: 1137-1144.

Knaus, K.J., Morillas, M., Swietnicki, W., Malone, M., Surewicz,W.K., and Yee, V.C. 2001. Crystal structure of the human prion protein reveals a mechanism for oligomerization. Nat. Struct. Biol. 8, 770-774.

Kramer, M.L., Kratzin, H.D., Schmidt, B., Römer, A., Windl, O., Liemann, S., Hornemann, S., and Kretzschmar, H. 2001. Prion protein binds copper within the physiological concentration range. J. Biol. Chem. 276: 16711-16719.

Kuwahara, C., Takeuchi, A.M., Nishimura, T., Haraguchi, K., Kubosaki, A., Matsumoto, Y., Saeki, K., Matsumoto, Y.,

Yokoyama, T., Itohara, S., and Onodera, T. 1999. Prions prevent neuronal cell line death. Nature. 400: 225-226.

Landriscina, M., Bagala, C., Mandinova, A., Soldi, R., Micucci, I., Bellum, S., Prudovsky, I., and Maciag, T. 2001. Copper induces the assembly of a multiprotein aggregate implicated in the release of fibroblast growth factor 1 in response to stress. J. Biol. Chem. 276: 25549-25557.

Lehmann, S. 2002. Metal ions and prion disease. Curr. Opin. Chem. Biol. 6:187-192.

Linder, M.C. 1991. Biochemistry of Copper. Plenum Press, New York.

Lovell, M.A., Robertson, J.D., Teesdale, W.J., Campbell, J.L. and Markesbery, W.R. 1998. Copper, iron and zinc in Alzheimer's disease senile plaques. J. Neurol. Sci. 158: 47-52.

Marklund, S.L. 1982. Human copper-containing superoxide dismutase of high molecular weight. Proc. Natl. Acad. Sci. U.S.A. 79: 7634-7638.

McKenzie, D., Bartz, J., Mirwald, J., Olander, D., Marsh, R., and Aiken J. 1998. Reversibility of scrapie inactivation is enhanced by copper. J. Biol. Chem. 273: 25545-25547.

McLoughlin, D.M., Standen, C.L., Lau, K.F., Ackerley, S., Bartnikas, T.P., Gitlin, J.D., and Miller, C.C. 2001. The neuronal adaptor protein X11alpha interacts with the copper chaperone for SOD1 and regulates SOD1 activity. J. Biol. Chem. 276: 9303-9307.

McMahon, H.E., Mange, A., Nishida, N., Creminon, C., Casanova, D., and Lehmann, S. 2001. Cleavage of the amino terminus of the prion protein by reactive oxygen species. J. Biol. Chem. 276: 2286-2291.

Miura, T., Hori-I, A., Mototani, H., and Takeuchi, H. 1999. Raman spectroscopic study on copper(II) binding mode of prion octapeptide and its pH dependence. Biochemistry. 38: 11560-11569.

Mueller, H.T., Borg, J.P., Margolis, B., and Turner, R.S. 2000. Modulation of amyloid precursor protein metabolism by X11alpha /Mint-1. A deletion analysis of protein-protein interaction domains. J. Biol. Chem. 275:39302-39306.

Multhaup, G., Schlicksupp, A., Hesse, L., Beher, D., Ruppert, T., Masters, C.L., and Beyreuther, K. 1996. The amyloid precursor protein of Alzheimer's disease in the reduction of copper(II) to copper(I). Science. 271: 1406-1409.

Multhaup, G., Dieter, H.H., Beyreuther, K., and Bayer, T.A. 2002. Role of copper and other transition metal ions in the pathogenesis of Parkinson's disease, prion diseases, familial amytrophic lateral sclerosis and Alzheimer's disease. In: Handbook of Copper and Pharmacology and Toxicology. E.J. Massaro, ed. Humana Press, New Jersey.

Naslavsky, N., Stein, R., Yanai, A., Friedlander, G., and Taraboulos, A. 1997. Characterization of detergent-insoluble complexes containing the cellular prion protein and its scrapie isoform. J. Biol. Chem. 272: 6324-6331.

Nunziante, M., Gilch, S., and Schatzl, H.M. 2003. Essential role of the prion protein N-terminus in subcellular trafficking and half-life of PrPc. J. Biol. Chem. 278: 3726-3734.

Ookawara, T., Imazeki, N., Matsubara, O., Kizaki, T., Oh-Ishi, S., Nakao, C., Sato, Y., and Ohno, H. 1998. Tissue distribution of immunoreactive mouse extracellular superoxide dismutase. Am. J. Physiol. 275: C840-7.

Owen, F., Poulter, M., Shah, T., Collinge, J., Lofthouse, R., Baker, H., Ridley, R., McVey, J., and Crow, T.J. 1990. An in-frame insertion in the prion protein gene in familial Creutzfeldt-Jakob disease. Brain Res. Mol. Brain Res. 7: 273-276.

Owen, F., Poulter, M., Collinge, J., Leach, M., Lofthouse, R., Crow, T.J., and Harding, A.E. 1992. A dementing illness associated with a novel insertion in the prion protein gene. Brain Res. Mol. Brain Res. 13: 155-157.

Paik, S.R., Shin, H.J., Lee, J.H., Chang, C.S., and Kim, J. 1999. Copper(II)-induced self-oligomerization of alpha-synuclein. Biochem. J. 340: 821-828.

Palmer, M.S., Mahal, S.P., Campbell, T.A., Hill, A.F., Sidle, K.C., Laplanche, J.L., and Collinge. J. 1993. Deletions in the prion protein gene are not associated with CJD. Hum. Mol. Genet. 2: 541-544.

Pan, K.M., Stahl, N., and Prusiner, S.B. 1992. Purification and properties of the cellular prion protein from Syrian hamster brain. Protein Sci. 1: 1343-1352.

Pattison, I.H., and Jebbett, J.N. 1971a. Histopathological similarities between scrapie and cuprizone toxicity in mice. Nature. 230:115-117.

Pattison, I.H., and Jebbett, J.N. 1971b. Clinical and histological observations on cuprizone toxicity and scrapie in mice. Res Vet Sci. 12: 378-380.

Pattison, I.H., and Jebbett, J.N. 1973. Unsuccessful attempts to produce disease with tissues from mice fed on a diet containing cuprizone. Res Vet Sci. 14: 128-130.

Pauly, P.C., and Harris, D.A. 1998. Copper stimulates endocytosis of the prion protein. J. Biol. Chem. 273: 33107-33110.

Priola, S.A., and Chesebro, B. 1998. Abnormal properties of prion protein with insertional mutations in different cell types. J. Biol. Chem. 273: 11980-11985.

Pruisner, S.B. 1982. Novel proteinaceous infectious particles cause scrapie. Science. 216: 136-144.

Puig, S., Thiele, D.J. 2002. Molecular mechanisms of copper uptake and distribution. Curr. Opin. Chem. Biol. 6: 171-180.

Purdey, M. 2000. Ecosystems supporting clusters of sporadic TSEs demonstrate excesses of the radical generating divalent cation manganese and deficiencies of antioxidant cofactors; Cu, Se, Fe, Zn. Does a foreign cation substitution at PrP's Cu domain initiate TSE? Med. Hypoth. 54: 278-306.

Qin, K., Yang, D.-S., Yang, Y., Christi, M.A., Meng, L.-J., Kretzschmar, H.A., Yip, C.M., Fraser, P.E., and Westaway, D. 2000. Copper(II)-induced conformational changes and protease resistance in recombinant and cellular PrP. J. Biol. Chem. 275: 19121-19131.

Qin, K., Yang, D.S., Mastrangelo, P., and Westaway, D. 2002. Mapping Cu(II) binding sites in the prion protein by diethyl pyrocarbonate modification and matrix-associate laser desorption ionisation-time of flight (MALDI-TOF) mass spectrometric footprinting. J. Biol. Chem. 277: 1981-1990.

Quaglio, E., Chiesa, R., and Harris, D.A. 2001 Copper converts the cellular prion protein into a protease-resistant species that is distinct from the scrapie isoform. J. Biol. Chem. 276: 11432-11438.

Ruiz, F.H., Silva, E., and Inestrosa, N.C. 2000. The N-terminal tandem repeat region of the human prion protein reduces Cu: the role of tryptophan residues. Biochem. Biophys. Res. Comm. 269: 491-495.

Salès, N., Rodolfo, K., Hassig, R., Faucheux, B., Di Giamberardino, L., and Moya, K.L. 1998. Cellular prion protein localization in rodent and primate brain. Eur. J. Neurosci. 10: 2464-2471.

Salès, N. Hässig, R., Rodolfo, K., Di Giamberardino, L., Traiffort, E., Ruat, M., Fre´tier, P., and Moya, K.L. 2002. Developmental expression of the cellular prion protein in elongating axons. Eur. J. Neurosci. 15 : 1163-1170.

Schafer, B.W., Fritschy, J.M., Murmann, P., Troxler, H., Durussel, I., Heizmann, C.W., and Cox, J.A. 2000. Brain S100A5 is a novel calcium-, zinc-, and copper ion-binding protein of the EF-hand superfamily. J. Biol. Chem. 275: 30623-30630.

Schätzl, H.M., Da Costa, M. Taylor, L., Cohen, F.E., and Pruisner, S.B.1995. Prion protein variation among primates. J. Mol. Biol. 245: 362-374.

Selvaggini, C., De Gioia, L., Cantu, L., Ghibaudi, E., Diomede, L., Passerini, F., Forloni, G., Bugiani, O., Tagliavini, F., and Salmona, R. 1993. Molecular characteristics of a protease resistano amyloidogenic and neurotoxic peptide homologous to residues 106-126 of the prion protein. Biochem. Biophys. Res. Comm. 194: 1380-1386.

Sendo, F., Suzuki, K., Watanabe, T., Takeda, Y., and Araki, Y. 1998. Modulation of leukocyte transendothelial migration by integrin-associated glycosyl phosphatidyl inositol (GPI)-anchored proteins. Inflamm. Res. 47 Suppl 3: S133-136.

Shiraishi, N., and Nishikimi, M. 1998. Suppression of copper-induced cellular damage by copper sequestration with S100b protein. Arch. Biochem. Biophys. 357: 225-230.

Shiraishi, N., Ohta, Y., and Nishikimi, M. 2000. The octarepeat region of prion protein binds Cu(II) in the redox inactive state. Biochem. Biophys. Res. Comm. 267: 398-402.

Shiraishi, N., and Nishikimi, M. 2002. Carbonyl formation on a copper bound prion protein fragment, PrP23-98, associated with its dopamine oxidase activity. FEBS Lett. 511: 118-122.

Shyng, S.L., Heuser, J.E., and Harris, D.A. 1994. A glycolipid-anchored prion protein is endocytosed via clathrin-coated pits. J. Cell Biol. 125:1239-1250.

Shyng, S.L., Moulder, K.L., Lesko, A., and Harris, D.A. 1995. The N-terminal domain of a glycolipid-anchored prion protein is essential for its endocytosis via clathrin-coated pits. J. Biol. Chem. 270: 14793-14800.

Strausak, D., Mercer, J.F., Dieter, H.H., Stremmel, W., and Multhaup, G. 2001. Copper in disorders with neurological symptoms: Alzheimer's, Menkes, and Wilson diseases. Brain Res Bull. 55: 175-185.

Rae, T.D., Schmidt, P.J., Pufahl, R.A., Culotta, V.C., and O'Halloran, T.V. 1999. Undetectable intracellular free copper: the requirement of a copper chaperone for superoxide dismutase. Science. 284: 805-808.

Requena, J.R., Groth, D., Legname, G., Stadtman, E.R., Prusiner, S.B., and Levine, R.L. 2001. Copper-catalyzed oxidation of the recombinant SHa(29-231) prion protein. Proc. Natl. Acad. Sci. U.S.A. 98: 7170-7175.

Riek, R., Hornemann, S., Wider, G., Billeter, M., Glockshuber, R., and Wurthrich, K. 1996. NMR structure of the mouse prion protein domain PrP(121-231). Nature. 382, 180-182.

Rodolfo, K., Hassig, R., Moya, K.L., Frobert, Y., Grassi, J., and Di Giamberardino, L. 1999. A novel cellular prion protein isoform present in rapid anterograde axonal transport. Neuroreport. 10: 3639-3644.

Rosenzweig, A.C. 2001. Copper Delivery by metallochaperone proteins. Acc. Chem. Res. 34: 119-128

Tanzi, R.E., Petrukhin, K., Chernov, I., Pellequer, J.L., Wasco, W., Ross, B., Romano, D.M., Parano, E., Pavone, L., Brzustowicz, L.M., Devoto,M., Peppercorn,J., Bush, A.I., Sternlieb, I., Pirastu, M., Gusella, J:F., Evgrafov, O., Penchaszadeh, G.K., Honig, B. Edelmann, I.S., Soares, M.B., Scheinberg,I.H., and Gilliam, T.C. 1993. The Wilson disease gene is a copper transporting ATPase with homology to the Menkes disease gene. Nat Genet. 5: 344-350.

Tobler, I., Gaus, S.E., Deboer, T., Achermann, P., Fischer, M., Rulicke, T., Moser, M., Oesch, B., McBride, P.A., and Manson, J.C. 1996. Altered cicadian rhythms and sleep in mice devoid of prion proteins. Nature. 380: 639-642.

Viles, J.H., Cohen, F.E., Pruisner, S.B., Goodin, D.B., Wright, P.E., and Dyson, H.J. 1997. Copper binding to the prion protein structural implications of four cooperative binding sites. Proc. Natl. Acad. Sci. U.S.A. 96: 2042-2047.

Vlachová, V., Zemkova, H., and Vyklickt Jr., L. 1996. Copper modulation of NMDA responses in mouse and rat cultured hippocampal neurons. Eur. J. Neurosci. 8: 2257-2264.

Wadsworth, J.D.F., Hill, A.F., Joiner, S., Jackson, G.S., and Clarke, A.R., Collinge, J. 1999. Strain-specific prion-protein conformation determined by metal ions. Nat. Cell. Biol. 1: 55-59.

Waggoner, D.J., Drisaldi, B., Bartnikas, T.B., Casareno, R.L.B., Prohaska, J.R., Gitlin, J.D. and Harris, D.A. 2000 Brain copper content and cuproenzyme activity do not vary with prion protein expression level. J. Biol. Chem. 275: 7455-7458.

White, A.R., Collins, S.J., Maher, F., Jobling, M.F., Stewart, L.R., Thyer,J.M., Beyreuther, K., Masters, C.L., and Cappai, R. 1999.

Prion protein deficient neurons reveal lower glutathione reductase activity and increased susceptability to hydrogen peroxide toxicity. Am. J. Pathol. 155: 1723-1730.

Whittal, R.M., Ball, H.L., Cohen, F.E., Burlingame, A.L., Pruisner, S.B. and Baldwin, M.A. 2000. Copper binding to octarepeat peptides of the prion protein monitered by mass spectrometry. Protein Sci. 9: 332-342.

Wildegger, G., Liemann, S., and Glockshuber, R. 1999. Extremely rapid folding of the C-terminal domain of the prion protein without kinetic intermediates. Nat. Struct. Biol. 6, 550-553.

Wong, B.S., Wang, H., Brown, D.R., and Jones, I.M. 1999. Selective oxidation of methionine residues in prion proteins. Biochem. Biophys. Res. Commun. 259: 352-355.

Wong, B.S., Venien-Bryan, C., Williamson, R.A., Burton, D.R., Gambetti, P., Sy, M.S., Brown, D.R., and Jones, I.M. 2000a. Copper refolding of prion protein. Biochem. Biophys. Res. Commun. 276: 1217-1224.

Wong, B.S., Pan, T., Liu, T., Li, R.L., Gambetti, P., and Sy, M.S. 2000b. Differential contribution of superoxide dismutase activity by prion protein *in vivo*. Biochem. Biophys. Res. Commun. 273: 136-139.

Wong, B.-S., Chen, S.G., Colucci, M., Xie, Z., Pan, T., Liu, T., Li, R., Gambetti, P., Sy, M.-S., and Brown, D.R. 2001a. Aberrant metal binding by prion protein in human prion disease. J. Neurochem. 78: 1400-1408.

Wong, B.-S., Brown, D. R., Pan, T., Whiteman, M., Liu, T., Bu, X., Li, R., Gambetti, P., Olesik, J., Rubinstein, R. and Sy, M.-S. 2001b. Oxidative impairment in scrapie-infected mice is associated with brain metal perturbations and altered ani-oxidantion activities. J.. Neurochem. 79: 689-698.

Wopfner, F., Weidenhofer, G., Schneider, R., von Brunn, A., Gilch, S., Schwarz, T.F., Werner, T., and Schatzl, H.M. 1999. Analysis of 27 mammalian and 9 avian PrPs reveals high conservation of flexible regions of the prion protein. J. Mol. Biol. 289: 1163-1178.

Zahn, R., Liu, A., Luhrs, T., Riek, R., von Schroetter, C., Lopez Garcia, F., Billeter, M., Calzolai, L.,Wider, G., and Wurthrich, K.. 2000. NMR solution structure of the human prion protein. Proc. Natl. Acad. Sci. U.S.A. 97: 145-150.

From: Prions and Prion Diseases: Current Perspectives
Edited by: Glenn C. Telling

Chapter 2

Binding and Conversion Reactions Between Prion Protein Isoforms

Byron Caughey, Jay R. Silveira,
Jonathan O. Speare and
Gerald S. Baron

Abstract

A fundamental process in the transmissible spongiform encephalopathies (TSE) or prion diseases is the conversion of PrP^C to abnormal and usually protease-resistant forms such as PrP^{Sc}. A variety of *in vitro* experiments have shown that TSE-associated forms of PrP can cause PrP^C to convert to forms that are similarly protease-resistant. These observations have shown that pathologic forms of PrP have at least limited capacity to propagate themselves, which would be necessary for them to be infectious proteins. However, it has not been shown that new infectivity is generated in any such reactions. Nonetheless, PrP conversion reactions are highly specific and may account, at least in part, for TSE species barriers and the propagation of strains. Such *in vitro* conversion systems have yielded insights into the molecular mechanisms of TSE disease and are being exploited as screens for anti-TSE drugs and as bases for diagnostic tests.

1. Introduction

Transmissible spongiform encephalopathies (TSEs) or prion diseases are critically dependent upon the conversion of PrP^C to neuropathologic forms. The primary difference between PrP^C and the scrapie-associated form, PrP^{Sc}, appears to be conformational. At the cellular level in the case of scrapie-infected cells, PrP^{Sc} is formed post-translationally after mature PrP^C is translocated to the cell surface (Caughey and Raymond, 1991; Harris, 2001). According to the popular protein-only or prion hypotheses, PrP^{Sc} is the infectious agent that propagates itself by causing the conversion of PrP^C to PrP^{Sc}. Although the mechanism and full neuropathologic implications of PrP^{Sc} formation *in vivo* remain uncertain, studies of PrP conversion under a variety of conditions have provided important insights into the ways in which PrP^C can adopt a PrP^{Sc}-like protease-resistant state upon interaction with PrP^{Sc}. Related studies have revealed thar PrP^C can also misfold spontaneously under certain circumstances (reviewed in (Glockshuber, 2001; Alonso and Daggett, 2001)). Sporadic and familial TSE diseases may be initiated by the spontaneous misfolding of PrP^C. In contrast, misfolding of PrP^C that is induced by preexisting PrP^{Sc} is likely to be fundamental to TSEs of infectious origin and to the propagation of spontaneously formed PrP^{Sc} in sporadic and familial TSE diseases. The present chapter will focus on recent developments in the study of PrP^{Sc}-induced misfolding of PrP^C and its relevance to TSE diseases.

Pathologic forms of PrP are often designated PrP^{Sc}, for PrP-scrapie, as if this were a single entity or state of PrP. However, abnormal TSE-linked forms of PrP vary considerably in terms of protease-resistance, association with infectivity, and apparent neurotoxicity. Here we will use the term PrP^{Sc} to refer to scrapie-associated PrP and PrP-res to designate PrP isoforms with the usual partial protease resistance of PrP^{Sc}, whether they are scrapie-associated or not. The term PrP^C will be used to refer to PrP in its normal structure and conformation and the term PrP-sen to refer generically to protease-sensitive forms of PrP, whether normal (i.e. PrP^C) or not (e.g., various recombinant forms).

2. PrP Binding and Conversion Reactions

PrP^C is usually found as a monomeric glycosylphosphatidylinositol (GPI)-linked membrane glycoprotein that is soluble in mild detergents. When incubated with PrP^{Sc} or other types of PrP-res, the PrP^C tends to bind to the PrP-res aggregate in a protease-sensitive state and convert to a protease-resistant state that, so far, has been indistinguishable from that of the original PrP-res (DebBurman *et al.*, 1997; Horiuchi *et al.*, 1999; Horiuchi *et al.*, 2000; Callahan *et al.*, 2001). This reaction can occur under a variety of conditions. The simplest, most biochemically defined reactions contain mixtures of largely purified PrP-sen and PrP-res preparations and can be stimulated by chaotropes, detergents and/ or chaperone proteins (Kocisko *et al.*, 1994; DebBurman *et al.*, 1997; Horiuchi *et al.*, 1999). In the absence of denaturants and detergents, conversion reactions between isolated PrP isoforms can be stimulated by sulfated glycans and/or elevated temperature (Wong *et al.*, 2001). Of intermediate complexity are conversion reactions between membrane-bound forms of PrP-sen and PrP-res (Baron *et al.*, 2002; Baron and Caughey, 2003). These cell-free reactions, as well as cellular systems in which the acute formation of PrP-res in intact cells is monitored after exposure to PrP^{Sc}-containing brain extracts (Korth *et al.*, 2000; Vorberg and Priola, 2002) provide new experimental models of the cell-to-cell propagation of PrP^{Sc} and infection. Conversion reactions have also been demonstrated in intact scrapie-infected brain slices (Bessen *et al.*, 1997), cellular extracts (Saborio *et al.*, 1999; Vorberg and Priola, 2002) and brain homogenates (Saborio *et al.*, 2001).

Given the uncertainties about the precise nature of the infectious agent (prion) and its propagation, it remains important to determine whether some sort of "conversion" of PrP^C alone or in combination with other molecules forms *bona fide* new TSE infectivity. In all but the most recent cell-free PrP conversion reactions, the yield of PrP-res conversion product has been less than that of the already infectious PrP-res used to drive the reaction, making it technically difficult to detect an increase in infectivity titer by bioassay in animals. To try to circumvent this problem, cross species conversion reactions have been performed whereby PrP-res (or "seed") from one species (e.g. hamster 263K scrapie) that does not itself induce illness in another species (e.g., mice), is used to drive conversion of chimeric PrP-sen molecules that should be compatible with both species (Scott *et al.*, 1992). In this scenario, any newly generated chimeric PrP-res should be capable of making the recipient species (mice) sick if PrP-res were, in fact, the infectious prion. However, attempts to demonstrate new infectivity under a variety of conditions have failed in our lab (unpublished results) and another (Hill *et al.*, 1999).

In recent attempts to amplify PrP-res in an *in vitro* conversion reaction, Saborio and Soto developed the protein misfolding cyclic amplification (PMCA) system (Saborio *et al.*, 2001). Detergent extracts of TSE-infected brain homogenate are mixed with vast excesses of similar extracts of PrP^C-containing normal brain tissue and subjected to repeated cycles of sonication and incubation. Such reactions were reported to yield >30-fold amplifications of the PrP-res. This tantalizing result has raised the possibility that the PMCA procedure might be useful for amplifying the detection of PrP-res in TSE diagnostic tests. Because the reported yield of newly formed PrP-res in this crude system was much higher than the yields observed in previously described conversion reactions between purified PrP isoforms, Saborio and Soto suggested that unidentified auxiliary factors provided in the crude normal brain homogenate might be important. More recently, it was reported that the PMCA-type of reaction can also work (>10-fold amplification) without denaturing detergents and sonication (Lucassen *et al.*, 2003). Furthermore, another method of brain homogenate-based conversion has been reported (Zou and Cashman, 2002). In this system, acidification and detergent treatment of PrP^C in normal brain homogenates promoted the formation of PrP^{Sc}-like, PK-resistant PrP species. In the presence of trace quantities of PrP^{Sc} derived from CJD brain homogenates, this technique amplified the signal of PK-resistant PrP 3-10 fold.

Given the levels of amplification afforded by the PMCA and related techniques, it should be possible to test whether or not the newly formed PrP-res is infectious; however, no such data has been published. In any case, the complexity of the whole brain extracts used in the current PMCA PrP-res amplification scheme will leave open the question of whether factors besides PrP are critical in the composition and/or formation of TSE infectivity. Ultimately, to fully understand the nature of the TSE agent and its propagation mechanism, it will be necessary to reconcile biochemically defined PrP conversion reactions, which so far have not proven to generate new TSE infectivity, and the TSE agent propagation that is known to occur readily in intact TSE-infected cells and animals.

Although none of the products of cell-free PrP conversion reactions are known to be infectious, correlations have been observed between infectivity and cell-free converting activity (Caughey *et al.*, 1997). Furthermore, a wealth of data from transgenic mice and TSE-infected cell culture, suggest that interactions and molecular compatibility between PrP-res and PrP^C are important in PrP-res formation and the transmission of TSE diseases (reviewed in (Priola, 2001; Asante and Collinge, 2001)). Hence, it remains important to define how the different PrP isoforms interact under various circumstances.

3. Selectivity and Membrane Dependence of Interactions Between PrP-sen and PrP-res

Selective binding of PrP-res to PrP-sen over a vast excess of other proteins has been observed directly in lysates or culture supernatants from ^{35}S-methionine-labeled cells expressing soluble PrP-sen lacking the GPI anchor (GPI⁻) (Horiuchi *et al.*, 1999). Recent studies of the interaction and conversion of membrane-bound PrP species did not detect conversion of membrane-anchored (GPI⁺) PrP^C by exogenous PrP-res (Baron *et al.*, 2002), suggesting that membrane attachment of PrP^C may limit accessibility to PrP-res. The inaccessibility may be due to the location of apparent initial PrP-res binding surfaces near the C-terminus of PrP-sen which may be blocked by attachment to the membrane via the C-terminal GPI anchor (Horiuchi *et al.*, 1999). Alternatively, direct interactions between membrane surfaces and the PrP^C polypeptide chain may block its access to PrP^{Sc} (Morillas *et al.*, 1999; Sanghera and Pinheiro, 2002), although data from cell-free studies argue against this possibility (Baron and Caughey, 2003).

4. Sequence Specificity of Interactions Between PrP Isoforms

Striking sequence specificity is often observed when 'heterologous' cell-free conversion reactions have been performed using PrP-res of one species/sequence and PrP-sen of another. The rank order of the efficiencies of such reactions seem to correlate roughly with relative susceptibilities of hosts to cross-species TSE infections (reviewed in Caughey *et al.*, 2001). Cell-free conversion assays have been used tentatively to gauge the relative susceptibilities of various hosts to

TSE agents from different source species or genotypes (Raymond *et al.*, 1997; Bossers *et al.*, 2000; Raymond *et al.*, 2000) when *in vivo* transmission data is unavailable. For instance, little is known about the transmissibility of CWD of deer and elk to noncervid species (Bartz *et al.*, 1998; Miller *et al.*, 2000). In cell-free conversion reactions, CWD-associated PrP-res (PrP^{CWD}) of cervids readily induces the conversion of cervid PrP-sen molecules to the protease-resistant state consistent with the known transmissibility of CWD between cervids (Raymond *et al.*, 2000). In contrast, PrP^{CWD} –induced conversions of human, bovine and ovine PrP-sen were 14-100 fold, 5-12 fold, and 2-3 fold less efficient, respectively, than the most efficiently converted cervid PrP-sen. This conversion incompatibility may limit the susceptibility of these non-cervid species to CWD. In addition, the data suggest that the rank order of susceptibilities to CWD, would be humans < cattle < sheep < cervids. However, although these interspecies *in vitro* conversion experiments are relatively rapid, they are only tentative surrogates for *in vivo* transmission and epidemiological data and should not be used as concrete evidence of species susceptibility or lack thereof.

Further mechanistic studies of the species specificity of conversion reactions have revealed that in some interspecies combinations the binding of PrP-sen to heterologous PrP^{Sc} occurs much more efficiently than the conversion to PrP-res (Horiuchi *et al.*, 2000). Thus, the species specificity of the conversion reaction may be determined more by the conversion step than the initial binding step. Nonetheless, the binding of heterologous, non-converting PrP-sen molecules to PrP^{Sc} can interfere with the conversion of homologous PrP-sen, arguing that the convertible and non-convertible PrP-sen molecules can compete for the same site. Such interference effects could explain reductions in PrP-res accumulation and increased incubation periods in hosts that co-express different PrP^{C} molecules such as humans who are heterozygous at PrP codon 129 (Prusiner, 1998; Priola, 1999). Studies using mouse/hamster chimeric PrP have shown that the central part of the PrP-sen molecule, which encompasses three amino acid substitutions at mouse/hamster residues 138/139, 154/155, and 169/170, is important in the conversion of PrP-sen to PrP-res (Prusiner, 1998; Priola, 1999; Priola *et al.*, 2001). Hence, critical interactions in the vicinity of these residues on PrP-sen and/or PrP^{Sc} may occur as part of the conversion step.

5. TSE strains and the Conformational Fidelity of PrP Conversions

Variations in PrP-res glycoform ratios, conformations, degrees of protease-resistance and aggregation states are associated with different TSE strains (reviewed in Horiuchi and Caughey, 1999). The mechanism by which these strain-dependent characteristics of PrP-res are maintained within individual host species in the absence of PrP sequence variation remains unclear. *In vitro* conversion reactions with largely purified PrP molecules have shown that different PrP^{Sc} "strains" can impose their different conformations on a single species of unglycosylated PrP-sen. This conformational fidelity provides evidence that the strain-specific conformers of PrP-res are faithfully propagated (or templated) through direct PrP^C-PrP-res interactions (Bessen *et al.*, 1995). More recent studies showed that mouse PrP^{Sc} strains can preferentially convert different PrP^C glycoforms from the complex pool of glycosylated PrP molecules that are produced by cells (Vorberg and Priola, 2002). Because the pool of PrP^C glycoforms can vary between cell types and within subcellular compartments, strain variations in the preferred cellular and subcellular sites of conversion can modulate the repertoire of conversion products. Thus, the propagation of TSE strain-dependent forms of PrP-res may be determined by multiple, self-propagating, and perhaps mutually exclusive, conformations or ordered aggregation states of PrP-res that preferentially select certain PrP^C glycoforms from various cellular pools. The binding of other ligands such as metal ions and sulfated glycosaminoglycans also might add to strain-associated PrP-res diversity. For instance, copper and zinc ions have been shown to alter the site at which PrP-res is cleaved *in vitro* by PK (Wadsworth *et al.*, 1999). The interplay between PrP-res conformation, glycoform selection and ligand binding may cause PrP-res strains to target and be most efficiently propagated within subsets of cells that can provide the preferred glycoforms and ligands. This cellular targeting, as well as differences in susceptibility of target cells to the toxic effects of PrP^{Sc} accumulation may, in turn, establish strain-dependent patterns of pathological lesions in the central nervous system (Bruce *et al.*, 1989; Caughey, 1991).

6. Mechanism of Conversion

Without full knowledge of the three-dimensional structure of PrP-res and the minimal essential conditions for continuous conversion, it is impossible to fully describe the PrP conversion mechanism. Nonetheless, mechanistic studies have shown that the PrP conversion is induced by PrP^{Sc} aggregates/polymers and not soluble, monomeric forms of PrP (Caughey *et al.*, 1995) (J. Silveira and B. Caughey, unpublished data) and that newly converted PrP molecules become bound to the polymers (Bessen *et al.*, 1997; Callahan *et al.*, 2001). The greatly increased β-sheet content in PrP-res, as well as the fact that conversion can be stimulated by a variety of treatments that affect protein conformation, suggest that the conversion involves major conformational changes in addition to the binding of PrP^C to PrP-res. PrP-sen appears to first bind to PrP-res via a localized site near in space to the C-terminal end of the third helix and then is more slowly converted to the protease-resistant state (Bessen *et al.*, 1997; DebBurman *et al.*, 1997; Horiuchi *et al.*, 1999). Modeling studies predicted a preeminent role of helix 1 of PrP^C in the conformational change (Morrissey and Shakhnovich, 1999). Experiments have since shown that mutations of the helix 1 aspartates (D144 and D147) to neutral residues, which eliminate stabilization by intrahelix salt bridges, can enhance conversion efficiencies by several fold (Speare *et al.*, 2003). Although it has been postulated that conversion involves breakage of the single PrP^C disulfide bond and formation of intermolecular disulfide-linked polymers (Welker *et al.*, 2001) or domain-swapped dimers (Knaus *et al.*, 2001), recent cell-free conversion reactions suggest that this need not be the case (Welker *et al.*, 2002). Taken together, these and various other observations, such as the formation of amyloid fibrils by PrP-res, are consistent with an autocatalytic, templated or seeded polymerization mechanism (reviewed in Caughey *et al.*, 2001). Seeded or autocatalytic protein polymerization mechanisms are well-precedented for yeast prion propagation and conventional amyloid fibril formation by other proteins and peptides (Caughey *et al.*, 2001; Wickner *et al.*, 2001; Serio and Lindquist, 2001). With TSE diseases, the preponderance of experimental evidence supports the idea that salient properties of PrP^{Sc}, i.e., partial protease-resistance, high β-sheet content, association with infectivity, and ability to cause PrP-sen to convert to PrP-res are dependent upon its being in an oligomeric state (reviewed in Caughey *et al.*, 2001). Full dissociation of PrP^{Sc} usually, if not always, results in the loss of these properties. PrP^{Sc} aggregates tend to be quite stable *in vitro* as indicated by the lack of dissociation of PrP monomers from PrP^{Sc} aggregates under nondenaturing conditions (Callahan *et al.*, 2001).

There are presumably two important stages in PrP-res formation, as there are in well-characterized seeded protein polymerizations (Jarrett and Lansbury, Jr., 1993; Lansbury, Jr. and Caughey, 1995). The spontaneous formation of PrP-res from mutant PrP^C is probably an important rate-determining step in familial TSE diseases. This step is analogous to a nucleation phase, which requires a rare or slow association of monomers to form a stable nucleus or seed. In TSEs of infectious origin, this step is bypassed altogether because of the introduction of PrP-res from an outside source. The most relevant mechanism in this case would then be PrP-res-induced conversion of PrP^C, analogous to the growth phase of seeded polymerizations. It should be borne in mind, however, that hosts may also have protective mechanisms that strongly modulate the rates of PrP-res initiation and accumulation.

7. PrP Conversion in Association with Membranes

Seeded protein polymerizations usually take the form of linear fibrils or bundles of fibrils. However, in the case of PrP-res formation it remains possible that the predominant mechanism involves a nonfibrillar autocatalytic assembly of GPI-anchored PrP molecules within the two-dimensional plane of membranes (Caughey, 2001). This has been suggested in part by the two-dimensional sheet-like arrays of PrP^{Sc} that have been observed in addition to fibrils in purified preparations of PrP^{Sc} (Wille *et al.*, 2002). It seems possible that such arrays, or short, laterally bundled protofibrils, might also be possible on membranes. Membranes might also stabilize monomeric forms of PrP^{Sc}; however such a species has not been clearly documented. Conversion reactions between GPI-anchored, membrane bound PrP isoforms revealed that conversion is not efficient when PrP-sen and PrP^{Sc} are attached to separate membrane vesicles. However, conversion can be detected when the vesicles are fused or PrP^C is detached from its GPI anchor (Baron *et al.*, 2002). PrP-sen can also bind to membrane vesicles in a GPI-independent mode (Morillas *et al.*, 1999; Sanghera and Pinheiro, 2002), in which case it is susceptible to conversion by PrP^{Sc} in separate vesicles (Baron and Caughey, 2003). These observations not only provide evidence that PrP conversion can occur within the context of a membrane, but also suggest that, in the process of infecting new cells, exogenous PrP^{Sc} must somehow be inserted into membranes of recipient cells in a manner that allows appropriate contacts between PrP^{Sc} and PrP^C molecules. These findings are consistent with evidence that contact between PrP^{Sc} and PrP^C is made via surfaces that are close in space to the C-terminal GPI-anchor in the folded PrP^C structure (Horiuchi *et al.*, 1999; Horiuchi *et al.*, 2001). Access to such surfaces by exogenous PrP molecules presumably would be limited by GPI anchoring to membranes (Baron and Caughey, 2003).

8. Solid Phase, High-throughput Conversion Reactions

Recently, we have developed solid-phase binding and conversion reactions in which PrP-res is fixed to the surface of 96-well plates and labeled PrP-sen is added in solution (Maxson *et al.*, 2003). These assays provide relatively rapid and high-throughput means of studying the PrP-res-PrP-sen interactions. One practical application of the solid phase reactions is the screening of inhibitors of conversion (Caughey *et al.*, 1998; Caughey *et al.*, 2003). Typically, we have first screened for inhibitors of PrP^{Sc} formation using scrapie-infected cell cultures. Since the cell culture experiments do not establish whether inhibitors act directly or indirectly on PrP conversion, we have used the solid-phase assays to look for direct inhibition of conversion. Given that the PrP-sen to PrP-res conversion is a major therapeutic and/or prophylactic target for TSE diseases, the availability of high throughput screens of conversion inhibitors should expedite the development of effective anti-TSE therapies.

References

Alonso, D.O. and Daggett, V. 2001. Simulations and computational analyses of prion protein conformations. Adv. Protein Chem. 57: 107-137.

Asante, E.A. and Collinge, J. 2001. Transgenic studies of the influence of the PrP structure on TSE diseases. Adv. Protein Chem. 57: 273-311.

Baron, G.S. and Caughey, B. 2003. Effect of glycosylphosphatidylinositol anchor-dependent and - independent prion protein association with model raft membranes on conversion to the protease-resistant isoform. J. Biol. Chem. 278: 14883-14892.

Baron, G.S., Wehrly, K., Dorward, D.W., Chesebro, B., and Caughey, B. 2002. Conversion of raft associated prion protein to the protease-resistant state requires insertion of PrP-res (PrP(Sc)) into contiguous membranes. EMBO J. 21: 1031-1040.

Bartz, J.C., Marsh, R.F., McKenzie, D.I., and Aiken, J.M. 1998. The host range of chronic wasting disease is altered on passage in ferrets. Virology. 251: 297-301.

Bessen, R.A., Kocisko, D.A., Raymond, G.J., Nandan, S., Lansbury, P.T., Jr., and Caughey, B. 1995. Nongenetic propagation of strain-specific phenotypes of scrapie prion protein. Nature. 375: 698-700.

Bessen, R.A., Raymond, G.J., and Caughey, B. 1997. *In situ* formation of protease-resistant prion protein in transmissible spongiform encephalopathy-infected brain slices. J. Biol. Chem. 272: 15227-15231.

Bossers, A., de Vries, R., and Smits, M.A. 2000. Susceptibility of sheep for scrapie as assessed by *in vitro* conversion of nine naturally occurring variants of PrP. J. Virol. 74: 1407-1414.

Bruce, M.E., McBride, P.A., and Farquhar, C.F. 1989. Precise targeting of the pathology of the sialoglycoprotein, PrP, and vacuolar degeneration in mouse scrapie. Neurosci. Lett. 102: 1-6.

Callahan, M.A., Xiong, L., and Caughey, B. 2001. Reversibility of scrapie-associated prion protein aggregation. J. Biol. Chem. 276: 28022-28028.

Caughey, B. 1991. Cellular metabolism of normal and scrapie-associated forms of PrP. Sem. Virol. 2: 189-196.

Caughey, B. 2001. Interactions between prion protein isoforms: the kiss of death? Trends Biochem. Sci. 26: 235-242.

Caughey, B., Kocisko, D.A., Raymond, G.J., and Lansbury, P.T. 1995. Aggregates of scrapie associated prion protein induce the cell-free conversion of protease-sensitive prion protein to the protease-resistant state. Chem. & Biol. 2: 807-817.

Caughey, B. and Raymond, G.J. 1991. The scrapie-associated form of PrP is made from a cell surface precursor that is both protease- and phospholipase-sensitive. J. Biol. Chem. 266: 18217-18223.

Caughey, B., Raymond, G.J., Callahan, M.A., Wong, C., Baron, G.S., and Xiong, L. 2001. Interactions and conversions of prion protein isoforms. Adv. Prot. Chem. 57: 139-169.

Caughey, B., Raymond, G.J., Kocisko, D.A., and Lansbury, P.T., Jr. 1997. Scrapie infectivity correlates with converting activity, protease resistance, and aggregation of scrapie-associated prion protein in guanidine denaturation studies. J. Virol. 71: 4107-4110.

Caughey, B., Raymond, L.D., Raymond, G.J., Maxson, L., Silveira, J., and Baron, G.S. 2003. Inhibition of protease-resistant prion protein accumulation *in vitro* by curcumin. J. Virol. 77: 5499-5502.

Caughey, W.S., Raymond, L.D., Horiuchi, M., and Caughey, B. 1998. Inhibition of protease-resistant prion protein formation by porphyrins and phthalocyanines. Proc. Natl. Acad. Sci. USA. 95: 12117-12122.

DebBurman, S.K., Raymond, G.J., Caughey, B., and Lindquist, S. 1997. Chaperone-supervised conversion of prion protein to its protease-resistant form. Proc. Natl. Acad. Sci. USA. 94: 13938-13943.

Glockshuber, R. 2001. Folding dynamics and energetics of recombinant prion proteins. Adv. Prot. Chem. 57: 83-105.

Harris, D.A. 2001. Biosynthesis and cellular processing of the prion protein. Adv. Prot. Chem. 57: 203-228.

Hill, A.F., Antoniou, M., and Collinge, J. 1999. Protease-resistant prion protein produced *in vitro* lacks detectable infectivity. J. Gen. Virol. 80: 11-14.

Horiuchi, M., Baron, G.S., Xiong, L.W., and Caughey, B. 2001. Inhibition of interactions and interconversions of prion protein isoforms by peptide fragments from the C-terminal folded domain. J. Biol. Chem. 276: 15489-15497.

Horiuchi, M. and Caughey, B. 1999. Prion protein interconversions and the transmissible spongiform encephalopathies. Structure. Fold. Des. 7: R231-40.

Horiuchi, M., Chabry, J., and Caughey, B. 1999. Specific binding of normal prion protein to the scrapie form via a localized domain initiates its conversion to the protease- resistant state. EMBO J. 18: 3193-3203.

Horiuchi, M., Priola, S.A., Chabry, J., and Caughey, B. 2000. Interactions between heterologous forms of prion protein: Binding, inhibition of conversion, and species barriers. Proc. Natl. Acad. Sci. USA. 97: 5836-5841.

Jarrett, J.T. and Lansbury, P.T., Jr. 1993. Seeding "one-dimensional crystallization" of amyloid: a pathogenic mechanism in alzheimer's disease and scrapie? Cell. 73: 1055-1058.

Knaus, K.J., Morillas, M., Swietnicki, W., Malone, M., Surewicz, W.K., and Yee, V.C. 2001. Crystal structure of the human prion protein reveals a mechanism for oligomerization. Nat. Struct. Biol. 8: 770-774.

Kocisko, D.A., Come, J.H., Priola, S.A., Chesebro, B., Raymond, G.J., Lansbury, P.T., and Caughey, B. 1994. Cell-free formation of protease-resistant prion protein. Nature. 370: 471-474.

Korth, C., Kaneko, K., and Prusiner, S.B. 2000. Expression of unglycosylated mutated prion protein facilitates PrPSc formation in neuroblastoma cells infected with different prion strains. J. Gen. Virol. 81: 2555-2563.

Lansbury, P.T., Jr. and Caughey, B. 1995. The chemistry of scrapie infection: implications of the 'ice 9' metaphor. Chem. & Biol. 2: 1-5.

Lucassen, R., Nishina, K., and Supattapone, S. 2003. *In vitro* amplification of protease-resistant prion protein requires free sulfhydryl groups. Biochemistry. 42: 4127-4135.

Maxson, L., Wong, C., Herrmann, L.M., Caughey, B., and Baron, G.S. 2003. A solid phase assay for identification of modulators of prion protein interactions. Anal. Biochem. 323: 54-64.

Miller, M.W., Williams, E.S., McCarty, C.W., Spraker, T.R., Kreeger, T.J., Larsen, C.T., and Thorne, E.T. 2000. Epidemiology of chronic wasting disease in free-ranging cervids. J. Wildl. Dis. 36: 676-690.

Morillas, M., Swietnicki, W., Gambetti, P., and Surewicz, W.K. 1999. Membrane environment alters the conformational structure of the recombinant human prion protein. J. Biol. Chem. 274: 36859-36865.

Morrissey, M.P. and Shakhnovich, E.I. 1999. Evidence for the role of PrP(C) helix 1 in the hydrophilic seeding of prion aggregates. Proc. Natl. Acad. Sci. USA 96: 11293-11298.

Priola, S.A. 1999. Prion protein and species barriers in the transmissible spongiform encephalopathies. Biomed. Pharmacother. 53: 27-33.

Priola, S.A. 2001. Prion protein diversity and disease in the transmissible spongiform encephalopathies. Adv. Protein Chem. 57: 1-27.

Priola, S.A., Chabry, J., and Chan, K. 2001. Efficient conversion of normal prion protein (PrP) by abnormal hamster PrP is determined by homology at amino acid residue 155. J. Virol. 75: 4673-4680.

Prusiner, S.B. 1998. Prions. Proc. Natl. Acad. Sci. USA. 95: 13363-13383.

Raymond, G.J., Bossers, A., Raymond, L.D., O'Rourke, K.I., McHolland, L.E., Bryant, P.K., III, Miller, M.W., Williams, E.S., Smits, M., and Caughey, B. 2000. Evidence of a molecular barrier limiting susceptibility of humans, cattle and sheep to chronic wasting disease. EMBO J. 19: 4425-4430.

Raymond, G.J., Hope, J., Kocisko, D.A., Priola, S.A., Raymond, L.D., Bossers, A., Ironside, J., Will, R.G., Chen, S.G., Petersen, R.B., Gambetti, P., Rubenstein, R., Smits, M.A., Lansbury, P.T., Jr., and Caughey, B. 1997. Molecular assessment of the transmissibilities of BSE and scrapie to humans. Nature. 388: 285-288.

Saborio, G.P., Permanne, B., and Soto, C. 2001. Sensitive detection of pathological prion protein by cyclic amplification of protein misfolding. Nature. 411: 810-813.

Saborio, G.P., Soto, C., Kascsak, R.J., Levy, E., Kascsak, R., Harris, D.A., and Frangione, B. 1999. Cell-lysate conversion of prion protein into its protease- resistant isoform suggests the participation of a cellular chaperone. Biochem. Biophys. Res. Commun. 258: 470-475.

Sanghera, N. and Pinheiro, T.J. 2002. Binding of prion protein to lipid membranes and implications for prion conversion. J. Mol. Biol. 315: 1241-1256.

Scott, M.R., Kohler, R., Foster, D., and Prusiner, S.B. 1992. Chimeric prion protein expression in cultured cells and transgenic mice. Protein Sci. 1: 986-997.

Serio, T.R. and Lindquist, S.L. 2001. The yeast prion [PSI+]: molecular insights and functional consequences. Adv. Protein Chem. 59: 391-412.

Speare, J.O., Rush, T.S., Bloom, M.E., and Caughey, B. 2003. The role of helix 1 aspartates and salt bridges in the stability and conversion of prion protein. J. Biol. Chem. 278: 12522-12529.

Vorberg, I. and Priola, S.A. 2002. Molecular basis of scrapie strain glycoform variation. J. Biol. Chem. 277: 36775-36781.

Wadsworth, J.D., Hill, A.F., Joiner, S., Jackson, G.S., Clarke, A.R., and Collinge, J. 1999. Strain-specific prion-protein conformation determined by metal ions. Nat. Cell Biol. 1: 55-59.

Welker, E., Raymond, L.D., Scheraga, H.A., and Caughey, B. 2002. Intramolecular versus intermolecular disulfide bonds in prion proteins. J. Biol. Chem. 277: 33477-33481.

Welker, E., Wedemeyer, W.J., and Scheraga, H.A. 2001. A role for intermolecular disulfide bonds in prion diseases? Proc. Natl. Acad. Sci. USA. 98: 4334-4336.

Wickner, R.B., Taylor, K.L., Edskes, H.K., Maddelein, M.L., Moriyama, H., and Roberts, B.T. 2001. Yeast prions act as genes composed of self-propagating protein amyloids. Adv. Protein Chem. 57: 313-334.

Wille, H., Michelitsch, M.D., Guenebaut, V., Supattapone, S., Serban, A., Cohen, F.E., Agard, D.A., and Prusiner, S.B. 2002. Structural studies of the scrapie prion protein by electron crystallography. Proc. Natl. Acad. Sci. USA. 99: 3563-3568.

Wong, C., Xiong, L.-W., Horiuchi, M., Raymond, L.D., Wehrly, K., Chesebro, B., and Caughey, B. 2001. Sulfated glycans and elevated temperature stimulate PrP^{Sc} dependent cell-free formation of protease-resistant prion protein. EMBO J. 20: 377-386.

Zou, W.Q. and Cashman, N.R. 2002. Acidic pH and detergents enhance *in vitro* conversion of human brain PrPC to a PrPSc-like form. J. Biol. Chem. 277: 43942-43947.

From: Prions and Prion Diseases: Current Perspectives
Edited by: Glenn C. Telling

Chapter 3

The Prion Protein in Cell Culture

Jeremy P. Brockes and Nnennaya Kanu

Abstract

Cell culture models have afforded an accessible context in which to study the biosynthesis and trafficking of the prion protein (PrP) with respect to its glycosylation, conformation, membrane topology, aggregation and disulfide bonding. These mechanisms may impinge directly or indirectly on the propensity of the normal form of PrP, referred to as PrP^C, to undergo conversion to the disease-associated isoform, PrP^{Sc}, during *de novo* infection of cells, or in a stably infected cell. Studies in cell culture have also delineated several mechanisms by which an infected cell might convert a neighbouring target cell, and have specifically suggested cell contact as an important aspect of this process.

1. Introduction

This chapter is concerned with recent developments in the study of prion mechanisms in cell culture. Cells in culture continue to provide an accessible context for investigating the basic molecular cell biology of PrP and its various mutants and conformers. Although much of this work has been focused on mouse N2a neuroblastoma cells and their scrapie infected derivatives, several other cell lines are also under investigation. Our account is divided into two sections, the first being concerned with various aspects of the behavior of PrP in a normal cell. The second addresses events in an infected cell both with respect to the regulation of intracellular conversion of PrP^C to PrP^{Sc}, and also the process whereby an infected cell is able to induce stable expression of PrP^{Sc} in a target cell. In view of space constraints we have had to omit several important and active areas of current interest. These include the use of cell culture assays to search for therapeutic or prophylactic agents, and the study of mechanisms of cytotoxicity on cultured neural cells.

2. Behavior of PrP in Uninfected Cells

2.1. Regulation of Glycosylation

Mouse PrP undergoes variable glycosylation at residues N180 and N196 (for hamster 181 and 197) (Endo *et al.*, 1989; Haraguchi *et al.*, 1989), giving rise to unglycosylated, monoglycosylated and diglycosylated species. The glycosylation of PrP is of interest from several perspectives. There is the basic issue of the regulation of glycosylation in a glycosylphosphatidylinositol (GPI)-anchored protein, a circumstance which exerts an additional level of regulation, at least for PrP. Glycosylation is a potential determinant with respect to endoplasmic reticulum associated degradation (ERAD), an aspect of the behavior of PrP which has received significant recent attention. It is also an important property of prion strains that PrP^{Sc} tends to exhibit a characteristic glycoform ratio, although the precise relationship between strain identity and the glycoforms remains unclear. Finally, glycosylation can be a significant influence on the conversion of PrP^C to PrP^{Sc}. There may well be additional issues whose significance will become clearer in the future, for example the effect of glycosylation and the extensive heterogeneity in the glycan structures on the biological activity of PrP (Rudd *et al.*, 1999; Stimson *et al.*, 1999).

Recent work has apparently established an important relationship between the C terminal anchoring of PrP and its glycosylation (Walmsley *et al.*, 2001). A construct expressing mouse PrP was transfected into human neuroblastoma cells and compared to constructs expressing PrP (a), anchored at its C terminus with a transmembrane region in place of a glycolipid or (b), anchored solely at its N terminus with a transmembrane region or (c), doubly anchored by a transmembrane region at the N terminus and a GPI anchor at the C terminus, or (d), a C terminal truncated construct missing the sequence that signals GPI transamidation. The constructs a-c were expressed appropriately at the cell surface as predicted by the primary sequences, and d was released into the medium. While b and d were predominantly unglycosylated, a and c were extensively N-glycosylated (Walmsley *et al.*, 2001). It is therefore noteworthy that C terminal anchorage, whether by a transmembrane region or by a glycolipid, appears to be necessary for substantial occupancy of the two sites. Although construct b was expressed at the cell surface, the N terminal anchorage resulted in a predominantly unglycosylated protein. While earlier results were generally consistent with a significant decrease in the glycosylation of C terminal truncated PrP molecules (Rogers *et al.*, 1993), the present study has underlined the constraints on endoplasmic reticulum (ER) glycosylation more clearly.

The mechanistic explanation for this dependence on C terminal anchorage is unclear. Although several GPI – anchored proteins appear to be glycosylated after their expression as truncated forms (Brown *et al.*, 1989; Caras *et al.*, 1989), a recent report indicates that Thy 1 shows a comparable dependence to PrP (Devasahayam *et al.*, 1999). The issue of sequon accessibility to the oligosaccharyltrans-ferase complex during translation has been addressed in a recent paper where glycosylation of the non-anchored deletion mutants of PrP was restored by increasing the distance between the sequons and the C terminus (Walmsley and Hooper, 2003). This suggests that the spacing is a more critical parameter for cotranslational glycosylation than anchorage, and its importance has been recognised in earlier studies of N glycosylation of model proteins (for references, see Walmsley and Hooper, 2003). Further investigation will be required to clarify the precise regulation of glycosylation in different cell types.

A second important constraint on the glycosylation of PrP is exerted by the redox state. The two N-linked glycans are located within the span of a disulfide bridge between residues 179 and 214. When disulfide bridge formation is blocked by the addition of dithiothreitol to the medium of M17 neuroblastoma cells, PrP becomes completely diglycosylated and is expressed on the cell surface (Capellari *et al.*, 1999). It is interesting that if disulfide bridge formation is blocked by mutation of a single cysteine residue then the resulting monothiol derivative is retained in the ER (Capellari *et*

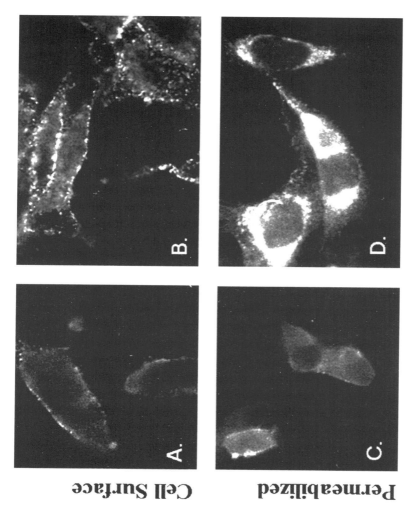

Cell Surface

Permeablized

Figure 1. Accumulation of PrP after treatment of cultured cells with a proteasome inhibitor. A and C show the surface and internal immunofluorescence staining for PrP in control CHO cells, while B and D show the corresponding images after treatment with inhibitor for 12 hours. Note the marked perinuclear accumulation in D. Reprinted with permission from Yedidia *et al.*, 2001.

al., 1999; Yanai *et al.*, 1999), while mutation of both residues leads to formation of diglycosylated PrP which reaches the surface but fails to recycle (Capellari *et al.*, 1999). The presence of the disulfide bond acts as a redox sensitive brake on glycosylation of the protein, and might indirectly influence the outcome of disease in certain contexts by the effect of glycosylation on the conversion to PrP^{Sc}, as mentioned later. It may be possible in future to dissect out the relative contributions of C terminal anchoring and redox state to the glycosylation pattern observed in various cell types.

2.2. ER Associated Degradation of PrP

Several recent studies have established the importance of ER associated degradation (ERAD) in the context of PrP biosynthesis (Zanusso *et al.*, 1999; Ma and Lindquist, 2001; 2002; Yedidia *et al.*, 2001; Ma *et al.*, 2002). Many proteins that are found in the ER may be translocated back into the cytosol, derivatized with ubiquitin, and degraded by proteasomes (Bonifacino and Weissman, 1998). If PrP-expressing cells are treated with proteasome inhibitors, there is a significant accumulation of PrP in the cytosol and this can be detected around the nucleus or centrosome after visualization by immunofluorescence (Figure 1) (Ma and Lindquist, 2001; 2002; Yedidia *et al.*, 2001; Ma *et al.*, 2002). The biochemical characterization of the accumulated PrP reveals a complex picture (Ma and Lindquist, 2001; 2002; Yedidia *et al.*, 2001; Ma *et al.*, 2002). Much of the PrP is apparently unglycosylated, although some higher molecular weight glycoforms are also observed. While there is some additional detergent soluble PrP, there is a marked accumulation of an aggregated form which is largely resistant to protease K digestion, and has an indistinguishable cleavage pattern from PrP^{Sc}. It is also possible to identify bands after Western blotting which react both with antibodies to ubiquitin and to PrP (Yedidia *et al.*, 2001). There is apparently considerable variability in the extent and nature of the accumulated species, depending on the precise inhibitor employed, the timing of treatment, and the extent of expression of PrP in individual cells following transient or stable transfection (Ma and Lindquist, 2001, 2002; Ma *et al.*, 2002); nonetheless it has been suggested that approximately 10% of nascent PrP molecules are normally diverted into the ERAD pathway (Yedidia *et al.*, 2001).

There are several interesting implications of ERAD for the biology of prion diseases. In a comparative analysis of the effects of accumulating PrP and presenilin-1 after proteasome inhibition, it was found that PrP appeared to induce cytotoxicity in neuroblastoma cells whereas presenilin did not (Ma *et al.*, 2002). This was also observed after expression in cultured cells and transgenic mice of an N and C terminal truncated species of PrP which does not enter the ER (Ma *et al.*, 2002). Furthermore, it has been shown that the D177N mutant form of PrP accumulates in the cytoplasm even without proteasome inhibitors (Ma and Lindquist, 2001). Hence it is possible that one route to disease might be the relative accumulation of misfolded forms in the ER, due to mutation or other factors, followed by their retrotranslocation into the cytoplasm and consequent cytotoxicity.

A second connection with disease has come from the proposal that the aggregated form, which accumulates in culture after proteasome inhibition, is able to initiate sustained PrP conversion in the same way as PrP^{Sc} (Ma and Lindquist, 2002). After transient treatment with inhibitor for 2 hours, the degree of accumulation of protease resistant PrP after 12 hours was consistent with conversion of newly synthesized PrP even after proteasome activity was restored. The time course presented for this effect was not an extended one, but it is possible that the conversion of retrotranslocated PrP to a PrP^{Sc}-like form might be a process that contributes to disease initiation and progression. In view of the absence of glycosylation after retrotranslocation, the demonstrable accumulation of glycosylated forms of PrP^{Sc} in infectious etiologies is indicative of other forms of propagation that are discussed later. It is worthwhile to note that the precise nature of the normal PrP species that is subject to retrotranslocation is unclear. In addition to misfolding, quality control mechanisms are capable of monitoring the trimming status of N-linked oligosaccharide chains (Cabral *et al.*, 2001), or the presence of short transmembrane domains (Bonifacino and Weissman, 1998). Since there are several examples where retrotranslocation is accompanied by deglycosylation, and a cytosolic N glycanase has been described (Suzuki *et al.*, 1994), it is also unclear if there is a preference for unglycosylated PrP to be retrotranslocated.

2.3. Transmembrane PrP

An important but unresolved hypothesis concerns the role of transmembrane forms of PrP, in particular the variant referred to as CtmPrP which has the C terminus on the lumenal side of the ER, and a transmembrane region from residues 111-134. This was originally identified as a minor product when PrP mRNA was translated *in vitro* with reticulocyte lysate supplemented with microsomes, but its preponderance can be increased by mutations in the vicinity of the transmembrane region (Hegde *et al.*, 1998; 1999). The Ctm form gives a characteristic product after proteolysis of the membrane associated translation reactions. Although much of the work on this form is based on studies in transgenic mice and hence falls outside the scope of this chapter (Hegde *et al.*, 1998; 1999), a recent study in cell culture has helped to clarify and also to challenge some aspects of the hypothesis that it is this form that mediates the pathogenic consequences of prion disease (Stewart *et al.*, 2001; Stewart and Harris, 2001). A detailed analysis of the CtmPrP *in vitro* translation product has revealed that the N terminal signal sequence escapes cleavage and hence remains intact. Furthermore the C terminal has a GPI anchor so that the molecule is doubly anchored in the membrane. A combination of a mutation in the signal sequence with the 3AV mutation in the transmembrane region yielded essentially all of the PrP in transfected cells in the Ctm form for the first time.

The CtmPrP expressed after transfection in baby hamster kidney (BHK) cells is quantitatively deglycosylated by endo H and hence does not transit beyond the mid Golgi stack where oligosaccharides become resistant to this enzyme (Stewart *et al.*, 2001). The authors point out that this result is in clear contrast to a previous analysis of the brains of transgenic mice expressing the 3AV mutation, which suggested that the Ctm form exited from the Golgi (Hegde *et al.*, 1998). They go on to show that proteasome inhibitors cause a marked accumulation of the mutant protein in transfected cells, indicating that CtmPrP is subject to ERAD. This raises the possibility of neurotoxicity by various mechanisms, for example by activating ER stress induced pathways, or by the mechanism associated with the appearance of cytosolic PrP discussed above. This incisive study (Stewart *et al.*, 2001) underlines the ability of cell culture systems to clarify basic mechanisms in the molecular cell biology of PrP.

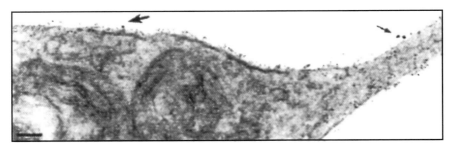

Figure 2. Transmission electron micrograph of immunogold labelling of Thy-1 and PrP on the surface of cultured neurites. The smaller arrow shows PrP within gaps in the Thy-1 labelling, while the larger arrow shows PrP in the midst of Thy-1 labelling. The scale bar is 100 nm. Reprinted with permission from Madore *et al.*, 1999.

2.4. PrP and Lipid Rafts

An important aspect of the cell biology of PrP is its location on the surface in the vicinity of lipid rafts. A variety of earlier work has underlined the importance of the glycolipid anchor in targeting the molecule to this location (Taraboulos *et al.*, 1995; Kaneko *et al.*, 1997). Furthermore, PrP that is expressed with a C terminal transmembrane region cannot be converted to PrP^{Sc} in an infected cell (Taraboulos *et al.*, 1995; Kaneko *et al.*, 1997); trafficking to rafts is also required for the dominant negative activity on conversion that is exhibited by PrP molecules substituted at various C terminal residues (Zulianello *et al.*, 2000). A recent study has compared both anatomical and biochemical aspects of the surface localization of PrP and the GPI-linked surface protein Thy 1 in cultured DRG neurons and brain membranes (Madore *et al.*, 1999). After solubilization in various non-ionic detergents followed by sucrose gradient flotation analysis, rafts containing PrP were more soluble than those containing Thy 1, yet at lower stringency of solubilization, where neuronal and Thy 1 domains remained intact, immunoaffinity purification of Thy 1 resulted in copurification of most of the PrP. Together with immunogold EM localization of the proteins on cultured neurites (Figure 2), this has led to a model in which ordered sphingolipid domains expressing Thy 1 have less ordered (and more soluble) domains containing PrP at their margins (Madore *et al.*, 1999). In future this kind of analysis may be combined with a detailed inventory of other proteins with which PrP interacts or co-fractionates.

One example of a study on such proteins is the identification of the coupling of PrP^{C} to the tyrosine kinase fyn after antibody crosslinking in the 1C11 cell line (Mouillet-Richard *et al.*, 2000). Antibodies to PrP induce a marked decrease in the phosphorylation level of the kinase and a consequent elevation in its activity. Interestingly the activation effect of antibodies seemed more marked on the neurites than on the cell bodies of differentiated 1C11 cells. Fyn is quite frequently implicated in studies of transduction by raft associated proteins, and this activity may be relevant to the functional roles of PrP.

The internalization of PrP is another important aspect of its surface localization, both with respect to turnover and possibly its functional role. Two recent studies have focused on the activity of the N terminal sequence in this process (Perera and Hooper 2001; Nunziante *et al.*, 2003). The human neuroblastoma line SY5 was stably transfected with mouse PrP constructs in which the octapeptide repeat region was deleted or mutated in the repeats, or additional repeats were introduced. In earlier work, cupric ions were found to stimulate endocytosis of chicken PrP (Pauly and Harris, 1998), and each of the three manipulations was found to prevent metal ion stimulated endocytosis (essentially all detectable endocytosis in this context) in these cells (Perera and Hooper, 2001). A related investigation of internalization, and also PrP trafficking to the cell surface, has come to somewhat different conclusions (Nunziante *et al.*, 2003). As before, truncation of the N terminus profoundly inhibited internalisation but addition of the N terminus of Xenopus PrP, a region which does not encode a copper binding repeat element, was able to restore the endocytic kinetics of wild type PrP. Thus while there is a consensus about the importance of the N terminal region for endocytosis, the role of metal ion binding in different cell contexts needs to be further addressed.

3. Intracellular Conversion of PrP in an Infected Cell in Culture

Although this is an area of obvious importance from a therapeutic perspective, we have selected only a subset of the issues currently being addressed. One is to return to the topic of glycosylation, in this case in relation to its effect on the conversion of PrP^C to PrP^{Sc}. A second relates to recent reports on new strategies for curing scrapie infected cells which illustrate different aspects of the underlying cell biology.

3.1. Effect of Glycosylation on Conversion

While earlier studies with tunicamycin revealed that unglycosylated PrP^C is a substrate for conversion (Taraboulos *et al.*, 1990; Lehmann and Harris, 1997), attempts to create a metabolically stable, double mutant unglycosylated form were unsuccessful (Rogers *et al.*, 1990; DeArmond *et al.*, 1997; Lehmann and Harris, 1997). This has now been accomplished by mutating the two critical Asn residues to Gln (Korth *et al.*, 2000). The resulting protein apparently passes the quality control mechanisms in the ER and is expressed on the cell surface after transient transfection of scrapie infected N2a cells. It is converted to PrP^{Sc} after expression in these cells, and is converted by mouse prions from brain homogenates after expression in normal N2a cells. A comparison with wild type PrP indicated that the N180Q, N196Q double mutant was more effectively converted (Korth *et al.*, 2000). Although several strains of mouse prions were equally effective, the authors remark that questions of strain identity and glycoform expression will be best addressed in mice expressing the double mutant PrP as a transgene.

Studies with *in vitro* conversion systems have addressed the issue of glycosylation and conversion both within and across species barriers but will not be considered further here (Priola and Lawson, 2001; Vorberg and Priola, 2002).

3.2. Curing of Infected Cells

Certain PrP-specific Fab antibodies have proved remarkably effective at curing infection after addition to the culture medium of ScN2a cells (Peretz *et al.*, 2001). One antibody called D18 eliminated 50% of the PrP^{Sc} from cells in 28 hours – a sufficiently rapid effect to suggest that it completely abolished formation of new PrP^{Sc}, while pre-existing PrP^{Sc} was cleared from the cells. The most straightforward explanation is that the antibody binds to PrP^C molecules in rafts and prevents the interaction with PrP^{Sc} that is critical for conversion. The binding epitope spans residues 132-156, possibly a critical region of the protein for transmission. A second study has used antibody 6H4, which recognises residues 144-152 of mouse PrP (Korth *et al.*, 1997), both to inhibit infection by scrapie infected brain homogenate and also to cure ScN2a cells (Enari *et al.*, 2001). The cured cells were tested as late as 6 weeks after treatment to ensure that they had been stably cured. In this study the removal of PrP^C from the cell surface by phosphatidyl inositol specific phospholipase C was also an effective curing strategy, presumably by removing the critical pool of precursor PrP^C, either in rafts or possibly after internalisation.

A rather different mechanism underlies the curing activity of the compound suramin, as revealed in a recent comprehensive analysis (Gilch *et al.*, 2001). Suramin, a bis-hexasulfonated napthylurea, is known to interfere with the oligomerization state of certain proteins. When N2a cells were treated with suramin, it led to the formation of aggregates of PrP^C which were sensitive to protease digestion and were not infectious. This activity was also detectable when assayed *in vitro* on detergent free lysates. Aggregation of PrP^C in cells was induced in the middle and late Golgi, and a fraction of PrP was also present in lysosomal vesicles. It is suggested from earlier work that suramin is able to access endosomes and the trans Golgi network, so it may interact at this location with folded and glycosylated PrP^C. The PrP aggregates are recognised as misfolded and marked for degradation. This re-routing of the aggregates prevents further targeting to the plasma membrane and leads to curing of scrapie infected N2a and hypothalamic GT1 cells. In experiments with scrapie infected mice, suramin significantly delayed the onset of disease when applied around the time of inoculation (Gilch *et al.*, 2001). This study has identified another point of 'quality control' for PrP and suggested a novel approach for new therapeutics or prophylactics.

4. Infection of Cultured Cells

In an experimental prion infection of an animal the initial inoculum is cleared quite rapidly relative to the incubation period. While the displacement of infectivity is likely to play some part in the progressive course of the disease, for example by anterograde and retrograde axoplasmic transport in the PNS and CNS, a critical aspect of progression is the ability of infected cells to propagate the infection by conversion of neighboring target cells to stable expression of abnormal forms of PrP. The mechanism of cell based infection has been unclear but it has recently received attention in dissociated cell culture, a context where it is significantly easier to evaluate the dependence on cell contact or extracellular mediators. Before considering these issues we will review some of the recent reports of cell lines that show enhanced sensitivity to infection in cell culture.

It has been difficult to identify cells or cell lines that are readily infectable and retain stable expression of PrP^{Sc}, and there has recently been significant progress in this area.

4.1. New Cell lines for Prion Infection

It has been possible to derive sublines from multiple clonal isolates of the stock N2a line (Bosque and Prusiner, 2000). About 20% of these sublines are readily susceptible to scrapie prion infection, about 20% resistant, and the remainder is intermediate. The initial stock apparently has a considerable level of variability which can be exploited to isolate susceptible sublines. In quantitative estimates of the susceptibility of sublines, it was observed that the cells were about 1000 fold more sensitive to homogenates of scrapie infected N2a cells than homogenates of infected brain (Bosque and Prusiner, 2000). In the former case the susceptible subline is sensitive to only 10 ID50 units. One of the aims of deriving infected/uninfected paired cell lines is to identify, for example by microarray analysis, the specific consequences of infection. This is particularly attractive for the cell lines where pathogenesis is effectively uncoupled from infection. Such a comparison seems more compelling for an uninfected/infected pair than an infected/cured pair because of the possible diverse effects of the curing agent.

An alternative approach to derivation of new lines is the overexpression of PrP in stable transfectants of N2a or hypothalamic GT1 cells (Nishida *et al.*, 2000). Such overexpressing lines were successfully infected and stably produced PrP^{Sc} after multiple passages in culture. Although primary neural cells are a desirable target for infection in the future, this has not yet been compellingly documented. A mouse Schwann cell line has been infected with the Chandler scrapie strain, and stably expresses PrP^{Sc} (Follet *et al.*, 2002). Since primary rodent Schwann cells can be propagated indefinitely (Mathon *et al.*, 2001) it seems worthwhile to test their susceptibility in future.

A third example concerns a new model for assay of sheep scrapie. In this case ovine PrP was placed under an inducible promoter and stably transfected into a rabbit epithelial cell line (Vilette *et al.*, 2001). The resulting Rov 9 cells were induced for high level ovine PrP expression and infected with sheep brain isolates of ovine prions. The infected cultures were also assayed by inoculation into mice carrying an ovine PrP transgene. This led to death of the mice after a defined incubation period with acute neurological disorders and concomitant accumulation of PrP^{Sc}. These cells seem a promising bioassay system and model for future studies on sheep scrapie and resistant PrP genotypes.

An alternative approach to derivation of a susceptible cell line is to begin with a stably infected cell, and to cure it of prion infection. If the properties that led to initial infection and to stable maintenance of infection are sustained after curing then the derivative line may be favorable for re-infection. This is illustrated by a report that begins with the curing of the SMB cell (Birkett *et al.*, 2001), which was originally established by culture from a brain of a mouse infected with scrapie. The SMB cell is sufficiently stable, both in its level of infectivity and its expression of PrP^{Sc}, that it provided the first compelling evidence that the scrapie agent could replicate in culture (Clarke and Haig, 1970; Haig and Clarke, 1971). SMB cells were cured by growth in the presence of pentosan sulfate (PS) to yield the clonally derived SMB-PS cell line which has lost detectable infectivity and expression of PrP^{Sc}. The SMB-PS cell could be readily reinfected with several mouse scrapie strains, and one of these strains, 22F, was selected for more extensive analysis (Birkett *et al.*, 2001). The strain characteristics exhibited by 22F after propagation in the SMB-PS cell were clearly different from the Chandler strain as analyzed in SMB cells. This was the case both for the pattern of histopathology after inoculation into mice, as well as the length of the incubation periods. While the possibility of variation in clonal isolates in culture must always be born in mind, and indeed is underlined by the work on N2a cells (Bosque and Prusiner, 2000), this study nonetheless provides a demonstration that two different strains of scrapie retain their identity after propagation in the same cell type.

4.2. Cell Based Infection in Dissociated Culture

It is possible to use live cells as a source of infectivity in culture rather than a subcellular prion preparation (Kanu *et al.*, 2002). SMB cells were plated along with a neomycin resistant line derived by transfection of mouse SMB-PS cells. The two cell types mix in closely apposed monolayers and after 7 days the cells were passaged and subjected to continuous selection with neomycin. The rigorous

(a) infection by brain homogenate

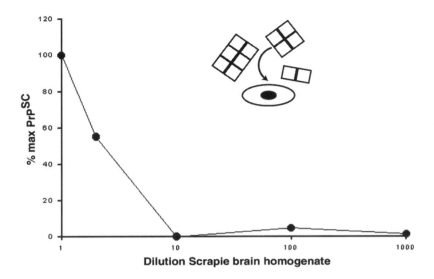

(b) infection by live cell

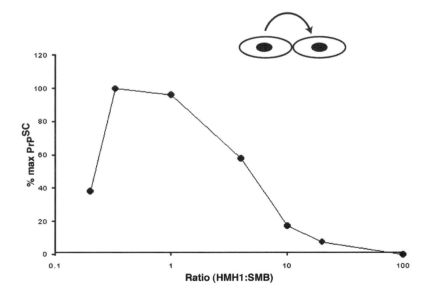

Figure 3. Infection of HMH target cells by a scrapie infected brain homogenate as compared to live SMB cells. (a), Concentration dependent infection by a 79A brain homogenate; HMH cells were exposed for eighteen hours to varying dilutions of a 2% brain homogenate and expanded by passaging for 17 days, prior to the preparation of PrP^{Sc} fractions and Western blotting. The dose response curve was constructed after quantitation of PrP^{Sc} expression by imager analysis. (b), Dependence of infection on the cell ratio in the co-culture with SMB cells. HMH cells (a constant number per dish) were cultured with SMB so as to vary the ratio over a range of 500 fold. After coculture the HMH cells were selected with neomycin over 14 days, processed as in (a) and analyzed by quantitative Western blotting. Note that conversion of the target cells is optimal at around a unitary ratio and decreases at higher SMB/HMH ratios. Although (a) and (b) are not strictly comparable in terms of the time of exposure to PrP^{Sc}, and the strain of scrapie (79A versus Chandler), they show the response of the same cell line in a quasilinear range. The PrP^{Sc} content of the brain homogenate and the SMB cells was compared by comparative Western blotting, and independently by dot blotting. After correction for the area of HMH cells exposed in the assays of (a) and (b), the amount of PrP^{Sc} required to give a comparable degree of conversion in the two cases was 2500 fold higher for (a) than (b). Details of the assays are given in Kanu *et al.*, 2002.

removal of the SMB cells was verified by labeling with cell tracker dyes, and by expression of the neo transcript in all surviving cells as evidenced by *in situ* hybridization. This procedure leads to conversion of the SMB-PS cells which can be propagated for at least 10-15 generations while maintaining stable expression of PrP^{Sc}. This was also analyzed by biosynthetic labeling of PrP^{Sc} after pulse chase in medium containing ^{35}S-methionine (Kanu *et al.*, 2002). The converted SMB-PS cells express PrP^{Sc} at 10-30% of the level of an equivalent number of SMB cells but in the absence of a reagent able to identify single cells that express PrP^{Sc}, it is unclear what percentage of cells become infected. The level of conversion is sensitive to the ratio of SMB to target cells and it is possible to make a semiquantitative comparison of the efficacy of infecting the same target cells. As detailed in Figure 3, infection with live cells requires 2500 fold less PrP^{Sc} than a homogenate of 79A scrapie-infected brain.

Although this comparison has some caveats (see Figure 3), the difference is sufficiently large that it suggests there are distinctive aspects to the mechanism of cell based conversion that need further elucidation. In view of the forthcoming evidence about the process of infection, we suggest that the difference of 1000-fold in the susceptibility of N2a cells to needle sheared homogenates of infected cells versus homogenates of infected brain (Bosque and Prusiner, 2000) also reflects the difference between presentation of PrP^{Sc} on a cell surface, versus presentation as prion rods derived from terminally infected brain. One obvious possibility for transfer is that the SMB cells release forms of PrP^Sc into the culture, but medium conditioned by dense cultures is completely inactive when assayed on the target cells (Kanu *et al.*, 2002). The two cell types can be confronted in close proximity but not contact by culturing them on alternative surfaces in dishes with an insert membrane carrying a high density of pores (0.4 micron diameter). Although such pores might retard larger species of PrP^Sc they would allow passage of membrane exovesicles which, as discussed below, are a possible agent for transfer of infectivity. Nonetheless transfer is only observed when cells are cultured together on the same surface, raising the possibility that contact is obligatory. This is supported by the activity of aldehyde fixed, killed SMB cells which have been cultured with live target cells and then separated by trypsinisation. Fixed SMB cells have approximately 10% of the activity of live cells assayed in parallel, but this level is readily detectable and strongly indicative of a contact dependent mechanism of transfer. Interestingly there is also clear time dependence for the target cells which must be cultured for 7-10 days in order to amplify the initial limited conversion of PrP^C to detectable levels. There is a marked dependence on the time of contact prior to trypsinization, since conversion is undetectable after one day (followed by 10 days of subsequent culture) and increases from 1-7 days (Kanu *et al.* 2002). This strongly indicates that infection does not depend on species produced by the action of trypsin on the fixed SMB cells, and underlines the importance of cell contact.

This study points to the activity of PrP^Sc presented on a surface, and similar conclusions have been reached in relation to the activity of stainless steel wires dipped into scrapie-infected brain homogenate and washed extensively (Zobeley *et al.*, 1999; Flechsig *et al.*, 2001; Weissmann *et al.*, 2002). Insertion of such wires into the brain proved to be at least as infectious as the injection of homogenates which contain far more protein. In an assay of infectivity in culture the prion-coated wires were placed onto a monolayer of target neuroblastoma cells for 14 days. The wire and its closely adherent cells were transferred to another dish, and cultures derived from the wire-bound cells proved to be stable producers of PrP^Sc, in contrast to cells in the original monolayer which were not converted by the wire (Weissmann *et al.*, 2002). The authors conclude that "intimate contact between the prion-carrying surface and susceptible cells greatly promotes infection or is prerequisite". One possibility for contact dependent infection is that PrP^Sc molecules on the surface of fixed cells or immobilized on a wire can act in *trans* to convert PrP^C molecules in rafts on a target cell surface, a conversion which is subsequently amplified by conversion of more PrP^C molecules after synthesis and trafficking to this location. In the simplest form of such a 'template' hypothesis, no PrP^{Sc} is transferred to the target (Figure 4a). This is consistent with observations that molecules of PrP^{Sc} are effectively crosslinked into aggregates by the aldehyde fixation (Kanu *et al.*, 2002), while bonding to the wire is sufficiently stable to withstand elution with 2M NaOH (Weissmann *et al.*, 2002). The possibility that small amounts of infectious material are transferred in these experiments cannot be ruled out at present. It is conceivable for example that membrane bound proteases could be activated by contact, and cleave off a fragment of PrPSc which acts locally (Figure 4d).

A recent study in cell culture that is relevant to these issues is concerned with the transfer of PrP^C between cells (Liu *et al.*, 2002). The donor cell is a stably transfected human neuroblastoma cell line called M17, while the recipient is a human erythroleukemia derivative, IA, which lacks PrP expression because of a defect in GPI anchor assembly. After co-culture PrP is not normally detected on the surface of IA cells but if either cell type, or better if both are pretreated with a protein kinase C activator, then PrP is transferred within 12 hours, and is found inserted along with its GPI anchor (Liu *et al.*, 2002). There are a number of examples of the transfer of GPI linked proteins between cells, including CD4 and CD59, a phenomenon sometimes referred to as 'painting' (Kooyman *et al.*, 1998). One plausible mechanism for transfer is the release of membrane exovesicles from the donor cell. Two recent examples of this process are the transfer of the CCR5 chemokine receptor between cells (Mack *et al.*, 2000), and the analysis of microvesicles released from *Drosophila* wing disc cells as visualized with a GPI linked version of Green Fluorescent Protein (Greco *et al.*, 2001). In the present case neither conditioned medium nor microvesicle preparations from M17 had a level of activity that could account for the observed transfer between cells. Furthermore an analysis with membrane insert dishes, analogous to that described above (Kanu *et al.*, 2002), indicated that transfer was dependent on cell contact, thus leading the authors to favour a mechanism based on transient and local fusion of the outer leaflets of the two cells and transfer of PrP (Figure 4b) (Liu *et al.*, 2002). While this would not operate in the context of fixed cells or immobilized prions, it could account for the approximately 90% of contact dependent conversion seen in live cells and not in fixed cells as described above. It is also

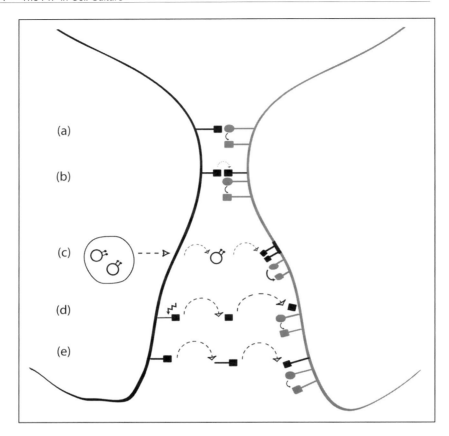

Figure 4. Schematic diagram of possible mechanisms for transfer of PrP^{Sc} between cells. The infected cell is depicted on the left in black, while the target cell on the right is in gray. Models (a) and (b) are by their nature strictly local and dependent on cell-cell contact, whereas (c), (d) and (e) involve diffusible species in the extracellular space. The membrane anchored form of PrP^{C} is represented by a line surmounted with an oval, while PrP^{Sc} is represented as a line and a square. (a), Template model; PrP^{Sc} on the infected cell acts in trans to convert PrP^{C} on the target cell (Kanu *et al.*, 2002; Weissmann *et al.*, 2002). (b), Contact dependent transfer model; local membrane fusion leads to transfer of PrP^{Sc} (Liu *et al.*, 2002). (c), Exovesicle model; membrane exovesicles arise in internal compartments and are discharged into the extracellular space leading to local insertion of PrP^{Sc} into the target cell (Mack *et al.*, 2000; Greco *et al.*, 2001; Baron *et al.*, 2002). (d) Cleavage model; PrP^{Sc} is cleaved by a protease (jagged line) and an active fragment diffuses to the target cell surface, leading to conversion. If protease activation were dependent on cell contact, this model would be in the category of (a) and (b) in that sense, while also involving extracellular diffusion. (e), Painting model; non-vesicular release of anchored form such that it can directly insert into the target membrane (Kooyman *et al.*, 1998).

very likely that the exovesicle possibility might operate in some cell types where transfer is not strictly dependent on cell-cell contact (Figure 4c).

A recent study of conversion in a cell free system provides a different perspective on the problem of intercellular infection (Baron *et al.*, 2002). The authors set up a system where the source of PrP^C was lipid rafts purified by detergent extraction of metabolically labeled N2a cells followed by flotation gradient centrifugation. The source of PrP^{Sc} was a microsomal membrane preparation from scrapie infected brains. The PrP^C was only converted to a protease resistant form if it was released from the membrane by phosphatidyl-inositol specific phospholipase C (PIPLC) treatment, or if the mixture was treated with a concentration of polyethylene glycol which promoted vesicle fusion. The authors conclude that conversion of raft associated PrP^C requires insertion of PrP^{Sc} into contiguous membranes and hence favor models for cell-cell transmission based on vesicle transfer or movement of GPI-anchored PrP (Figure 4e) (Baron *et al.*, 2002). Neither is necessarily dependent on cell contact but this could be accommodated by a contact dependent process of activation. It would be helpful to clarify the nature of the PrP^{Sc} in the microsomal preparations which causes conversion and the precise effect of the polyethylene glycol treatment. The various possibilities for intercellular conversion are summarized schematically in Figure 4 and the accompanying legend. In the absence of detailed studies in various cell contexts, it seems appropriate to regard them all as plausible candidates. Note that model (a) is the only one that does not involve transfer of PrP^{Sc}.

5. Conclusions

This chapter has reviewed some of the considerable progress in understanding the intricate mechanisms regulating the biosynthesis and trafficking of PrP. This progress serves to underline a future direction already proposed in the earlier volume (Harris, 1999) – that is to identify the key molecules with which PrP interacts at various locations in the cell. These include the ER and Golgi components mediating quality control in relation to conformation, topological identity, aggregation state, glycosylation, disulfide bonding and other variables. The molecules associated with lipid rafts that mediate or regulate the conversion to PrP^{Sc}, internalization, vesiculation, signal transduction and other activities of PrP are also an important target.

There has been rather limited progress in extending cell culture studies to primary cells such as neurons and glia – another goal identified earlier (Harris, 1999) – but it is likely that this will soon be a prominent direction. A defined culture system for prion infection of primary neurons seems desirable, for example to explore the link between infection and pathogenesis. In view of the limited number of such cells, it would be very helpful to have selective cell labeling reagents to distinguish infected cells. It is clear overall that progress in understanding the normal cell biology of PrP goes hand in hand with understanding the events of infection and the consequences for a stably infected cell.

Acknowledgements

We thank Tim Landy for helpful comments on the manuscript, Anoop Kumar for help, Roger Morris for providing the micrograph of Figure 2, and the Biotechnology and Biological Sciences Research Council for support. JPB is a MRC Non-clinical Research Professor.

References

Baron, G. S., Wehrly, K., Dorward, D. W., Chesebro, B. and Caughey, B. 2002. Conversion of raft associated prion protein to the protease-resistant state requires insertion of PrP-res (PrP(Sc)) into contiguous membranes. EMBO J. 21: 1031-1040.

Birkett, C. R., Hennion, R. M., Bembridge, D. A., Clarke, M. C., Chree, A., Bruce, M. E. and Bostock, C. J. 2001. Scrapie strains maintain biological phenotypes on propagation in a cell line in culture. EMBO J. 20: 3351-3358.

Bonifacino, J. S. and Weissman, A. M. 1998. Ubiquitin and the control of protein fate in the secretory and endocytic pathways. Annu. Rev. Cell. Dev. Biol. 14: 19-57.

Bosque, P. J. and Prusiner, S. B. 2000. Cultured cell sublines highly susceptible to prion infection. J. Virol. 74: 4377-4386.

Brown, D. A., Crise, B. and Rose, J. K. 1989. Mechanism of membrane anchoring affects polarized expression of two proteins in MDCK cells. Science. 245: 1499-1501.

Cabral, C. M., Liu, Y. and Sifers, R. N. 2001. Dissecting glycoprotein quality control in the secretory pathway. Trends Biochem. Sci. 26: 619-624.

Capellari, S., Zaidi, S. I., Urig, C. B., Perry, G., Smith, M. A. and Petersen, R. B. 1999. Prion protein glycosylation is sensitive to redox change. J. Biol. Chem. 274: 34846-34850.

Caras, I. W., Weddell, G. N. and Williams, S. R. 1989. Analysis of the signal for attachment of a glycophospholipid membrane anchor. J. Cell. Biol. 108: 1387-1396.

Clarke, M. C. and Haig, D. A. 1970. Evidence for the multiplication of scrapie agent in cell culture. Nature. 225: 100-101.

DeArmond, S. J., Sanchez, H., Yehiely, F., Qiu, Y., Ninchak-Casey, A., Daggett, V., Camerino, A. P., Cayetano, J., Rogers, M., Groth, D., Torchia, M., Tremblay, P., Scott, M. R., Cohen, F. E. and Prusiner, S. B. 1997. Selective neuronal targeting in prion disease. Neuron. 19: 1337-1348.

Devasahayam, M., Catalino, P. D., Rudd, P. M., Dwek, R. A. and Barclay, A. N. 1999. The glycan processing and site occupancy of recombinant Thy-1 is markedly affected by the presence of a glyc osylphosphatidylinositol anchor. Glycobiology. 9: 1381-1387.

Enari, M., Flechsig, E. and Weissmann, C. 2001. Scrapie prion protein accumulation by scrapie-infected neuroblastoma cells abrogated by exposure to a prion protein antibody. Proc. Natl. Acad. Sci. USA. 98: 9295-9299.

Endo, T., Groth, D., Prusiner, S. B. and Kobata, A. 1989. Diversity of oligosaccharide structures linked to asparagines of the scrapie prion protein. Biochemistry. 28: 8380-8388.

Flechsig, E., Hegyi, I., Enari, M., Schwarz, P., Collinge, J. and Weissmann, C. 2001. Transmission of scrapie by steel-surface-bound prions. Mol. Med. 7: 679-684.

Follet, J., Lemaire-Vieille, C., Blanquet-Grossard, F., Podevin-Dimster, V., Lehmann, S., Chauvin, J. P., Decavel, J. P., Varea, R., Grassi, J., Fontes, M. and Cesbron, J. Y. 2002. PrP expression and replication by Schwann cells: implications in prion spreading. J. Virol. 76: 2434-2439.

Gilch, S., Winklhofer, K. F., Groschup, M. H., Nunziante, M., Lucassen, R., Spielhaupter, C., Muranyi, W., Riesner, D., Tatzelt, J. and Schatzl, H. M. 2001. Intracellular re-routing of prion protein prevents propagation of PrP(Sc) and delays onset of prion disease. EMBO J. 20: 3957-3966.

Greco, V., Hannus, M. and Eaton, S. 2001. Argosomes: a potential vehicle for the spread of morphogens through epithelia. Cell. 106: 633-645.

Haig, D. A. and Clarke, M. C. 1971. Multiplication of the scrapie agent. Nature. 234: 106-107.

Haraguchi, T., Fisher, S., Olofsson, S., Endo, T., Groth, D., Tarentino, A., Borchelt, D. R., Teplow, D., Hood, L., Burlingame, A. and et al. 1989. Asparagine-linked glycosylation of the scrapie and cellular prion proteins. Arch. Biochem. Biophys. 274: 1-13.

Harris, D. A. 1999 Cell. biological studies of the prion protein. In: Prions Molecular and Cellular Biology. D. A. Harris, ed. Horizon Scientific Press, Norfolk, UK. p. 53-65.

Hegde, R. S., Mastrianni, J. A., Scott, M. R., DeFea, K. A., Tremblay, P., Torchia, M., DeArmond, S. J., Prusiner, S. B. and Lingappa, V. R. 1998. A transmembrane form of the prion protein in neurodegenerative disease. Science. 279: 827-834.

Hegde, R. S., Tremblay, P., Groth, D., DeArmond, S. J., Prusiner, S. B. and Lingappa, V. R. 1999. Transmissible and genetic prion diseases share a common pathway of neurodegeneration. Nature. 402: 822-826.

Kaneko, K., Vey, M., Scott, M., Pilkuhn, S., Cohen, F. E. and Prusiner, S. B. 1997. COOH-terminal sequence of the cellular prion protein directs subcellular trafficking and controls conversion into the scrapie isoform. Proc. Natl. Acad. Sci. USA. 94: 2333-2338.

Kanu, N., Imokawa, Y., Drechsel, D. N., Williamson, R. A., Birkett, C. R., Bostock, C. J. and Brockes, J. P. 2002. Transfer of scrapie prion infectivity by cell contact in culture. Curr. Biol. 12: 523-530.

Kooyman, D. L., Byrne, G. W. and Logan, J. S. 1998. Glycosyl phosphatidylinositol anchor. Exp. Nephrol. 6: 148-151.

Korth, C., Kaneko, K. and Prusiner, S. B. 2000. Expression of unglycosylated mutated prion protein facilitates PrP(Sc) formation in neuroblastoma cells infected with different prion strains. J. Gen. Virol. 81: 2555-2563.

Korth, C., Stierli, B., Streit, P., Moser, M., Schaller, O., Fischer, R., Schulz-Schaeffer, W., Kretzschmar, H., Raeber, A., Braun, U., Ehrensperger, F., Hornemann, S., Glockshuber, R., Riek, R., Billeter, M., Wuthrich, K. and Oesch, B. 1997. Prion (PrPSc)-specific epitope defined by a monoclonal antibody. Nature. 390: 74-77.

Lehmann, S. and Harris, D. A. 1997. Blockade of glycosylation promotes acquisition of scrapie-like properties by the prion protein in cultured cells. J. Biol. Chem. 272: 21479-21487.

Liu, T., Li, R., Pan, T., Liu, D., Petersen, R. B., Wong, B. S., Gambetti, P. and Sy, M. S. 2002. Intercellular transfer of the cellular prion protein. J. Biol. Chem. 277: 47671-47678.

Ma, J. and Lindquist, S. 2001. Wild-type PrP and a mutant associated with prion disease are subject to retrograde transport and proteasome degradation. Proc. Natl. Acad. Sci. USA. 98: 14955-14960.

Ma, J. and Lindquist, S. 2002. Conversion of PrP to a Self-Perpetuating PrPSc-like Conformation in the Cytosol. Science. 298: 1785-1788.

Ma, J., Wollmann, R. and Lindquist, S. 2002. Neurotoxicity and neurodegeneration when PrP accumulates in the cytosol. Science. 298: 1781-1785.

Mack, M., Kleinschmidt, A., Bruhl, H., Klier, C., Nelson, P. J., Cihak, J., Plachy, J., Stangassinger, M., Erfle, V. and Schlondorff, D. 2000. Transfer of the chemokine receptor CCR5 between cells by membrane-derived microparticles: a mechanism for cellular human immunodeficiency virus 1 infection. Nat. Med. 6: 769-775.

Madore, N., Smith, K. L., Graham, C. H., Jen, A., Brady, K., Hall, S. and Morris, R. 1999. Functionally different GPI proteins are organized in different domains on the neuronal surface. EMBO J. 18: 6917-6926.

Mathon, N. F., Malcolm, D. S., Harrisingh, M. C., Cheng, L. and Lloyd, A. C. 2001. Lack of replicative senescence in normal rodent glia. Science. 291: 872-875.

Mouillet-Richard, S., Ermonval, M., Chebassier, C., Laplanche, J. L., Lehmann, S., Launay, J. M. and Kellermann, O. 2000. Signal transduction through prion protein. Science. 289: 1925-1928.

Nishida, N., Harris, D. A., Vilette, D., Laude, H., Frobert, Y., Grassi, J., Casanova, D., Milhavet, O. and Lehmann, S. 2000. Successful transmission of three mouse-adapted scrapie strains to murine neuroblastoma cell lines overexpressing wild-type mouse prion protein. J. Virol. 74: 320-325.

Nunziante, M., Gilch, S. and Schatzl, H. M. 2003. Essential role of the prion protein N-terminus in subcellular trafficking and half-life of PrPc. J. Biol. Chem. 278: 3726-3734.

Pauly, P. C. and Harris, D. A. 1998. Copper stimulates endocytosis of the prion protein. J. Biol. Chem. 273: 33107-33110.

Perera, W. S. and Hooper, N. M. 2001. Ablation of the metal ion-induced endocytosis of the prion protein by disease-associated mutation of the octarepeat region. Curr. Biol. 11: 519-523.

Peretz, D., Williamson, R. A., Kaneko, K., Vergara, J., Leclerc, E., Schmitt-Ulms, G., Mehlhorn, I. R., Legname, G., Wormald, M. R., Rudd, P. M., Dwek, R. A., Burton, D. R. and Prusiner, S. B. 2001. Antibodies inhibit prion propagation and clear cell cultures of prion infectivity. Nature. 412: 739-743.

Priola, S. A. and Lawson, V. A. 2001. Glycosylation influences cross-species formation of protease-resistant prion protein. EMBO J. 20: 6692-6699.

Rogers, M., Taraboulos, A., Scott, M., Groth, D. and Prusiner, S. B. 1990. Intracellular accumulation of the cellular prion protein after mutagenesis of its Asn-linked glycosylation sites. Glycobiology. 1: 101-109.

Rogers, M., Yehiely, F., Scott, M. and Prusiner, S. B. 1993. Conversion of truncated and elongated prion proteins into the scrapie isoform in cultured cells. Proc. Natl. Acad. Sci. USA. 90: 3182-3186.

Rudd, P. M., Endo, T., Colominas, C., Groth, D., Wheeler, S. F., Harvey, D. J., Wormald, M. R., Serban, H., Prusiner, S. B., Kobata, A. and Dwek, R. A. 1999. Glycosylation differences between the normal and pathogenic prion protein isoforms. Proc. Natl. Acad. Sci. USA. 96: 13044-13049.

Stewart, R. S., Drisaldi, B. and Harris, D. A. 2001. A transmembrane form of the prion protein contains an uncleaved signal peptide and is retained in the endoplasmic reticulum. Mol. Biol. Cell. 12: 881-889.

Stewart, R. S. and Harris, D. A. 2001. Most pathogenic mutations do not alter the membrane topology of the prion protein. J. Biol. Chem. 276: 2212-2220.

Stimson, E., Hope, J., Chong, A. and Burlingame, A. L. 1999. Site-specific characterization of the N-linked glycans of murine prion protein by high-performance liquid chromatography/electrospray mass spectrometry and exoglycosidase digestions. Biochemistry. 38: 4885-4895.

Suzuki, T., Seko, A., Kitajima, K., Inoue, Y. and Inoue, S. 1994. Purification and enzymatic properties of peptide:N-glycanase from C3H mouse-derived L-929 fibroblast cells. Possible widespread occurrence of post-translational remodification of proteins by N-deglycosylation. J. Biol. Chem. 269: 17611-17618.

Taraboulos, A., Rogers, M., Borchelt, D. R., McKinley, M. P., Scott, M., Serban, D. and Prusiner, S. B. 1990. Acquisition of protease resistance by prion proteins in scrapie-infected cells does not require asparagine-linked glycosylation. Proc. Natl. Acad. Sci. USA. 87: 8262-8266.

Taraboulos, A., Scott, M., Semenov, A., Avrahami, D., Laszlo, L., Prusiner, S. B. and Avraham, D. 1995. Cholesterol depletion and modification of COOH-terminal targeting sequence of the prion protein inhibit formation of the scrapie isoform. J. Cell. Biol. 129: 121-132.

Vilette, D., Andreoletti, O., Archer, F., Madelaine, M. F., Vilotte, J. L., Lehmann, S. and Laude, H. 2001. *Ex vivo* propagation of infectious sheep scrapie agent in heterologous epithelial cells expressing ovine prion protein. Proc. Natl. Acad. Sci. USA. 98: 4055-4059.

Vorberg, I. and Priola, S. A. 2002. Molecular basis of scrapie strain glycoform variation. J. Biol. Chem. 277: 36775-36781.

Walmsley, A. R. and Hooper, N. M. 2003. Distance of sequons to the C-terminus influences the cellular N-glycosylation of the prion protein. Biochem. J. 370: 351-355.

Walmsley, A. R., Zeng, F. and Hooper, N. M. 2001. Membrane topology influences N-glycosylation of the prion protein. EMBO J. 20: 703-712.

Weissmann, C., Enari, M., Klohn, P. C., Rossi, D. and Flechsig, E. 2002. Transmission of prions. Proc. Natl. Acad. Sci. USA. 99 Suppl 4: 16378-16383.

Yanai, A., Meiner, Z., Gahali, I., Gabizon, R. and Taraboulos, A. 1999. Subcellular trafficking abnormalities of a prion protein with a disrupted disulfide loop. FEBS Lett. 460: 11-16.

Yedidia, Y., Horonchik, L., Tzaban, S., Yanai, A. and Taraboulos, A. 2001. Proteasomes and ubiquitin are involved in the turnover of the wild-type prion protein. EMBO J. 20: 5383-5391.

Zanusso, G., Petersen, R. B., Jin, T., Jing, Y., Kanoush, R., Ferrari, S., Gambetti, P. and Singh, N. 1999. Proteasomal degradation and N-terminal protease resistance of the codon 145 mutant prion protein. J. Biol. Chem. 274: 23396-23404.

Zobeley, E., Flechsig, E., Cozzio, A., Enari, M. and Weissmann, C. 1999. Infectivity of scrapie prions bound to a stainless steel surface. Mol. Med. 5: 240-243.

Zulianello, L., Kaneko, K., Scott, M., Erpel, S., Han, D., Cohen, F. E. and Prusiner, S. B. 2000. Dominant-negative inhibition of prion formation diminished by deletion mutagenesis of the prion protein. J. Virol. 74: 4351-4360.

From: Prions and Prion Diseases: Current Perspectives
Edited by: Glenn C. Telling

Chapter 4

PrP Deletion Mutants

Surachai Supattapone
and Judith R. Rees

Abstract

A number of investigators have successfully used deletion mutagenesis of the prion protein (PrP) to investigate the mechanisms of prion propagation, PrP folding, and PrP^{Sc}-induced neurotoxicity. Several of these studies have shown that PrP is comprised of at least three distinct domains. The N-terminal domain spanning residues 23-88 facilitates efficient prion propagation. The central, hydrophobic region spanning residues 89-140 is amyloidogenic and mediates neurotoxicity. The C-terminal domain is structured and helps prevent PrP^C from misfolding. Other deletion mutagenesis studies have shown that a 106 amino acid molecule, PrP (Δ23-88,Δ141-176) can form infectious miniprions in transgenic (Tg) mice with an artificial transmission barrier. *In vitro* translation experiments demonstrated that deletion of the signal sequence spanning residues 1-22 increases production of the pathogenic topologic variant ^{Ctm}PrP, indicating that ^{Ctm}PrP is generated by post-translational translocation into endoplasmic reticulum membranes.

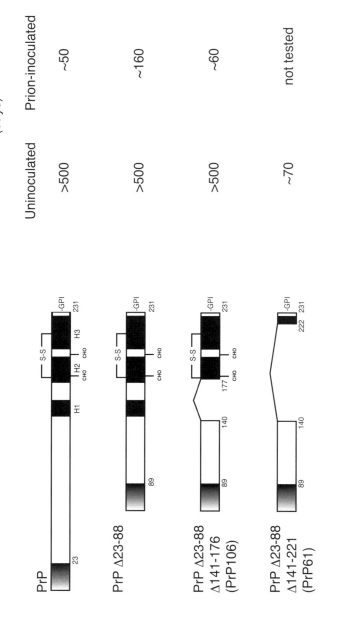

Figure 1. Lifespan and prion transmission data for Tg mice expressing PrP deletion mutants. Black areas on drawings correspond to α-helices of recombinant hamster PrP 90-231 determined by NMR spectroscopy (James *et al.*, 1997), and gradated gray areas represent the signal sequence (residues 1-22). Lifespan and scrapie transmission data for transgenic mice expressing each PrP deletion mutant are given in columns on the right. The inoculum for Tg(PrP)Prnp[0/0] and Tg(PrP,Δ23-88)Prnp[0/0] mice was RML brain homogenate containing full-length or N-terminally truncated prions. The inoculum for Tg(PrP106)Prnp[0/0] was RML106 brain homogenate containing PrP[Sc]106.

1. Introduction

For the past two decades, the major focus in prion research has been to determine the biochemical composition of the infectious agent. Extensive studies support the unorthodox hypothesis that infectious prions lack informational nucleic acid and instead contain a pathogenic conformation of a normal, host-encoded glycoprotein (Prusiner, 1982; 2000). The protein-only hypothesis is now widely accepted, and the central focus of prion research has shifted to determining the mechanism by which the normal glycoprotein, PrP^C, is converted into its pathogenic conformation, PrP^{Sc}. The strategy of deletion mutagenesis has been used by several groups of investigators to study the molecular basis of prion propagation, PrP conformational change, and prion disease pathogenesis. The results of these studies have yielded several important insights, and these are summarized in this chapter.

1.1. The N-Terminus of PrP Facilitates Efficient Prion Propagation

A number of independent deletion mutagenesis studies have shown that the N-terminus of PrP^C facilitates prion propagation and PrP^{Sc} formation. Supattapone and co-workers discovered that $Tg(MoPrP,\Delta23-88)Prnp^{0/0}$ mice propagated prions less efficiently than $Tg(MoPrP)Prnp^{0/0}$ mice. After inoculation with either full-length or truncated murine prions, $Tg(MoPrP,\Delta23-88)Prnp^{0/0}$ mice developed scrapie with incubation periods ~3 times longer than $Tg(MoPrP)Prnp^{0/0}$ mice (Figure 1) (Supattapone *et al.*, 2001b). Similarly, Flechsig and co-workers discovered that $Tg(MoPrP,\Delta32-93)Prnp^{0/0}$ mice had prolonged scrapie incubation times and prion titers ~30-fold lower than in wild-type mice (Flechsig *et al.*, 2000). In contrast, Fisher and co-workers found that $Tg(MoPrP,\Delta32-80)Prnp^{0/0}$ mice retaining one octarepeat propagate prions efficiently, with scrapie incubation times similar to those of mice expressing full length MoPrP (Fischer *et al.*, 1996). Taken together, these transmission data indicate that PrP^C requires at least one octarepeat sequence to act as an efficient substrate for conversion to PrP^{Sc}. In agreement with the transmission data from transgenic mice, cell culture studies showed that epitope-labeled $MHM2,\Delta23-88$ produced less protease-resistant PrP^{Sc} in RML scrapie-infected ScN2a neuroblastoma cells than full-length MHM2 (Supattapone *et al.*, 1999; 2001a). Furthermore, removal of residues 34-94 decreased the yield of protease-resistant PrP in an *in vitro* radiolabel conversion assay (Lawson *et al.*, 2001). These cell culture and *in vitro* conversion results suggest that the N-terminus of PrP, particularly the octarepeat sequences, plays an important role in facilitating efficient prion propagation. This region is unstructured (Donne *et al.*, 1997; James *et al.*, 1997; Liu *et al.*, 1999; Riek *et al.*, 1996; 1997), but contains all five octarepeat sequences that have been shown to bind copper (Hornshaw *et al.*, 1995a; 1995b; Stockel *et al.*, 1998; Whittal *et al.*, 2000).

The mechanism(s) by which the N-terminus of PrP facilitates prion propagation remains undetermined. One possibility is that copper catalyzes the pathogenic conformational change from PrP[C] to PrP[Sc]. In support of this concept, McKenzie *et al.* found that addition of copper to chemically denatured PrP[Sc] facilitated its renaturation into infectious PrP[Sc] in some but not all experiments (McKenzie *et al.*, 1998). Another possibility is that the N-terminus may participate in either intra- or inter-molecular interactions that destabilize PrP[C] structure. Donne and co-workers provided evidence in favor of the hypothesis that the N-terminus of PrP participates in intra-molecular interactions. They found that residues 23-120 of full-length *Escherichia coli*-derived, recombinant PrP influenced the structure of the C-terminal α-helices, as judged by chemical shift index analysis and nuclear Overhauser effect data obtained through NMR spectroscopy (Donne *et al.*, 1997). Zulianello and co-workers provided data that the N-terminus of PrP may participate in inter-molecular interactions. These investigators discovered that deletion of residues 23-88 from a dominant negative PrP molecule destroyed the ability of the resulting molecule (PrP,Δ23-88, Q218K) to inhibit conversion of wild type PrP (Zulianello *et al.*, 2000). It is possible that the N-terminus of PrP facilitates prion propagation in several different ways, but the critical mechanism(s) have yet to be elucidated.

1.2. A 106 Amino Acid PrP Molecule Forms Infectious Miniprions Efficiently

The smallest deletion mutant capable of propagating infectious prions was identified in 1999, and the characterization of this mutant has provided insight into the mechanism of prion conversion (Supattapone *et al.*, 1999). Prnp[0/0] mice expressing a mutant PrP molecule with 2 separate deletions, PrP(Δ23-88,Δ141-176), propagated infectious prions as efficiently as Prnp$^{0/0}$ mice expressing full length PrP (Figure 1). This molecule was named PrP106 because it contains only 106 amino acids in its processed sequence, compared to 208 for full length PrP. The ability of PrP106 to form infectious miniprions efficiently was surprising because this short molecule lacks all five N-terminal octarepeat sequences. To investigate why PrP106 is an efficient conversion template, the biophysical properties of PrP106 molecules expressed *in vivo* as well as purified recombinant and chemically synthesized PrP106 molecules refolded *in vitro* were characterized (Baskakov *et al.*, 2000; Bonetto *et al.*, 2002; Supattapone *et al.*, 1999). These studies showed that PrP106 molecules expressed in uninfected Tg mice and N2a neuroblastoma cells have an intrinsic propensity to fold into a conformation that is soluble in non-denaturing detergents and is resistant to digestion by proteinase K (PK) at a total protein: enzyme ratio of 70:1 for 30 min at 37°C. *E. coli*-derived, recombinant and chemically synthesized PrP106 molecules spontaneously fold into structures rich in β-sheet and aggregate into soluble oligomers in a concentration-dependent manner, whereas PrP(Δ23-88) molecules refold into structures that are predominantly α-helical even at high concentrations. These results suggest that removal of the internal residues 141-176 destabilizes PrP106 structure and facilitates prion formation. In support of this hypothesis, NMR structures of long recombinant PrP molecules indicate that this region contains α-helix H1 (Donne *et al.*, 1997; James *et al.*, 1997; Liu *et al.*, 1999; Riek *et al.*, 1996; 1997). Removal of H1 may leave PrP106 poised to form prions even though it lacks the N-terminal octarepeat region. Consistent with this hypothesis, Eberl and Glockshuber found that specific removal of H1 from an *E. coli-derived*, recombinant PrP polypeptide containing the C-terminal residues 121-231 thermodynamically destabilized the resulting structure, PrP(Δ23-120,Δ135-160), and produced a free energy of folding close to zero, as judged by CD spectroscopy (Eberl and Glockshuber, 2002). In the future, it will be interesting to create Tg(PrP,Δ141-176)Prnp$^{0/0}$ mice expressing PrP molecules that contain a disruptive internal deletion but retain the N-terminus. If destabilization of PrP structure caused by the internal deletion supersedes the action of the N-terminus, then Tg(PrP,Δ141-176)Prnp$^{0/0}$ mice would propagate prions as efficiently as Tg(PrP106)Prnp$^{0/0}$ mice. On the other hand, if the conversion-promoting effects of internal deletion the N-terminus were additive rather than complementary, Tg(PrP,Δ141-176)Prnp$^{0/0}$ mice would be expected propagate prions more efficiently than Tg(PrP106)Prnp$^{0/0}$ mice.

An interesting feature that was discovered during animal transmission and cell culture studies was that PrP106 possesses an artificial prion transmission barrier relative to full length PrP (Supattapone *et al.*, 1999). In wild type animals, the ability to transmit prion disease from one animal species to another is often inefficient and correlates with the degree of homology between the PrP molecules in the inoculum and the host animal. An analogous effect was observed in Tg(PrP106)Prnp$^{0/0}$ mice inoculated with full-length prions: whereas Tg(MoPrP)Prnp$^{0/0}$ mice had a scrapie incubation time of ~50 days after inoculation with full-length prions, Tg(PrP106)Prnp$^{0/0}$ mice had a scrapie incubation time of 300 days. The artificial prion transmission barrier between full-length PrP and PrP106 was reciprocal: whereas Tg(PrP106)Prnp$^{0/0}$ mice had a scrapie incubation time of 66 days after inoculation with RML106 prions, Tg(MoPrP)Prnp$^{0/0}$ mice did not develop disease even after 500 days. The artificial prion transmission barrier between full-length PrP and PrP molecules with internal deletions has also been observed in cell culture and *in vitro* conversion assays. Supattapone and co-workers found that ScN2a murine neuroblastoma cells infected with full-length RML prions did not convert PrP106 into PrP^{Sc}106 (Supattapone *et al.*, 1999; 2001a). Similarly, Vorberg and co-workers found that a PrP deletion mutant lacking α-helix H1 did not acquire protease-resistance when expressed in ScN2a cells or when mixed with full-length RML *in vitro* (Vorberg *et al.*, 2001). The creation of an artificial prion transmission barrier by deletion mutagenesis provides strong additional support for the prion hypothesis (Prusiner, 1982) because it is difficult to envision how the deliberate mutation of PrP sequence could alter the host range of the resulting prion inoculum unless PrP is a central component of the infectious agent. The knowledge that deletion mutagenesis creates artificial transmission barriers also has implications for the designs of future experiments involving new PrP deletion mutants. For example, we predict that infectious miniprions such as PrP^{Sc}106 would be better inocula than full-length PrP^{Sc} for animal, cell culture, and biochemical experiments involving PrP deletion mutants because full-length prions are inefficient at driving the conversion of such mutants, and may yield false-negative results.

In addition to providing insight into the mechanism of prion formation, studies of PrP106 also generated a miniprion useful for comparative structural studies. An analysis of the difference between the X-ray crystallographic structures of PrP27-30 and PrP^{Sc}106 revealed the location of the region encompassing residues 141-176 within the PrP^{Sc} crystal structure (Wille *et al.*, 2002). This region was found to reside within the core of a PrP^{Sc} hexamer.

1.3. PrP Deletion Mutants Causing Spontaneous Neurodegeneration

Experiments with transgenic mice expressing PrP deletion mutants have also provided insights into the mechanisms of PrP-mediated neurodegeneration. Three different neurodegenerative disease models have been generated in transgenic mice by expression of PrP deletion mutants (Muramoto *et al.*, 1997; Shmerling *et al.*, 1998; Supattapone *et al.*, 2001a). (1) Supattapone *et al.* identified a 61 amino acid deletion mutant, PrP(Δ23-88,Δ141-221), designated PrP61, that folds spontaneously into a conformation resembling PrP^{Sc} (Supattapone *et al.*, 2001a). PrP61molecules expressed in uninfected N2a neuroblastoma cells and Tg mice are protease-resistant and insoluble, and synthetic PrP61 peptide forms amyloid rods *in vitro*. Low-level expression of PrP61 in Tg mice causes a fatal, spontaneous neurodegenerative disease characterized by apoptosis of hippocampal and cortical neurons accompanied by reactive gliosis (Figure 1). (2) Shmerling *et al.* found that expression of PrP(Δ32-121) and PrP(Δ32-134) in Tg mice caused cell death limited to the granule layer of the cerebellum (Shmerling *et al.*, 1998). Interestingly, the neuronal death caused by these truncated molecules could be prevented by co-expression of full-length PrP. This phenomenon suggests that intermolecular interactions between truncated and full-length PrP molecules may prevent protein misfolding. (3) Muramoto *et al.* discovered that expression of PrP(Δ23-88,Δ177-200) or PrP(Δ23-88,Δ201-217) in Tg mice created heritable disorders resembling neuronal storage diseases (Muramoto *et al.*, 1997). These mutant PrP molecules, which lack either of the C-terminal α-helices, were sensitive to protease digestion but insoluble in non-denaturing detergents. These interesting results raise the possibility that some naturally occurring hereditary neuronal storage diseases may be caused by PrP mutations.

An important observation in all of these studies was that each misfolded PrP deletion mutant that causes neurodegeneration also demonstrates abnormal trafficking. Misfolded PrP61, PrP(Δ23-88,Δ177-200), and PrP(Δ23-88,Δ201-217) molecules all accumulate within inclusion bodies in the cytoplasm of neurons, as judged by immunocytochemistry. GFP-tagged PrP(Δ32-121) and PrP(Δ32-134) molecules appear to accumulate at the plasma membrane and fail to undergo copper-stimulated endocytosis in SN56 neuronal cells (Lee *et al.*, 2001). It is likely that the abnormal trafficking of misfolded PrP deletion mutants plays an important role in the pathogenesis of neurodegeneration.

1.4. Interactions Between PrP Deletion Mutants and Full Length PrP Molecules

Radiation-inactivation, x-ray crystallographic, and theoretical studies suggest that PrP molecules may dimerize or oligomerize, and that dimer/oligomer formation may play a critical role in prion propagation (Bellinger-Kawahara *et al.*, 1987; Hardy, 1991; Knaus *et al.*, 2001; Priola *et al.*, 1995; Tompa *et al.*, 2002). Several investigators co-expressed PrP deletion mutants with full-length PrP molecules to investigate whether heterologous PrP molecules can interact in *trans*. (1) Supattapone and co-workers showed that, whereas Tg(MHM2,Δ23-88)Prnp$^{0/0}$ mice expressing only N-terminally truncated, epitope-tagged PrP could not propagate infectious prions, Tg(MHM2,Δ23-88)Prnp$^{+/0}$ mice co-expressing full-length MoPrP propagated RML prions with an incubation time of ~165 days and generated MHM2(Δ23-88)PrP^{Sc} (Supattapone *et al.*, 1999). Most likely, MoPrP facilitated conversion of MHM2(Δ23-88) by providing a complementary N-terminus in *trans*. (2) Shmerling and co-workers found that the spontaneous neurodegenerative disease of Tg(PrP,Δ32-121)Prnp$^{0/0}$ and Tg(PrP,Δ32-134)Prnp$^{0/0}$ mice could be rescued by restoring one allele of full-length MoPrP (Shmerling *et al.*, 1998). (3) Holscher and co-workers showed that expression of PrP (Δ114-121), a nonconvertible deletion mutant, in ScN2a cells led to *trans*-dominant inhibition of endogenous full-length PrP^{Sc} accumulation (Holscher *et al.*, 1998). Taken together, these results suggest that PrP molecules normally exist as oligomers, and that several artificially redacted PrP molecules contain the domains needed to interact with wild type PrP molecules within such oligomers.

1.5. Sorting Signal Deletions

The open reading frame (ORF) of PrP contains 254 amino acids. After processing, the mature PrP polypeptide sequence contains 208 amino acids encompassing residues 23-231 (Oesch *et al.*, 1985). Residues

1-22 constitute the signal sequence that targets PrP into the lumen of the endoplasmic reticulum (ER) before cleavage by signal peptidase (Robakis *et al.*, 1986). Residues 232-254 constitute the glycophosphatidylinositol (GPI)-addition signal that is cleaved within the ER and directs the attachment of a GPI anchor to the C-terminus of residue 231 (Stahl *et al.*, 1987). The C-terminal GPI anchor causes mature PrP to traffic to cholesterol-rich domains within the plasma membrane (Kaneko *et al.*, 1997; Naslavsky *et al.*, 1997; Taraboulos *et al.*, 1995; Vey *et al.*, 1996).

Both the signal sequence and the GPI-addition signal of PrP have been deleted experimentally. Holscher and co-workers examined the effect of selectively deleting the N-terminal signal sequence from the PrP ORF (Holscher *et al.*, 2001). When MHM2(Δ1-22) was translated *in vitro* using a reticulocyte lysate, ~7% of the mature PrP molecules successfully translocated across the ER membrane despite the lack of an N-terminal signal sequence. The investigators found that these MHM2(Δ1-22) molecules used the C-terminal GPI-addition signal as an alternative signal sequence to translocate across the ER membrane after translation. Interestingly, the translocated MHM2(Δ1-22) molecules were predominantly oriented with ^{Ctm}PrP topology. The ^{Ctm}PrP topological form, which is characterized by a single transmembrane domain and a luminal C-terminus, may play a role in mediating the neurodegeneration of inherited prion diseases (Hegde *et al.*, 1998). In contrast, deletion of the GPI-addition signal results in a 5-fold reduction of ^{Ctm}PrP formation during *in vitro* translation experiments (Holscher *et al.*, 2001). Attempts to express PrP(Δ232-254) in Tg mice have thus far been unsuccessful, probably because PrP unattached to the membrane would either be subject to quality control mechanisms within the ER or secreted into the extracellular space.

1.6. Artificial PrP^{Sc} Analogs

An unexpected byproduct of PrP deletion mutagenesis studies has been the creation of artificial PrP^{Sc} analogs. As mentioned previously, PrP106 folds spontaneously into a conformation with biophysical properties that lie in between those of PrP^C and PrP^{Sc}. Further modification of the PrP106 backbone either by extension of the internal deletion or by addition of positively charged affinity tags yields PrP molecules with biophysical properties resembling PrP^{Sc}.

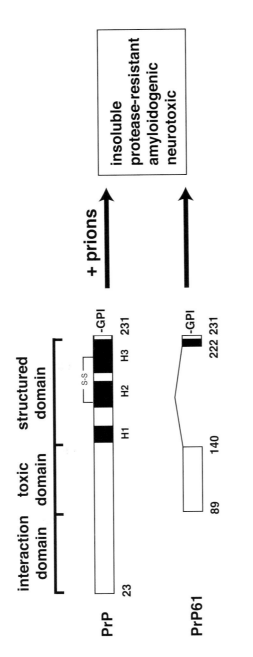

Figure 2. PrP domains identified by deletion mutagenesis. Schematic summary of experiments with PrP deletion mutants. PrP61 is a deletion mutant that spontaneously folds into protease-resistant conformation, whereas full-length PrP^C requires infectious prions to convert into protease-resistant PrP^{Sc}.

Extension of the internal deletion of PrP106 yielded surprising results. As mentioned above, PrP106 expressed in uninfected cells and animals spontaneously folds into a conformation that is resistant to mild PK digestion at a total protein: enzyme ratio of 70:1 for 30 min at 37°C (Supattapone *et al.*, 1999). Unlike PrP^{Sc}106 and full-length PrP^{Sc} molecules, PrP106 in uninfected cells and animals is soluble in non-denaturing detergents and is sensitive to harsh PK digestion at a total protein: enzyme ratio of 25:1 for 1 hr at 37°C. To investigate the molecular basis of miniprion folding, Supattapone and co-workers mutated the PrP106 backbone by creating progressively longer internal deletions and expressed the resulting constructs in uninfected N2a and scrapie-infected ScN2a cells (Supattapone *et al.*, 2001a). These experiments revealed that extension of the internal deletion to 141-186 created a PrP molecule that was sensitive to mild PK digestion, but further extension of the internal deletion to 141-205 or 141-221 yielded molecules that were detergent-insoluble and resistant to harsh PK digestion in uninfected as well as infected cells. These results suggest that (1) PrP spontaneously folds into a conformation resistant to mild PK digestion when α-helix H1 is disrupted and α-helix H2 is intact, and (2) PrP spontaneously folds into a conformation that is resistant to harsh PK digestion when all three α-helices are disrupted. Simultaneous removal of the N-terminal interaction domain and the C-terminal α-helices generated PrP61 bound to the membrane by the GPI anchor. When PrP61 was expressed at low levels in Tg mice, neurodegeneration accompanied by apoptosis was observed (Figure 1). Synthetic PrP61 covalently attached to myristic acid at the C-terminus spontaneously folded into amyloid fibrils rich in β-sheet structure in the presence of physiological salt concentrations. These fibrils were detergent-insoluble and resistant to protease digestion, but did not cause prion disease when injected intracerebrally into wild type or Tg(PrP106)Prnp$^{0/0}$ mice. Because full-length PrP^C requires infectious prions to catalyze conversion into protease-resistant PrP^{Sc} but PrP61 spontaneously folds into its protease-resistant conformation (Figure 2), it is likely that the structured domain 140-222 provides the thermodynamic barrier to PrP misfolding during prion propagation.

Addition of positively charged affinity tags to PrP106 also creates insoluble molecules resistant to harsh PK digestion at a total protein: enzyme ratio of 25:1 for 1 hr at 37°C (Supattapone *et al.*, 2000). The maximal yield of protease-resistant PrP106 was obtained with addition of a six-histidine (6His) or six-lysine tag. Using shorter, longer, hydrophobic, and negatively charged tags reduced the recovery of protease-resistant PrP106 molecules. The location of the affinity tag within PrP106 also influenced the biophysical properties of the resulting molecules. Replacement of the internal deletion of PrP106 with 6His generates a protein that is protease-resistant and partially soluble in 0.5% NP40, 0.5% deoxycholate. Placement of the 6His tag near the C-terminus of PrP106 between residues 225 and 226 yields a protease-resistant molecule that is not soluble in non-denaturing detergents. Addition of the 6His tag to the N-terminus of PrP106 at residue 89 yielded a molecule that was sensitive to PK digestion at 25:1 total protein: enzyme ratio. Interestingly, addition of a disease-associated mutation (E200K) to PrP106(140/6His) increased the yield of protease-resistant PrP >5 fold, raising the possibility that PrP106(140/6His) could be used as a sentinel molecule to screen for potentially pathogenic PrP mutations.

Both PrP61 and PrP106(140/6His,E200K) could be purged from N2a cells by the prion-clearing compound polypropyleneimine (PPI) generation 4.0, demonstrating in principle the usefulness of these PrP^{Sc} analogs in drug screening (Supattapone *et al.,* 2001a; 2000). These artificial analog molecules offer several advantages over *bona fide* PrP^{Sc} as experimental reagents. Both PrP61 and PrP106(140/6His,E200K) can be synthesized chemically, and therefore can be generated in large quantities and linked covalently to useful adducts such as biotin or fluorescent probes. Both molecules fold spontaneously into a conformation resembling PrP^{Sc}, but neither is infectious for Tg(HuPrP)Prnp$^{0/0}$ or Tg(MoPrP)Prnp$^{0/0}$ mice. Thus, drug-screening assays using PrP61 and PrP106(140/6His,E200K) molecules would pose less biohazard risk than assays using bona fide PrP^{Sc} preparations.

2. Future Prospects

The general strategy of deletion mutagenesis has already yielded many important insights into the mechanisms of prion propagation, PrP folding, and PrP^{Sc}-mediated neurotoxicity. PrP deletion mutagenesis studies have also generated a number of molecules that will be useful tools for future biological, structural, and pharmacological studies. A few examples of how PrP deletion mutants could be used in the future are provided below.

1) Tg(PrP106)Prnp$^{0/0}$ mice can be used to investigate the structural basis of strain diversity. Strain diversity is an interesting and incompletely explained feature of prion biology that is characterized by the ability of distinct infectious isolates to create distinguishable diseases, with unique clinical symptoms, pathology, prion incubation times, and PrP^{Sc} biophysical properties (Dickinson *et al.*, 1968; Kimberlin, 1990). Several lines of evidence suggest that prion strain diversity is encoded by subtle conformational differences in PrP^{Sc} molecules (Bessen *et al.*, 1995; Collinge *et al.*, 1996; Telling *et al.*, 1996). Because PrP106 lacks ~50% of the full-length polypeptide sequence and is likely to have a constrained structure, it would be interesting to test whether PrP106 is capable of supporting prion strain diversity. A variety of established wild type murine prion strains could be passaged through Tg(PrP106)Prnp$^{0/0}$ mice to generate different isolates of infectious miniprions. These miniprions could then be assayed to determine whether they remain phenotypically distinct or whether their properties converge as a result of the structural constraints imposed by PrP106.

2) Infectious RML106 prions can be used to help create smaller miniprions. The ability of PrP106 to support prion propagation suggests that PrP residues 23-88 and 141-176 are not essential for the propagation process. It would simplify our analysis of the conformational change process to determine the minimal structure capable of forming PrP^{Sc}. Because miniprions possess artificial prion transmission barriers in relation to full-length prions, the most logical approach to creating smaller miniprions would be to extend the deletions of PrP106 gradually. Each smaller molecule could be tested for its ability to propagate prions in Tg mice by using the smallest infectious miniprion available as an inoculum, starting with RML106.

3) Synthetic PrP106 can be used to determine the structure of a folding intermediate. In uninfected cells, PrP106 appears to adopt a conformation that may resemble an intermediate on the folding pathway from PrP^{C} to PrP^{Sc}. Unglycosylated PrP106 purified from *E. coli* or synthesized chemically spontaneously folds into soluble oligomers rich in β-sheet structure (Baskakov *et al.*, 2000; Bonetto *et al.*, 2002). Such oligomers have a PK-resistant core comprised of residues 134-215. Because PrP106 is soluble, it may be possible to determine its three-dimensional molecular

structure either by X-ray crystallography or NMR spectroscopy, and thereby gain insight into the possible structure of a prion conversion intermediate. Because it has a destabilized structure, PrP106 may also be a good substrate for attempts to form PrP^{Sc} *in vitro*.

4) PrP61 and affinity-tagged PrP106 molecules can be used as PrP^{Sc} analogs. Supattapone and co-workers found that PrP61 and some affinity-tagged PrP106 molecules share biophysical properties with PrP^{Sc} and cause neuronal death when expressed in Tg mice (Supattapone *et al.*, 2001a; 2000). Such molecules could be used as convenient and easily manipulable PrP^{Sc} analogs for studies of PrP^{Sc}-induced cellular toxicity or therapeutic drug screening. In particular, PrP61 could be expressed in cells or animals using an inducible promoter as a tool to control PrP-induced apoptosis experimentally.

5) PrP(Δ1-22) can be used to study the pathogenic relevance of ^{Ctm}PrP. *In vitro* translation studies showed that PrP(Δ1-22) molecules lacking the normal signal sequence generate >5-fold more ^{Ctm}PrP than full-length PrP (Holscher *et al.*, 2001). PrP(Δ1-22) molecules containing wild type and disease-associated mutations can be expressed in Tg mice to test the hypothesis that ^{Ctm}PrP is responsible for neurodegeneration in inherited prion diseases (Hegde *et al.*, 1998). If the hypothesis that ^{Ctm}PrP causes neurodegeneration is true, increasing ^{Ctm}PrP by deleting the signal sequence should lead to more rapid neurodegeneration.

3. Conclusion

In conclusion, the results obtained from studies of deletion mutants suggest that full-length PrP contains three separate functional domains (Figure 2). The N-terminus appears to facilitate prion propagation either by intra- or inter-molecular interactions. The central, hydrophobic sequence 90-140, which forms the core peptide sequence of PrP61, appears to be amyloidogenic and neurotoxic. The C-terminus appears to be a structured domain, and disruption of this domain facilitates the formation of infectious PrP106 miniprions and other misfolded, neurotoxic PrP isoforms. Studies of PrP deletion mutants have provided many important insights into the mechanisms of prion propagation, PrP folding, and PrP^{Sc}-mediated neurotoxicity. Several PrP deletion mutants will serve as useful tools for future research.

4. Acknowledgements

S.S. is a recipient of the Burroughs Wellcome Fund Career Development Award, a Hitchcock Foundation Award, and an NIH Clinical Investigator Development Award (K08 NS02048-02).

References

Baskakov, I. V., Aagaard, C., Mehlhorn, I., Wille, H., Groth, D., Baldwin, M. A., Prusiner, S. B., and Cohen, F. E. 2000. Self-assembly of recombinant prion protein of 106 residues. Biochemistry. 39: 2792-2804.

Bellinger-Kawahara, C., Cleaver, J. E., Diener, T. O., and Prusiner, S. B. 1987. Purified scrapie prions resist inactivation by UV irradiation. J. Virol. 61: 159-166.

Bessen, R. A., Kocisko, D. A., Raymond, G. J., Nandan, S., Lansbury, P. T., and Caughey, B. 1995. Non-genetic propagation of strain-specific properties of scrapie prion protein. Nature. 375: 698-700.

Bonetto, V., Massignan, T., Chiesa, R., Morbin, M., Mazzoleni, G., Diomede, L., Angeretti, N., Colombo, L., Forloni, G., Tagliavini, F., and Salmona, M. 2002. Synthetic miniprion PrP106. J. Biol. Chem. 277: 31327-31334.

Collinge, J., Sidle, K. C., Meads, J., Ironside, J., and Hill, A. F. 1996. Molecular analysis of prion strain variation and the aetiology of 'new variant' CJD. Nature. 383: 685-690.

Dickinson, A. G., Meikle, V. M., and Fraser, H. 1968. Identification of a gene which controls the incubation period of some strains of scrapie agent in mice. J. Comp. Pathol. 78: 293-299.

Donne, D. G., Viles, J. H., Groth, D., Mehlhorn, I., James, T. L., Cohen, F. E., Prusiner, S. B., Wright, P. E., and Dyson, H. J. 1997. Structure of the recombinant full-length hamster prion protein PrP(29- 231): the N terminus is highly flexible. Proc. Natl. Acad. Sci. USA. 94: 13452-13457.

Eberl, H., and Glockshuber, R. 2002. Folding and intrinsic stability of deletion variants of PrP(121-231), the folded C-terminal domain of the prion protein. Biophys. Chem. 96: 293-303.

Fischer, M., Rulicke, T., Raeber, A., Sailer, A., Moser, M., Oesch, B., Brandner, S., Aguzzi, A., and Weissmann, C. 1996. Prion protein (PrP) with amino-proximal deletions restoring susceptibility of PrP knockout mice to scrapie. EMBO J. 15: 1255-1264.

Flechsig, E., Shmerling, D., Hegyi, I., Raeber, A. J., Fischer, M., Cozzio, A., von Mering, C., Aguzzi, A., and Weissmann, C. 2000. Prion protein devoid of the octapeptide repeat region restores susceptibility to scrapie in PrP knockout mice. Neuron. 27: 399-408.

Hardy, J. 1991. Prion dimers: a deadly duo? Trends Neurosci. 14: 423-424.

Hegde, R. S., Mastrianni, J. A., Scott, M. R., DeFea, K. A., Tremblay, P., Torchia, M., DeArmond, S. J., Prusiner, S. B., and Lingappa, V. R. 1998. A transmembrane form of the prion protein in neurodegenerative disease. Science. 279: 827-834.

Holscher, C., Bach, U. C., and Dobberstein, B. 2001. Prion protein contains a second endoplasmic reticulum targeting signal sequence located at its C terminus. J. Biol. Chem. 276: 13388-13394.

Holscher, C., Delius, H., and Burkle, A. 1998. Overexpression of nonconvertible PrPc delta114-121 in scrapie-infected mouse neuroblastoma cells leads to trans-dominant inhibition of wild-type PrP(Sc) accumulation. J. Virol. 72: 1153-1159.

Hornshaw, M. P., McDermott, J. R., and Candy, J. M. 1995a. Copper binding to the N-terminal tandem repeat regions of mammalian and avian prion protein. Biochem. Biophys. Res. Commun. 207: 621-629.

Hornshaw, M. P., McDermott, J. R., Candy, J. M., and Lakey, J. H. 1995b. Copper binding to the N-terminal tandem repeat region of mammalian and avian prion protein: structural studies using synthetic peptides. Biochem. Biophys. Res. Commun. 214: 993-999.

James, T. L., Liu, H., Ulyanov, N. B., Farr-Jones, S., Zhang, H., Donne, D. G., Kaneko, K., Groth, D., Mehlhorn, I., Prusiner, S. B., and Cohen, F. E. 1997. Solution structure of a 142-residue recombinant prion protein corresponding to the infectious fragment of the scrapie isoform. Proc. Natl. Acad. Sci. USA. 94: 10086-10091.

Kaneko, K., Vey, M., Scott, M., Pilkuhn, S., Cohen, F. E., and Prusiner, S. B. 1997. COOH-terminal sequence of the cellular prion protein directs subcellular trafficking and controls conversion into the scrapie isoform. Proc. Natl. Acad. Sci. USA. 94: 2333-2338.

Kimberlin, R. H. 1990. Scrapie and possible relationships with viroids. Semin. Virol. 1: 153-162.

Knaus, K. J., Morillas, M., Swietnicki, W., Malone, M., Surewicz, W. K., and Yee, V. C. 2001. Crystal structure of the human prion protein reveals a mechanism for oligomerization. Nat. Struct. Biol. 8: 770-774.

Lawson, V. A., Priola, S. A., Wehrly, K., and Chesebro, B. 2001. N-terminal truncation of prion protein affects both formation and conformation of abnormal protease-resistant prion protein generated in vitro. J. Biol. Chem. 276: 35265-35271.

Lee, K. S., Magalhaes, A. C., Zanata, S. M., Brentani, R. R., Martins, V. R., and Prado, M. A. 2001. Internalization of mammalian fluorescent cellular prion protein and N- terminal deletion mutants in living cells. J. Neurochem. 79: 79-87.

Liu, H., Farr-Jones, S., Ulyanov, N. B., Llinas, M., Marqusee, S., Groth, D., Cohen, F. E., Prusiner, S. B., and James, T. L. 1999. Solution structure of Syrian hamster prion protein rPrP(90-231). Biochemistry. 38: 5362-5377.

McKenzie, D., Bartz, J., Mirwald, J., Olander, D., Marsh, R., and Aiken, J. 1998. Reversibility of scrapie inactivation is enhanced by copper. J. Biol. Chem. 273: 25545-25547.

Muramoto, T., DeArmond, S. J., Scott, M., Telling, G. C., Cohen, F. E., and Prusiner, S. B. 1997. Heritable disorder resembling neuronal storage disease in mice expressing prion protein with deletion of an alpha-helix. Nat. Med. 3: 750-755.

Naslavsky, N., Stein, R., Yanai, A., Friedlander, G., and Taraboulos, A. 1997. Characterization of detergent-insoluble complexes containing the cellular prion protein and its scrapie isoform. J. Biol. Chem. 272: 6324-6331.

Oesch, B., Westaway, D., Walchli, M., McKinley, M. P., Kent, S. B., Aebersold, R., Barry, R. A., Tempst, P., Teplow, D. B., Hood, L. E., *et al*. 1985. A cellular gene encodes scrapie PrP 27-30 protein. Cell. 40: 735-746.

Priola, S. A., Caughey, B., Wehrly, K., and Chesebro, B. 1995. A 60-kDa prion protein (PrP) with properties of both the normal and scrapie-associated forms of PrP. J. Biol. Chem. 270: 3299-3305.

Prusiner, S. B. 1982. Novel proteinaceous infectious particles cause scrapie. Science. 216: 136-144.

Prusiner, S. B. 2000. Prion Biology and Diseases. Cold Spring Harbor, N.Y., Cold Spring Harbor Laboratory Press.

Riek, R., Hornemann, S., Wider, G., Billeter, M., Glockshuber, R., and Wuthrich, K. 1996. NMR structure of the mouse prion protein domain PrP(121-321). Nature. 382: 180-182.

Riek, R., Hornemann, S., Wider, G., Glockshuber, R., and Wuthrich, K. 1997. NMR characterization of the full-length recombinant murine prion protein, mPrP(23-231). FEBS Lett. 413: 282-288.

Robakis, N. K., Sawh, P. R., Wolfe, G. C., Rubenstein, R., Carp, R. I., and Innis, M. A. 1986. Isolation of a cDNA clone encoding the leader peptide of prion protein and expression of the homologous gene in various tissues. Proc. Natl. Acad. Sci. USA. 83: 6377-6381.

Shmerling, D., Hegyi, I., Fischer, M., Blattler, T., Brandner, S., Gotz, J., Rulicke, T., Flechsig, E., Cozzio, A., von Mering, C., *et al.* 1998. Expression of amino-terminally truncated PrP in the mouse leading to ataxia and specific cerebellar lesions. Cell. 93: 203-214.

Stahl, N., Borchelt, D. R., Hsiao, K., and Prusiner, S. B. 1987. Scrapie prion protein contains a phosphatidylinositol glycolipid. Cell. 51: 229-240.

Stockel, J., Safar, J., Wallace, A. C., Cohen, F. E., and Prusiner, S. B. 1998. Prion protein selectively binds copper(II) ions. Biochemistry. 37: 7185-7193.

Supattapone, S., Bosque, P., Muramoto, T., Wille, H., Aagaard, C., Peretz, D., Nguyen, H. O., Heinrich, C., Torchia, M., Safar, J., *et al.* 1999. Prion protein of 106 residues creates an artifical transmission barrier for prion replication in transgenic mice. Cell. 96: 869-878.

Supattapone, S., Bouzamondo, E., Ball, H. L., Wille, H., Nguyen, H. O., Cohen, F. E., DeArmond, S. J., Prusiner, S. B., and Scott, M. 2001a. A protease-resistant 61-residue prion peptide causes neurodegeneration in transgenic mice. Mol. Cell. Biol. 21: 2608-2616.

Supattapone, S., Muramoto, T., Legname, G., Mehlhorn, I., Cohen, F. E., DeArmond, S. J., Prusiner, S. B., and Scott, M. R. 2001b. Identification of two prion protein regions that modify scrapie incubation time. J. Virol. 75: 1408-1413.

Supattapone, S., Nguyen, H. O., Muramoto, T., Cohen, F. E., DeArmond, S. J., Prusiner, S. B., and Scott, M. 2000. Affinity-tagged miniprion derivatives spontaneously adopt protease-resistant conformations. J. Virol. 74: 11928-11934.

Taraboulos, A., Scott, M., Semenov, A., Avrahami, D., Laszlo, L., Prusiner, S. B., and Avraham, D. 1995. Cholesterol depletion and modification of COOH-terminal targeting sequence of the prion protein inhibit formation of the scrapie isoform. J. Cell. Biol. 129: 121-132.

Telling, G. C., Parchi, P., DeArmond, S. J., Cortelli, P., Montagna, P., Gabizon, R., Mastrianni, J., Lugaresi, E., Gambetti, P., and Prusiner, S. B. 1996. Evidence for the conformation of the pathologic isoform of the prion protein enciphering and propagating prion diversity. Science. 274: 2079-2082.

Tompa, P., Tusnady, G. E., Friedrich, P., and Simon, I. 2002. The role of dimerization in prion replication. Biophys. J. 82: 1711-1718.

Vey, M., Pilkuhn, S., Wille, H., Nixon, R., DeArmond, S. J., Smart, E. J., Anderson, R. G., Taraboulos, A., and Prusiner, S. B. 1996. Subcellular colocalization of the cellular and scrapie prion proteins in caveolae-like membranous domains. Proc. Natl. Acad. Sci. USA. 93: 14945-14949.

Vorberg, I., Chan, K., and Priola, S. A. 2001. Deletion of beta-strand and alpha-helix secondary structure in normal prion protein inhibits formation of its protease-resistant isoform. J. Virol. 75: 10024-10032.

Whittal, R. M., Ball, H. L., Cohen, F. E., Burlingame, A. L., Prusiner, S. B., and Baldwin, M. A. 2000. Copper binding to octarepeat peptides of the prion protein monitored by mass spectrometry. Protein. Sci. 9: 332-343.

Wille, H., Michelitsch, M. D., Guenebaut, V., Supattapone, S., Serban, A., Cohen, F. E., Agard, D. A., and Prusiner, S. B. 2002. Structural studies of the scrapie prion protein by electron crystallography. Proc. Natl. Acad. Sci. USA. 99: 3563-3568.

Zulianello, L., Kaneko, K., Scott, M., Erpel, S., Han, D., Cohen, F. E., and Prusiner, S. B. 2000. Dominant-negative inhibition of prion formation diminished by deletion mutagenesis of the prion protein. J. Virol. 74: 4351-4360.

From: Prions and Prion Diseases: Current Perspectives
Edited by: Glenn C. Telling

Chapter 5

Targeting the Murine PrP Gene

Rona M. Barron and Jean C. Manson

Abstract

The prion protein (PrP) is known to be central to the Transmissible
Spongiform Encephalopathies (TSE) and point mutations and
polymorphisms in the PrP gene have a major effect in defining the
incubation time of TSE disease and the susceptibility of the host
to TSE infection. In mice, polymorphisms at amino acids 108 and
189 of PrP have been implicated in the control of TSE incubation
time. In humans, point mutations in the PrP protein are linked to
the occurrence of familial forms of TSE disease, such as P102L
Gerstmann-Straussler-Scheinker Syndrome (P102L GSS). In order
to study the role of these point mutations and polymorphisms in PrP
in TSE disease, we have developed several transgenic lines in which
specific mutations have been introduced into the endogenous murine
PrP protein by gene targeting. These transgenic mice express PrP at
the same level and under the same regulatory controls as wild type
mice, and all transgenic lines are directly comparable as the mutations
have been introduced on the same genetic background. Inoculation
of these mice has shown that amino acids 108 and 189 in murine PrP
do indeed control scrapie incubation time in mice, and that the 101L
mutation alters the susceptibility of the mice to several strains of TSE
agent from different species.

1. Introduction

The study of TSE diseases has been aided in recent years by the use of transgenic mice. Standard transgenic technology has been used to produce a large number of different transgenic lines expressing varying levels of PrP, specific PrP polymorphisms and mutations, PrP from different species, and chimeric proteins combining the PrP sequences of two different species. The varying expression levels and genetic backgrounds of these lines make direct comparison of lines difficult. Therefore a controlled system producing transgenic mice with wild type expression levels and tissue specificity would allow a more detailed study of the effects of specific mutations and polymorphisms in PrP in TSE disease. This can be achieved by replacing the endogenous murine gene with a mutated gene by gene targeting.

2. Production of Gene Targeted Transgenic Mice

The production of murine transgenic models by gene targeting results in the creation of lines either heterozygous or homozygous for the desired transgene, which is present in the correct location in the mouse genome under the correct transcriptional controls. These mice are inbred to produce transgenic mice which are identical to wild type mice, with the exception of the specific mutation introduced in the transgene, removing the complications encountered with site of insertion and expression level in standard transgenic lines. In gene targeting, the endogenous murine gene in embryonic stem (ES) cells is replaced by homologous recombination with a mutated gene, or the equivalent gene from another species. Successfully targeted ES cell clones containing the mutant gene are injected into mouse embryos (3.5 day old blastocysts), where they are capable of contributing towards the developing embryo. Any pups born which are derived from both embryo and stem cell line are called chimeric mice, and are identified by the presence of two coat colors (one from the stem cell line, the other from the blastocyst into which the ES cells were injected). If the ES cells contribute to the germ line of the chimera, it can be inbred to produce a transgenic line, differing from the wild type by only the inserted mutation. All lines produced by this method will express the mutated gene under the correct transcriptional and translational controls, allowing the direct comparison of transgenic lines encoding different mutations in the same gene. We have used this technology to produce a number of different lines of transgenic mice with specific alterations in the PrP gene. The identical genetic background, PrP protein levels and tissue specificity in these lines allows us to directly compare the effect of different mutations in TSE disease.

3. Gene Targeted Transgenic Lines

3.1. Murine 108/189 PrP Polymorphisms

Studies of TSE transmissions in congenic mice identified a gene/locus, which appeared to control the incubation time of mouse scrapie, called *Sinc* or *Prni* (Dickinson *et al.*, 1968; Carlson *et al.*, 1986). $Sinc^{s7}/Prni^N$ mice were identified by their short incubation times after inoculation with ME7 and Chandler mouse scrapie strains, while $Sinc^{p7}/Prni^I$ mice showed prolonged incubation times with these strains of agent. The isolation of PrP from purified hamster infectivity, and the subsequent identification of its gene (*Prnp*) provided a candidate for the gene controlling scrapie incubation time. Restriction-fragment length polymorphism (RFLP) analysis suggested that *Sinc/Prni* and *Prnp* were linked (Carlson *et al.*, 1986; Hunter *et al.*, 1992), and the two alleles of *Prnp* (*Prnp^a* and *Prnp^b*) were shown to segregate with $Sinc^{s7}/Prni^N$ and $Sinc^{p7}/Prni^I$ (Westaway *et al.*, 1987). However, classical genetic analysis was not able to prove formally that *Sinc/Prni and Prnp* were the same gene (Hunter *et al.*, 1992; Carlson *et al.*, 1993).

The PrPs encoded by *Prnp^a* and *Prnp^b* were shown to contain two amino acid polymorphisms at codons 108 and 189. The *Prnp^a* allele encoded 108L/189T (PrP A), while *Prnp^b* PrP contained 108F/189V (PrP B). All other positions in the two proteins were identical. Gene targeting was used to introduce 108F and 189V into the endogenous murine *Prnp^a* gene of 129/Ola mice ($Prnp^{a(108F/189V)}$), giving rise to transgenic mice (FV/FV) which expressed PrP B (Moore *et al.*, 1998), but were in all other ways identical to wild type 129/Ola mice. On challenge with the $Sinc^{p7}$ passaged BSE strain 301V, dramatically reduced incubation times were observed in FV/FV mice, proving that the polymorphisms at 108/189 do control TSE incubation times and that *Sinc, Prni and Prnp* are the same gene (Moore *et al.*, 1998). Further transmissions of mouse scrapie strains ME7, 79A, 139A, 301C, and 22A to the FV/FV transgenic mice (Barron *et al.*, 2003)

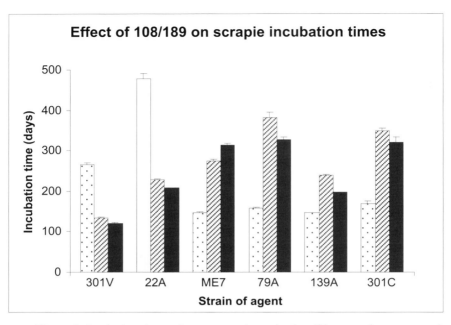

Figure 1. Incubation times of mouse scrapie strains in wild type and gene targeted FV/FV transgenic mice. Dotted bars, control 129/Ola mice (*Prnp^a*); hatched bars, gene targeted FV/FV mice (*Prnp^b* on 129/Ola background); solid bars, VM mice (*Prnp^b*).

have confirmed these observations, and show that when introduced into the *Prnp^a* allele, the 108F/189V polymorphisms cause shortening or lengthening of incubation time with each scrapie strain in the same manner as PrP B (Figure 1). However, differences in the length of incubation period were evident when comparing FV/FV mice with other lines of mice carrying the *Prnp^b* allele (Figure 1), suggesting that although 108/189 exerts the major control on scrapie incubation time, other genetic factors in the different mouse lines also have an effect on survival. This is also evident in different lines of *Prnp^a* mice, where BSE incubation times in C57BL mice are approximately 100 days longer than in RIII mice, despite the fact that both lines express PrP A (Fraser *et al.*, 1992).

Gene targeted models have also been produced in which the 108 and 189 polymorphisms have been introduced separately. Mice both homozygous and heterozygous for the *Prnp^{a(108L/189V)}* and *Prnp^{a(108F/189T)}* alleles have been challenged with several well characterized mouse scrapie strains. The outcome of these experiments will identify the contribution of each of these polymorphisms, and if both are involved will identify whether they are acting co-operatively or independently to control incubation time of disease.

3.2. Human P102L Mutation

Familial forms of human TSE disease, which include Fatal Familial Insomnia (FFI) and Gerstmann-Straussler-Scheinker Syndrome (GSS), appear to be inherited in an autosomal dominant manner and account for approximately 15% of human TSEs. These diseases are diagnosed by differing clinical presentations and pathological profiles, and are associated with specific mutations in the host PrP gene. These mutations are thought to destabilize PrP, making it more prone to conversion into the abnormal disease associated isoform, PrP^{Sc}, causing the development of a 'spontaneous' TSE disease in the absence of any exogenous infectious agent. One specific mutation, which has been shown to be directly linked to the development of GSS, is a single base mutation in the human PrP gene that results in the substitution of leucine for proline at amino acid 102 (P102L GSS). This mutation has been shown to display 100% penetrance (Hsiao *et al.*, 1989), and many geographically distinct families have been identified which carry this mutation and show a family history of neurological disease (Kretzschmar *et al.*, 1992; Goldhammer *et al.*, 1993; Hainfellner *et al.*, 1995). Investigating the mechanism by which this mutation results in the development of TSE disease is difficult in humans. Therefore the production of transgenic mice expressing the equivalent mutation at amino acid 101 in murine PrP should provide a model of spontaneous TSE disease with which to investigate the mechanism by which the mutation results in the development of disease.

Transgenic mice containing a high copy number (64x) of murine *Prnp*^{a101L} transgenes were shown to develop a spontaneous neurological disease between 150 to 300 days of age (Hsiao *et al.*, 1990; 1991; 1994; Telling *et al.*, 1996). The neurological disorder showed spongiform change, astrogliosis and PrP amyloid plaques in the brain, although no proteinase K resistant (PK-res) PrP was detected by western analysis. Low copy number (2x) transgenic mice expressing approximately wild type levels of mutant 101L PrP did not show signs of CNS disorder until after 600 days of age, and this only occurred in a small number of the animals (Kaneko *et al.*, 2000). However, approximately 40% of these low copy number transgenic mice developed neurological disease before 600 days when inoculated with brain material from sick, high copy number transgenic mice (Hsiao *et al.*, 1994; Telling *et al.*, 1996). Hence, in this particular transgenic model, overexpression of the P101L transgene appeared to lead to a spontaneous neurological disease that could be transmitted to 101L transgenic mice and hamsters, but not to wild type mice (Hsiao *et al.*, 1994; Telling *et al.*, 1996).

3.2.1. Gene Targeting the 101L Mutation

In order to examine the precise effect of the 101L mutation in the absence of any possible phenotype caused by transgene overexpression, the 101L mutation was introduced into the endogenous murine PrP gene by gene targeting. This technique produced a line of transgenic mice which differed from the wild type 129/Ola mice (101PP) by only the single amino acid mutation at codon 101 in PrP (Moore and Melton, 1995; Moore *et al.*, 1995; Manson *et al.*, 1999). Inbred transgenic mice homozygous for $Prnp^{a101L}$ (101LL) were viable, and were shown by PCR and sequence analysis to contain the DNA polymorphism responsible for the proline to leucine substitution. Southern blot analysis confirmed that no other alterations had been made to the mice during gene targeting. Northern blot analysis detected similar levels of PrP mRNA in both 101LL and wild type (101PP) mice; however, the steady state protein levels initially detected by western analysis were found to be marginally lower in 101LL mice (Manson *et al.*, 1999). These differences in protein expression levels observed have subsequently been shown to be due to differences in antibody affinity between the two proteins, since analysis with a number of different antibodies has shown that the level of PrP in the brain of 101LL mice is identical to that of the wild type protein.

3.2.2. Does the 101L Mutation Cause Spontaneous TSE Disease?

During the course of our studies with the P101L line, several hundred homozygous (101LL) and heterozygous (101PL) inbred mice have been allowed to age, but no clinical signs other than normal ageing phenotypes were observed in these mice. The brains of all mice were analyzed for signs of early TSE pathology, but no vacuolation was detected. In addition, no PrP deposition has been observed by immunocytochemical analysis, and no protease resistant PrP^{Sc} was detected by western analysis of brain tissue from several mice (Manson *et al.*, 1999). Homogenates of brain and of spleen from 101LL mice over 600 days old have been bioassayed for the presence of infectivity by inoculation in 101LL and 101PP mice, but no infectivity was detectable in these tissues.

In contrast to the standard 101LL transgenic mice (Hsiao *et al.*, 1990; Telling *et al.*, 1996), the gene targeted 101LL mice did not develop spontaneous TSE disease in their lifespan, and could not transmit neurological disease to other mice expressing the transgene. These two conflicting observations may simply reflect the different expression levels of PrP in each model. Human P102L GSS can take over 40 years to develop, and the lifespan of a mouse expressing normal levels of 101L PrP (600-800 days) may not be sufficient for the specific event to occur which triggers the conversion and accumulation of PrP in the brain. The overexpression of 101L PrP may therefore allow this event to occur more rapidly, and cause disease. However, neurological disease was observed in some low copy number P101L microinjected transgenic mice over 600 days old (Kaneko *et al.*, 2000), suggesting that the wild type expression levels in the gene targeted 101LL transgenic mice should have been sufficient to allow the development of some neurological disease.

Despite the similar levels of 101L PrP protein in the brain, the two P101L mouse models appear to differ in respect to phenotype observed. It is also possible that the different genetic backgrounds of these mice might lead to the different phenotypes. The differences between these lines of mice will be addressed in experiments which will passage brain homogenates from the overexpressing 101L mice to the gene targeted line of mice.

3.3. Transmission of TSE Disease to Gene Targeted 101LL Mice

In the absence of spontaneous neurological disease in gene targeted 101LL mice, the possibility that the mutation may instead alter the susceptibility of the mice to TSE disease was examined. If the 101L

Table 1. Transmission of murine scrapie strains to wild type and 101LL mice

Strain of Agent	*Prnp* Genotype of agent source	Incubation Time (days ± SEM)		TSE Disease in 101LL Mice*
		129/Ola	101LL	
ME7	a/a	161 ± 2	338 ± 8	18/18
139A	a/a	147 ± 2	306 ± 7	16/16
79A	a/a	139 ± 2	298 ± 3	22/22
22A	b/b	493 ± 6	527 ± 28	9/16
79V	b/b	234 ± 6	377 ± 10	14/15
301V	b/b	253 ± 2	181 ± 1	19/19

101LL and control 129/Ola mice were inoculated intracerebrally with 5 different strains of mouse passaged scrapie and a strain of mouse passaged BSE to assess the effect of a single amino acid mutation in the N-terminal region of murine PrP on TSE incubation time.

*Number of 101LL mice with TSE disease/number inoculated. All 129/Ola mice developed disease with each strain of mouse scrapie, hence data not shown. Data taken from Barron *et al* 2001, Barron *et al.*, 2003, and Manson *et al.*, 1999.

mutation does, as suggested, make PrP unstable and more prone to forming PrP^Sc, then these mice may be more susceptible than wild type (101PP) mice to any source of TSE infectivity.

3.3.1. Transmission of Murine Scrapie

Inoculation of 101LL mice with the mouse scrapie strains ME7, 139A, 79A, 79V and 22A (Table 1) resulted in prolonged incubation times when compared to wild type mice. Conversely, *Prnp^a* incubation times with the mouse passaged BSE agent 301V were found to be reduced in 101LL mice (Manson *et al.*, 1999; Barron *et al.*, 2001; Barron *et al.*, 2003). These results suggested that 101L PrP is not more prone to conversion to PrP^Sc than 101P PrP, and that incubation times in 101LL mice are altered in a strain specific manner independent of *Prnp* genotype of the source of infectivity. A similar prolongation of incubation time has been observed previously in standard transgenic mice over-expressing murine PrP with the 3F4 epitope (108M/111M). Over-expression of the mutated 108M/111M PrP caused prolonged incubation times (between 1-3 fold) for each of the scrapie strains investigated (RML, ME7, 22A and 301V), independent of the *Prnp* genotype of the donor of infectivity (Supattapone *et al.*, 2001). The precise effect of the 108M/111M mutations is difficult to quantify in these experiments due to the differences in *Prnp* gene copy number and genetic background between control and mutant mice. However, this 1-3 fold increase in incubation time with ME7, 22A and 301V was not observed on inoculation of 101LL mice (Table 1), suggesting that the mechanism by which the 101L mutation controls incubation time is distinct from that of the 108M/111M mutations. In addition, the 101L mutation controls incubation time in a strain specific manner independent of *Prnp* genotype, indicating that this mechanism must also be distinct from that of the 108/189 polymorphisms.

3.3.2. Cross Species Transmissions

Transmission of TSE disease is most efficient when the donor and recipient of infectivity are of the same PrP genotype (Prusiner *et al.*, 1990). Transmission occurs rapidly, with tight, predictable incubation times that can be re-produced on subsequent sub-passage in the same species. When attempting to transmit disease between different species, incubation times are usually prolonged and occur over a

Table 2. Incubation times of non-murine TSE agents in wild type (101PP) and 101LL mice

Strain of Agent	*Prnp* Genotype of Agent Source	*Prnp* Genotype of Mouse	Incubation Time (days ± SEM)	TSE Disease*
Human GSS	102PL	101LL	288 ± 4	15/15
		101PP	456	1/6
Human vCJD	102PP	101LL	556 ± 12	8/18
		101PP	373 ± 8	24/24
Hamster 263K	101PP	101LL	374 ± 6	4/4
		101PP	707	1/4
Sheep SSBP/1	105 PP	101LL	346 ± 5	14/16
		101PP	535 ± 3	12/14

101LL and control 129/Ola (101PP) mice were inoculated intracerebrally with four different TSE agents isolated from humans, hamsters and sheep to assess the effect of a single amino acid mutation in the N-terminal region of murine PrP on the transmission of TSE disease between species.
*Number with TSE disease/number inoculated.
Data taken from Barron *et al.,* 2001, and Manson *et al.,* 1999.

wide range. Very often, these incubation times exceed the life-span of the animal, and no clinical signs of disease are ever detected. This phenomenon is known as the 'species barrier', and is thought to be due to the differences in PrP sequence found in different species. *In vitro* experiments have shown that the efficiency of conversion of PrP^C to PrP^{Sc} is greatest when both proteins share identical amino acid sequence (Horiuchi *et al.,* 2000; Raymond *et al.,* 2000). The presence of heterologous PrP from another species appears to block this conversion reaction, causing extended incubation times (Horiuchi *et al.,* 2000). We have shown that the 101L mutation can alter the incubation times of murine scrapie strains despite the presence of homologous PrP sequences in donor and recipient of infectivity. It is therefore possible that the 101L mutation can also affect transmission of TSE disease between species.

The 101LL mice were inoculated with infectivity derived from human (P102L GSS and vCJD), hamster (263K) and sheep (SSBP/1) sources (Table 2). All of these agents, with the exception of vCJD, have previously proved difficult to transmit to wild type (101PP) mice. In contrast to the results obtained with murine scrapie strains, decreased incubation times were observed in all cross species transmissions, with the exception of vCJD (Table 2). Although P102L GSS inoculum contained PrP^{Sc} with the 102L mutation, none of the other agents contained a mutation at this position. Yet P102L GSS, 263K and SSBP/1 transmitted more efficiently to 101LL mice than 101PP mice, despite the overall PrP sequence incompatibility.

In contrast, vCJD had longer incubation times in the 101LL mice than in the 101PP mice. Wild type mice showed 100% incidence of disease with vCJD at approximately 370 days, whereas 101LL mice showed increased incubation times of 556 days and only 45% incidence of disease (Table 2). This result shows that the 101LL mice are not more susceptible to all cross species transmissions, and that the effect is strain specific. The reduced susceptibility of 101LL mice to vCJD is in complete contrast with results obtained on transmission of the BSE derived mouse strain 301V, which showed reduced incubation times in 101LL mice compared with wild type 101PP mice (Barron *et al.*, 2003). vCJD has been demonstrated to be the human form of BSE (Bruce *et al.*, 1997; Hill *et al.*, 1997; Scott *et al.*, 1999), and the strain characteristics of this particular TSE agent in wild type mice have been show to remain remarkably consistent after transmission through several different species (Bruce *et al.*, 1997). However, passage of this strain of agent through a human produced increased incubation times and altered lesion profiles in 101LL mice compared to wild type mice (Barron *et al.*, 2001), while passage through a *Prnp^{b}* mouse (301V) produced shortened incubation times in 101LL mice (Barron *et al.*, 2003). It would appear that adaptation to mouse with this strain of agent increases the efficiency of transmission to 101LL mice, and this may in some way be related to why this human TSE agent alone transmits efficiently to wild type mice (Tateishi, 1996; Bruce *et al.*, 1997).

The results obtained from these cross species transmissions were not anticipated, and showed that neither sequence compatibility nor identity at amino acid 101 were critical for the efficient transmission of TSE disease, but that substitutions at that position could have a major influence on incubation time of disease. This phenomenon also appeared to be strain specific rather than species specific, as two human sources of infectivity (GSS and vCJD) showed the ability to decrease and increase incubation times respectively, despite the fact that both strains of agent were propagated in the same species.

4. Gene Replacements

The use of gene targeting is not restricted to the production of transgenic lines containing point mutations. This technique can be used also for gene replacement studies, where transgenic mice are produced which express PrP protein from another species. Two lines of transgenic mice expressing human PrP have been produced by gene targeting (Kitamoto *et al.*, 1996; 2002). These lines express either full length human (Hu-Ki) PrP (Kitamoto *et al.*, 1996), or a chimeric human/mouse (Ki-ChM) PrP where amino acids 23-188 are human, and 189-231 are murine (Kitamoto *et al.*, 2002). Although no data is available for transmissions to Hu-Ki transgenic mice, several CJD isolates transmitted to Ki-ChM transgenic mice with high efficiency, producing incubation times ranging from 140-180 days (Kitamoto *et al.*, 2002). These mice may prove to be a sensitive model for the detection of human TSE infectivity. However, the presence of C-terminal murine PrP sequence in this model suggests that the efficiency of transmission of TSE disease does not rely on PrP sequence compatibility alone and that other cellular components with species specificity may also be involved.

5. Future Developments

5.1. PrP Mutations/Polymorphisms

Results obtained from gene targeting point mutations in the murine PrP gene have shown distinct differences to those obtained from equivalent overexpressing transgenic lines. These differences may simply be due to the different expression levels in each transgenic line, but may also be due to site of integration, tissue specificity, post translational processing of the transgene product, or to the genetic background of the mice. We therefore aim to use gene targeting to study further disease associated mutations in PrP, and also mutations in specific structural regions of the protein to assess their effect on disease transmission. These studies should provide an important insight into the role of PrP in the TSE diseases, and the importance of different areas of the protein in disease transmission.

5.2. Species Barrier

Standard transgenic lines overexpressing non-murine PrP proteins have been produced as models to study TSE disease from a range of species. The overexpression of PrP in transgenic mice may provide highly sensitive models for assaying TSE infectivity from different species although the sensitivity of these models is still to be assessed. In contrast gene targeted lines of transgenic mice expressing human, bovine, ovine or cervid PrP maintained on an inbred genetic background express equivalent levels of PrP with the same tissue specificity as wild type mice. These lines will be used to study the role of PrP sequence in host susceptibility to disease and the development of disease in both the periphery and CNS.

Gene targeting is a powerful technique with which to study the TSE diseases, and further models are currently being produced in our laboratory to address the mechanisms of human familial TSE disease, the role of different regions of PrP in controlling disease susceptibility, and the role of PrP in transmission of disease both within and between species. As these models all express equivalent levels of PrP on the same genetic background we can directly compare the effects of different mutations and polymorphisms on these aspects of disease pathogenesis in order to gain a more detailed insight into the role of PrP in TSE disease.

6. Acknowledgements

The authors would like to acknowledge Professor David Melton and Dr Richard Moore for their major contribution in the production of the gene targeted mice, and V. Thomson, F. Robertson E. Murdoch and S.Dunlop for care and scoring of the animals. This work was supported by a grant from BBSRC.

References

Barron, R.M., Thomson, V., Jamieson, E., Melton, D.W., Ironside, J., Will, R. and Manson, J.C. 2001. Changing a single amino acid in the N-terminus of murine PrP alters TSE incubation time across three species barriers. EMBO J. 20: 5070-5078.

Barron, R.M., Thomson, V., King, D., Shaw, J., Melton, D.W. and Manson, J.C. 2003. Transmission of murine scrapie to P101L transgenic mice. J. Gen. Virol. 84: 3165-3172.

Bruce, M.E., Will, R.G., Ironside, J.W., McConnell, I., Drummond, D., Suttie, A., McCardle, L., Chree, A., Hope, J., Birkett, C., Cousens, S., Fraser, H. and Bostock, C.J. 1997. Transmissions to mice indicate that 'new variant' CJD is caused by the BSE agent. Nature. 389: 498-501.

Carlson, G.A., Ebeling, C., Torchia, M., Westaway, D. and Prusiner, S.B. 1993. Delimiting the location of the scrapie prion incubation-time gene on chromosome-2 of the mouse. Genetics. 133: 979-988.

Carlson, G.A., Kingsbury, D.T., Goodman, P.A., Coleman, S., Marshall, S.T., DeArmond, S., Westaway, D. and Prusiner, S.B. 1986. Linkage of prion protein and scrapie incubation time genes. Cell. 46: 503-511.

Dickinson, A.G., Meikle, V.M. and Fraser, H. 1968. Identification of a gene which controls the incubation period of some strains of scrapie agent in mice. J. Comp. Pathol. 78: 293-299.

Fraser, H., Bruce, M.E., Chree, A., McConnell, I. and Wells, G.A. 1992. Transmission of bovine spongiform encephalopathy and scrapie to mice. J. Gen. Virol. 73: 1891-1897.

Goldhammer, Y., Gabizon, R., Meiner, Z. and Sadeh, M. 1993. An Israeli family with Gerstmann-Straussler-Scheinker disease manifesting the codon-102 mutation in the prion protein gene. Neurology. 43: 2718-2719.

Hainfellner, J.A., Brantnerinthaler, S., Cervenakova, L., Brown, P., Kitamoto, T., Tateishi, J., Diringer, H., Liberski, P.P., Regele, H., Feucht, R., Mayr, N., Wessely, P., Summer, K., Seitelberger, F. and Budka, H. 1995. The original Gerstmann-Straussler-Scheinker family of Austria - Divergent clinicopathological phenotypes but constant PrP genotype. Brain Pathol. 5: 201-211.

Hill, A.F., Desbruslais, M., Joiner, S., Sidle, K.C.L., Gowland, I., Collinge, J., Doey, L.J. and Lantos, P. 1997. The same prion strain causes vCJD and BSE. Nature. 389: 448-450.

Horiuchi, M., Priola, S.A., Chabry, J. and Caughey, B. 2000. Interactions between heterologous forms of prion protein: binding, inhibition of conversion, and species barriers. Proc. Natl. Acad. Sci. U. S. A. 97: 5836-5841.

Hsiao, K., Baker, H.F., Crow, T.J., Poulter, M., Owen, F., Terwilliger, J.D., Westaway, D., Ott, J. and Prusiner, S.B. 1989. Linkage of a prion protein missense variant to Gerstmann-Straussler Syndrome. Nature. 338: 342-345.

Hsiao, K., Scott, M., Foster, D., DeArmond, S.J., Groth, D., Serban, H. and Prusiner, S.B. 1991. Spontaneous neurodegeneration in transgenic mice with prion protein codon 101 proline----leucine substitution. Ann. N. Y. Acad. Sci. 640: 166-170.

Hsiao, K.K., Groth, D., Scott, M., Yang, S.L., Serban, H., Raff, D., Foster, D., Torchia, M., Dearmond, S.J. and Prusiner, S.B. 1994. Serial transmission in rodents of neurodegeneration from transgenic mice expressing mutant prion protein. Proc. Natl. Acad. Sci. U. S. A. 91: 9126-9130.

Hsiao, K.K., Scott, M., Foster, D., Groth, D.F., DeArmond, S.J. and Prusiner, S.B. 1990. Spontaneous neurodegeneration in transgenic mice with mutant prion protein. Science. 250: 1587-1590.

Hunter, N., Dann, J.C., Bennett, A.D., Somerville, R.A., McConnell, I. and Hope, J. 1992. Are Sinc and the PrP gene congruent? Evidence from PrP gene analysis in Sinc congenic mice. J. Gen. Virol. 73: 2751-2755.

Kaneko, K., Ball, H.L., Wille, H., Zhang, H., Groth, D., Torchia, M., Tremblay, P., Safar, J., Prusiner, S.B., DeArmond, S.J., Baldwin, M.A. and Cohen, F. 2000. A synthetic peptide initiates Gerstmann-Straussler-Scheinker (GSS) disease in transgenic mice. J. Mol. Biol. 295: 997-1007.

Kitamoto, T., Mohri, S., Ironside, J.W., Miyoshi, I., Tanaka, T., Kitamoto, N., Itohara, S., Kasai, N., Katsuki, M., Higuchi, J., Muramoto, T. and Shin, R.W. 2002. Follicular dendritic cell of the knock-in mouse provides a new bioassay for human prions. Biochem. Biophys. Res. Commun. 294: 280-286.

Kitamoto, T., Nakamura, K., Nakao, K., Shibuya, S., Shin, R.W., Gondo, Y., Katsuki, M. and Tateishi, J. 1996. Humanized prion protein knock-in by cre-induced site-specific recombination in the mouse. Biochem. Biophys. Res. Commun. 222: 742-747.

Kretzschmar, H.A., Kufer, P., Riethmuller, G., DeArmond, S., Prusiner, S.B. and Schiffer, D. 1992. Prion protein mutation at codon 102 in an Italian family with Gerstmann-Straussler-Scheinker syndrome. Neurology. 42: 809-810.

Manson, J.C., Jamieson, E., Baybutt, H., Tuzi, N.L., Barron, R., McConnell, I., Somerville, R., Ironside, J., Will, R., Sy, M.S., Melton, D.W., Hope, J. and Bostock, C. 1999. A single amino acid alteration (101L) introduced into murine PrP dramatically alters incubation time of transmissible spongiform encephalopathy. EMBO J. 18: 6855-6864.

Moore, R.C., Hope, J., McBride, P.A., McConnell, I., Selfridge, J., Melton, D.W. and Manson, J.C. 1998. Mice with gene targetted prion protein alterations show that Prnp, Sinc and Prni are congruent. Nat. Genet. 18: 118-125.

Moore, R.C. and Melton, D.W. 1995. Models of human-disease through gene targeting. Biochem. Soc. Trans. 23: 398-403.

Moore, R.C., Redhead, N.J., Selfridge, J., Hope, J., Manson, J.C. and Melton, D.W. 1995. Double replacement gene targeting for the production of a series of mouse strains with different prion protein gene alterations. Biotechnology. (N. Y). 13: 999-1004.

Prusiner, S.B., Scott, M., Foster, D., Pan, K.-M., Groth, D., Mirenda, C., Torchia, M., Yang, S.-L., Serban, D., Carlson, G.A., Hoppe, P.C., Westaway, D. and DeArmond, S.J. 1990. Transgenetic studies implicate interactions between homologous PrP isoforms in scrapie prion replication. Cell. 63: 673-686.

Raymond, G.J., Bossers, A., Raymond, L.D., O'Rourke, K.I., McHolland, L.E., Bryant, P.K., Miller, M.W., Williams, E.S., Smits, M. and Caughey, B. 2000. Evidence of a molecular barrier limiting susceptibility of humans, cattle and sheep to chronic wasting disease. EMBO J. 19: 4425-4430.

Scott, M.R., Will, R., Ironside, J., Nguyen, H.O.B., Tremblay, P., DeArmond, S.J. and Prusiner, S.B. 1999. Compelling transgenetic evidence for transmission of bovine spongiform encephalopathy prions to humans. Proc. Natl. Acad. Sci. U. S. A. 96: 15137-15142.

Supattapone, S., Muramoto, T., Legname, G., Mehlhorn, I., Cohen, F.E., DeArmond, S.J., Prusiner, S.B. and Scott, M.R. 2001. Identification of two prion protein regions that modify scrapie incubation time. J. Virol. 75: 1408-1413.

Tateishi, J. 1996. Transmission Of Human Prion Diseases to Rodents. Seminars In Virology. 7: 175-180.

Telling, G.C., Haga, T., Torchia, M., Tremblay, P., Dearmond, S.J. and Prusiner, S.B. 1996. Interactions between wild-type and mutant prion proteins modulate neurodegeneration in transgenic mice. Genes Dev. 10: 1736-1750.

Westaway, D., Goodman, P.A., Mirenda, C.A., McKinley, M.P., Carlson, G.A. and Prusiner, S.B. 1987. Distinct prion proteins in short and long scrapie incubation period mice. Cell. 51: 651-662.

From: Prions and Prion Diseases: Current Perspectives
Edited by: Glenn C. Telling

Chapter 6

Transgenic Mouse Models
of Prion Diseases

Karah Nazor and Glenn C. Telling

Abstract

Manipulation of prion protein (PrP) genes by transgenesis in mice has provided important insights into mechanisms of prion propagation and the molecular basis of prion strains and species barriers. Despite these advances, our understanding of these unique pathogens is far from complete. Transgenic approaches will doubtless remain one of the cornerstones of investigations into the prion diseases in the coming years which will include mechanistic studies of prion pathogenesis and prion transmission barriers. Transgenic models will also be important tools for the evaluation of potential therapeutic agents.

1. Transgenic Expression Vectors

The majority of transgenic studies of prion diseases have involved the incorporation of wild type or mutant PrP genes from different species into the genome of fertilized mouse embryos by DNA microinjection. Various chimeric gene constructs incorporating PrP gene sequences from mouse and other species have also been used to produce transgenic mice. The seminal transgenic experiments utilized cosmid clones containing PrP gene sequences isolated from Syrian hamster

(SHa) and the I/lnJ strain of mice (Scott *et al.*, 1989; Westaway *et al.*, 1991) and this approach was also used to produce transgenic mice expressing sheep PrP (Westaway *et al.*, 1994). The cos.Tet vector is a modification of the SHa cosmid vector and contains a 43 kb DNA fragment encompassing the PrP gene and approximately 24 and 6 kb of 5' and 3' flanking sequences respectively (Scott *et al.*, 1992). The vector is designed to allow the convenient insertion of PrP coding sequences. A plasmid expression vector based upon the PrP gene derived from the I/lnJ PrP cosmid, referred to as phgPrP or the 'half-genomic' construct, has also been used to produce transgenic mice (Fischer *et al.*, 1996). A modification of the 'half-genomic' construct, referred to as MoPrP.Xho that allows for convenient insertion of cDNA coding sequences has been extensively used to overexpress a variety of genes in the central nervous system (CNS) including amyloid precursor protein (APP) and human presenilin 1 (Borchelt *et al.*, 1996). Smaller, plasmid-based expression vectors based on the mouse PrP gene (*Prnp*) (Campbell *et al.*, 2000; Browning, S. and Telling, G., unpublished data) and human PrP gene (*PRNP*) (Asante *et al.*, 2002a) control elements have also been produced for transgenetic modeling of human and animal prion diseases. SHa and mouse PrP gene constructs in which all intron sequences are removed, so-called minigene constructs, completely fail to express PrP in the CNS demonstrating the requirement for at least the smaller intron for efficient expression (Fischer *et al.*, 1996; Scott *et al.*, 1989).

2. Transgenic Studies of PrP Function

Some of the most compelling evidence to date for the so-called 'protein-only hypothesis' of prion replication derives from experiments with knockout transgenic mice. Since PrP^C is the source of PrP^{Sc}, the model predicted that elimination of PrP^C would abolish prion replication. To test this, the mouse PrP gene, referred to as *Prnp*, was disrupted by homologous recombination with a genetically modified version in embryonic stem cells. Stem cells containing the disrupted *Prnp* gene were introduced into mouse blastocysts and knockout mice were established (Büeler *et al.*, 1992; Manson *et al.*, 1994b; Sakaguchi *et al.*, 1995). Unlike wild type mice, the resultant homozygous null mice (*Prnp$^{0/0}$*) fail to develop the characteristic clinical and neuropathological symptoms of scrapie after inoculation with mouse prions and do not propagate prion infectivity (Büeler *et al.*, 1993; Manson *et al.*, 1994a; Prusiner *et al.*, 1993; Sakaguchi *et* *al.*, 1995), while mice that are hemizygous for PrP gene ablation have prolonged incubation times (Büeler *et al.*, 1994; Manson *et al.*, 1994a; Prusiner *et al.*, 1993).

Since $Prnp^{0/0}$ mice developed normally and suffered no gross phenotypic defects this raised the possibility that adaptive changes occur during the development of $Prnp^{0/0}$ mice that compensate for the loss of PrP^C function. To test this hypothesis transgenic mice were produced in which expression of transgene-expressed PrP^C could be controlled at will. Using the tetracycline gene-response system, mice were produced which co-express a tetracycline-responsive transactivator, referred to as tTA, and a tTA-responsive promoter that drives PrP expression (Tremblay *et al.*, 1998). Repressing PrP^C expression by oral administration of doxycycline was not deleterious to adult mice. However, since doxycycline treatment did not completely inhibit PrP^C expression in these mice it is not clear whether this residual expression masks the true phenotype of $Prnp^{0/0}$ mice.

An alternative approach to creating a conditional post-natal PrP knockout utilized a Cre-*lox*P system (Mallucci *et al.*, 2002). Double transgenic mice were produced by crossing mice in which the PrP gene is flanked by lox-P sites with mice in which cre recombinase is expressed in neurons under the control of the neurofilament promoter. Double transgenic offspring express PrP up until 10 weeks of age when adult-onset activation of the neurofilament controlled Cre recombinase occurs. Mice remained free of any abnormal physical phenotype up to 15 months post-knockout, but had decreased afterhyperpolarization potentials in hippocampal CA1 neurons, suggestive of an electrophysiological function of PrP, perhaps associated with Fyn-mediated PrP neuronal signaling (Mouillet-Richard *et al.*, 2000). These data are in support of other studies that demonstrate impairment of GABA$_A$ receptor-mediated fast inhibition and long-term potentiation in hippocampal slices from $Prnp^{0/0}$ mice (Collinge *et al.*, 1994; Whittington *et al.*, 1995) although in other studies, no electrophysiological defects could be identified (Herms *et al.*, 1995; Lledo *et al.*, 1996). Several other phenotypic defects have also been reported in $Prnp^{0/0}$ mice including altered circadian rhythms and sleep patterns (Tobler *et al.*, 1996), alterations in superoxide dismutase activity (SOD-1) (Brown *et al.*, 1997a) and defects in copper metabolism (Brown *et al.*, 1997b).

While *Prnp*$^{0/0}$ mice independently created in Zurich and Edinburgh developed normally and showed no overt phenotypic defects, a third line of gene-targeted *Prnp*$^{0/0}$ mice generated in Nagasaki showed progressive ataxia and cerebellar Purkinje cell degeneration at about 70 weeks (Sakaguchi *et al.*, 1996). In addition to the PrP coding sequence, ~900 nucleotides from the second intron and 450 nucleotides of the 3' non-coding sequence of *Prnp* is deleted in these mice. The molecular basis of the phenotypic differences in these *Prnp*$^{0/0}$ lines derived from the discovery of a PrP-like gene termed doppel (Prnd) located 16 kb downstream of *Prnp*. Expression of doppel was upregulated in Nagasaki *Prnp*$^{0/0}$ and a fourth *Prnp*$^{0/0}$ line termed Rcm0 which also develops late-onset ataxia and Purkinje cell loss (Moore *et al.*, 1999). Doppel expression results from an alternative intergenic splicing event ensuing from the different molecular strategies used to disrupt the PrP gene (Moore *et al.*, 1999; Rossi *et al.*, 2001). Doppel expression is upregulated in the Nagasaki *Prnp*$^{0/0}$ and Rcm0 *Prnp*$^{0/0}$ mice due to a diversion from the normal *Prnp* 5' splice acceptor site to a downstream acceptor, leading to inclusion of the doppel exon. In constructs used to create the Zurich and Edinburgh PrP null mice, which do not develop ataxia and Purkinje cell loss, the neomycin resistance gene cassette was placed downstream of the PrP 5' acceptor site leaving it intact and therefore preventing transcription of doppel.

Correlation of doppel expression with a neurological phenotype resembling clinical prion disease suggested that doppel could play a key role in prion pathogenicity. However, reintroduction of wild type PrP resulted in offspring lacking an ataxic phenotype (Moore *et al.*, 2001). Another knockout line, referred to as Zurich II (Rossi *et al.*, 2001), in which the entire PrP-encoding exon and its flanking regions are recombined between lox P sites, was created to further address this issue. Zurich II mice express doppel and develop ataxia and when crossed to Zurich I mice, heterozygous offspring expressed half as much doppel as Zurich II mice and had a 6-month delay in appearance of ataxia. Introduction of a single wild type PrP allele blocked the ataxic phenotype of Zurich II mice, indicating that PrP plays a role in protecting against doppel-induced neurotoxicity. To further investigate the protective mechanism of PrP from doppel-induced toxicity, various PrP transgenes were introduced into Nagasaki *Prnp*$^{0/0}$ mice (Atarashi *et al.*, 2003). All transgenes except for one construct lacking amino acid residues 23-88 rescued the ataxic phenotype and Purkinje cell death. Expression of N-terminally truncated PrP directed to Purkinje cells of Zurich I mice, which normally remain healthy, also resulted in ataxia and Purkinje cell loss (Flechsig *et al.*, 2003). Again, introduction of a single PrP allele rescued this phenotype. It is plausible that introduction of a single PrP allele rescues the ataxic phenotype through a neuroprotective signal elicited by PrP, possibly as a result of the Cu/Zn SOD activity of the copper-binding octapeptide repeat region which protects nerve cells from reactive oxygen species (ROS) -induced injury. The biology of doppel is extensively reviewed by Westaway in this volume.

2.1. Structure-Function Studies of PrP

The finding that the introduction of PrP transgenes into Zurich I *Prnp*$^{0/0}$ mice restores susceptibility to scrapie opened the possibility for assessing whether a modified PrP^C molecule remains functional, at least insofar as it continues to be eligible for supporting prion propagation. While spontaneous neuromyopathy has been reported in aged mice expressing high levels of mouse PrP-B, SHaPrP and sheep PrP (Fischer *et al.*, 1996; Westaway *et al.*, 1994), this phenotype has not been observed in mice that overexpress other wild type PrP transgenes. A series of amino terminal deletions between codons 69-84, 32-80, 32-93 or 32-106 of the PrP coding sequence was able to restore susceptibility to scrapie in *Prnp*$^{0/0}$ mice (Fischer *et al.*, 1996; Shmerling *et al.*, 1997). In contrast, deletions between codons 32-121 or 32-134 caused ataxia and degeneration of the granular layer of the cerebellum within 2-3 months after birth (Shmerling *et al.*, 1997). This defect was overcome by the co-expression of wild type MoPrP suggesting that the truncated PrP might compete with a functionally similar non-PrP molecule for a common ligand.

A series of PrP coding sequence deletions based on putative regions of secondary structure were also expressed in *Prnp*$^{0/0}$ mice (Muramoto *et al.*, 1997). These deletions were engineered in a modified PrP construct that lacks amino acid residues 23-88, the residues that are removed from the amino terminus of PrP^{Sc} by limited proteolysis. Transgenic mice with deletions between codons 95-107, 108-121 and 141-176 remained healthy. In contrast, transgenic mice with deletions at the carboxyl terminus between codons 177-190 and 201-217, which disrupted the penultimate and last α helix, showed neuronal cytoplasmic inclusions of PrP-derived deposits and spontaneously developed fatal CNS illnesses similar to neuronal storage diseases at 90 to 227 days of age.

Two of these deletion constructs were further characterized in transgenic mice with respect to their ability to support prion replication (Supattapone *et al.*, 1999). Surprisingly, transgenic mice in which residues 23-88 were deleted remained resistant to infection and this block to prion propagation was alleviated by further deleting residues 141-176. In both cases, the block to prion propagation was overcome by co-expression of wild type MoPrP. In contrast to these findings, introduction of PrP devoid of amino acids 32-93 to $Prnp^{0/0}$ mice restored susceptibility to mouse-adapted prions (Flechsig *et al.*, 2000) albeit with protracted incubation times and 30-fold less accumulation of PrP^{Sc} compared to wild type mice. The discrepancy between these two models presumably results from differences in transgene design that preserves residues 23-31 within the latter deletion mutant. Transgenic approaches to study structure-function relationships of PrP are reviewed more extensively by Supattapone in this volume.

3. Transgenic Models of Inherited Prion Diseases

Approximately 10-20% of human prion disease is inherited with an autosomal dominant mode of inheritance. To date, 20 different missense and insertion mutations that segregate with dominantly inherited neurodegenerative disorders have been identified in the coding sequence of *Prnp*. Five of these mutations are genetically linked to loci controlling familial Creutzfeldt Jakob disease (CJD), Gerstmann Straussler Scheinker (GSS) syndrome and fatal familial insomnia (FFI) which are inherited human prion diseases that can be transmitted to experimental animals.

Transgenic mice that express a proline to leucine mutation at codon 101 of mouse PrP, equivalent to the human GSS P102L mutation, referred to as Tg(MoPrP-P101L), spontaneously developed clinical and neuropathological symptoms similar to mouse scrapie between 150 and 300 days of age (Hsiao *et al.*, 1990; Telling *et al.*, 1996a). After crossing the mutant transgene onto the $Prnp^{0/0}$ background, the resulting Tg(MoPrP-P101L) $Prnp^{0/0}$ mice displayed a highly synchronous onset of illness at ~145 days of age which shortened to ~85 days upon breeding to homozygosity for the transgene array. In addition Tg(MoPrP-P101L) $Prnp^{0/0}$ mice had increased numbers of PrP plaques and more severe spongiform degeneration. In contrast, transgenic mice over-expressing wild type mouse PrP at equivalent levels did not spontaneously develop neurodegenerative disease although they had highly reduced mouse scrapie incubation times after inoculation with mouse prions (Telling *et al.*, 1996a).

Importantly, the serial propagation of infectivity from the brains of spontaneously sick Tg(MoPrP-P101L) mice to indicator mice expressing low levels of mutant protein which otherwise do not get sick (Tg196 mice), demonstrated the production of infectious prions in the brains of these spontaneously sick mice (Hsiao *et al.*, 1994; Telling *et al.*, 1996a). A 55-residue synthetic peptide comprising mouse PrP residues 89-103 containing the P101L mutation refolded into a beta-sheet conformation induced prion disease in Tg196 mice after 360 days. Mice displayed GSS-like neuropathological changes in their brains similar to mice which developed spontaneous disease. Mice inoculated with peptide in a non beta-sheet conformation remained healthy for more than 600 days (Kaneko *et al.*, 2000). Prion infectivity from brain extracts of humans expressing the P102L GSS mutation was also propagated in transgenic mice expressing a chimeric mouse-human PrP gene with the P101L mutation (Telling *et al.*, 1995). Other inherited human prion diseases have also transmitted to transgenic mice expressing human and chimeric mouse-human PrP (Collinge *et al.*, 1995; Telling *et al.*, 1995; 1996b).

In contrast to Tg(MoPrP-P101L) mice, transgenic mice overexpressing an mutant mouse PrP gene with a glutamate to lysine mutation at codon 199, equivalent to the codon 200 mutation linked to familial CJD (E200K), did not spontaneously develop neurologic disease (Telling *et al.*, 1996a). A third human prion disease mutation associated with GSS in which the Tyr residue at codon 145 was mutated to a stop codon was modeled in transgenic mice, designated Tg(MoPrP144#). However, no PrP expression was detected in high copy number lines and neither uninoculated Tg(MoPrP144#) mice or mice inoculated with mouse RML scrapie developed symptoms of neurodegenerative disease (Muramoto *et al.*, 1997). Expression of a mouse PrP version of a nine octapeptide insertion, designated Tg(PG14) mice, associated with human prion dementia produced a slowly progressive neurological disorder in transgenic mice (Chiesa *et al.*, 1998). Upon breeding to homozygosity, Tg(PG14) mice developed ataxia and clinical illness at ~ 65 days compared to 240 days for hemizygous mice (Chiesa *et al.*, 2000) and mice accumulated proteinase K (PK)-resistant and detergent-insoluble PrP which increased 20-80 fold from birth and appeared to be associated with apoptotic loss of granule cells in the cerebellum. Partially PK-resistant, detergent insoluble mutant PrP accumulated in spinal cord, skeletal muscle and heart and accumulation was associated with primary skeletal muscle myopathy (Chiesa *et al.*, 2001).

4. Transgenic Studies of Prion Species Barriers

The species barrier describes the difficulty with which prions from one species can cause disease in a different species. In experimental studies, the initial passage of prions between species is associated with prolonged incubation times with only a few animals developing illness. On subsequent passage in the same species, all the animals become ill after greatly shortened incubation times. As detailed below, prion species barriers have been abrogated in transgenic mice by expressing PrP genes from other species or artificially engineered chimeric PrP genes. Experiments designed to probe the molecular basis of the species barrier have also provided important clues about the mechanism of prion propagation involving association and conformational conversion of PrP^C into PrP^{Sc}.

4.1. Transgenic Studies of Rodent-Adapted Scrapie Prions

As a result of the species barrier, wild type mice are normally resistant to infection with SHa prions. Seminal experiments by Prusiner and colleagues demonstrated that expression of SHa PrP^C in transgenic mice, referred to as Tg(SHaPrP) mice, rendered them susceptible to SHa prions and produced CNS pathology similar to that found in Syrian hamsters with prion disease (Scott *et al.*, 1989). Expression levels of SHa PrP were inversely correlated with the incubation period of SHa prions (Prusiner *et al.*, 1990). Furthermore, inoculation with mouse prions resulted in propagation of prions pathogenic for mice, while inoculation with SHa prions resulted in the propagation of prions pathogenic for hamsters. This work strongly implied that a direct protein/protein interaction between PrP molecules was involved in prion propagation, and that for optimum progression of the disease, the interacting species should be identical in primary structure (Prusiner *et al.*, 1990). Chimeric SHa/mouse PrP transgenes produced prions with new properties. The MH2M transgene carries 5 amino acid substitutions found in SHaPrP lying between codons 94 and 188. Tg(MH2M) mice generated prions with an artificial host range such that infectivity produced by inoculation with SHa prions could be passaged from Tg(MH2M) mice to wild type mice and infectivity produced by inoculation with mouse prions could be passaged from Tg(MH2M) mice to Syrian hamsters (Scott *et al.*, 1993).

4.2. Transgenic Studies of Human Prion Diseases

The infrequent transmission of human prion disease to rodents is also an example of the species barrier. Based on the results with Tg(SHaPrP) mice, it was expected that the species barrier to human prion propagation would be abrogated in transgenic mice expressing human PrP. However, transmission of human prion disease was generally no more efficient in transgenic mice expressing human PrP^C on a wild type background than in non-transgenic mice. In contrast, propagation of human prions was highly efficient in transgenic mice expressing a chimeric mouse-human PrP gene, referred to as Tg(MHu2M) in which the region of the mouse gene between codons 94 and 188 were replaced with human PrP sequences (Telling *et al.*, 1994). These mice became ill with an average incubation time of 200 days after inoculation with brain homogenates from patients dying of CJD. These studies made possible the rapid and relatively inexpensive transmission of human prion diseases for the first time. With the aim of decreasing the incubation time of disease in Tg(MHu2M) mice, transgenes were constructed in which one or more of the nine human residues of the human insert of MHu2M were changed to mouse. Mice expressing a double substitution of two C-terminal residues (M165V and E167Q), designated Tg22372 mice, which express the transgene at 1-2 X level of wildtype PrP, became ill with a reduced incubation time of 106-114 days after challenge with sporadic CJD prions (Korth *et al.*, 2003). Following the success of the Tg(MHu2M) model, mice expressing chimeric human/mouse PrP were created using a knock-in strategy (Kitamoto *et al.*, 2002). Following inoculation with sporadic, familial and iatrogenic human CJD brain homogenates, mice succumbed to disease with incubation times between 151-200 days. Seventy-five days post inoculation, follicular dendritic cells (FDC) from all lymphoid organs contained converted chimeric PrP^{Sc}, with the splenic FDC proving infectious upon serial transmission. This model holds promise as a relatively rapid bioassay for human prions.

The barrier to CJD transmission in Tg(HuPrP) mice was abolished by expressing HuPrP on a *Prnp$^{0/0}$* background demonstrating that mouse PrP^C inhibited the transmission of prions to transgenic mice expressing human PrP^C but not to those expressing chimeric PrP (Telling *et al.*, 1995). To explain these and other data, it was proposed that the most likely mediator of this inhibition is an auxiliary non-PrP molecule, provisionally designated protein X, which participates in the formation of prions by interacting with the carboxy-terminal region of PrP^C to facilitate conversion to PrP^{Sc}.

4.3.   Transgenic   Studies   of   Bovine   Spongiform Encephalopathy and Scrapie

Because the bioassay for bovine prions in non-transgenic mice is relatively insensitive (Fraser *et al.*, 1992) and the expense and long bovine spongiform encephalopathy (BSE) incubation times in cattle make them unsuitable for bioassay studies, it seems likely that transgenic approaches will offer a more accurate and convenient means of determining of BSE titers. Based on the success of Tg(MHu2M) mice, transgenic mice expressing a similar chimeric mouse-bovine PrP construct referred to as MBo2M were produced. While transgenic mice expressing bovine PrP developed disease after inoculation with BSE, albeit with long incubation times between 250 and 300, Tg(MBo2M) mice did not develop disease after challenge with BSE (Scott *et al.*, 1997). Similar results were independently obtained by another group (Buschmann *et al.*, 2000).

Transgenic mice expressing chimeric sheep/mouse or bovine/mouse PrP gene resembling the Tg(MHu2M) model   were intracerebrally inoculated with three scrapie isolates (Gombojav *et al.*, 2003). On the *Prnp* $^{+/+}$ background, minimal amounts of chimeric sheep/mouse PrP converted to PrP^{Sc} with a disease incubation time of 493-555 days. No chimeric bovine/mouse PrP^{Sc} was detected and these mice had an even longer incubation period of 598-701 days. To address the issue that the wild type allele could inhibit the conversion of chimeric PrP, the transgenes were also expressed on a *Prnp$^{0/0}$* background which resulted in prolonged incubation times for the sheep chimeric mice and no transmission in bovine chimeric mice, again without evidence of conversion of chimeric PrP.

A recently developed bovine transgenic mouse model was created using the MoPrP.Xho expression vector and *Prnp$^{0/0}$* mice (Castilla *et al.*, 2003). After intracerebral inoculation with BSE, transgenic mice overexpressing bovine PrP^C 6-fold that of wild type mice displayed PK resistant PrP^{Sc}, PrP immunopositive plaques in the hippocampus and spongiform degeneration in the midbrain at 196 days post inoculation and a BSE incubation time of 287 ±12 days (± SEM). These mice were also susceptible to sheep scrapie with an extended incubation time of 495 ± 16 days. Seventy percent of these mice harbored PK resistant PrP^{Sc} in their brains.

Transgenic mice expressing the ovine PrP coding sequence with alanine, arginine and glutamine (ARQ) at codons 136, 154, and 171, under the control of a neuron-specific enolase promoter on a *Prnp$^{0/0}$* background were developed to study transmission of natural sheep scrapie (Crozet *et al.*, 2001a; 2001b). Mice inoculated with two different scrapie strain isolates became ill with an incubation period of 238-290 days and developed clinical signs, spongiform changes, and accumulation of PK resistant PrP^{Sc} in their brains, whereas non-transgenic mice remained scrapie free for more than 700 days. When inoculated with brain homogenate from sheep experimentally infected with BSE, these mice became ill within 300 days with prominent florid pathology as found in vCJD cases (Crozet *et al.*, 2001a). Transgenic mice overexpressing ovine PrP with valine, arginine and glutamine (VRQ) at codons 136, 154, and 171, under the control of *Prnp* regulatory sequences developed clinical illness ~70 days after inoculation with scrapie (Vilotte *et al.*, 2001).

5. Transgenic Studies of Prion Strains

Because of their unprecedented mode of replication, explaining the mechanism by which prions propagate strain information has posed a major challenge to the prion hypothesis. Two strains of transmissible mink encephalopathy (TME), produced different clinical symptoms and incubation periods in Syrian hamsters and showed different resistance to proteinase K digestion and altered amino-terminal proteinase K cleavage sites (Bessen and Marsh, 1992) suggesting that different strains might represent different conformational states of PrP^{Sc}. Evidence supporting this concept emerged from transmission studies of inherited human prion diseases in transgenic mice. Expression of mutant prion proteins in patients with FFI and familial CJD (fCJD) result in variations in PrP conformation, reflected in altered proteinase K cleavage sites which generate PrP^{Sc} molecules with molecular weights of 19 kDa in FFI and 21 kDa in fCJD(E200K) (Telling *et al.*, 1996b). Extracts from the brains of FFI and fCJD(E200K) patients transmitted disease to Tg(MHu2M) mice after about 200 days on first passage and induced formation of 19 kDa PrP^{Sc} and 21 kDa PrP^{Sc}, respectively. Upon second passage in Tg(MHu2M) mice these characteristic molecular sizes remained constant but the incubation times for FFI and fCJD(E200K) prions diverged. These results indicated that PrP^{Sc} conformers function as templates in directing the formation of nascent PrP^{Sc} and provide a mechanism to explain strains of prions where diversity is enciphered in the tertiary structure of PrP^{Sc}. The strain characteristics of fCJD caused by the V210I mutation were also defined following transmission to Tg(MHu2M) mice (Mastrianni *et al.*, 2001).

A sporadic form of fatal insomnia, referred to as sFI, has also been described (Mastrianni *et al.*, 1997). While patients with sFI have symptoms and neuropathological profiles indistinguishable from patients with FFI they do not express the D178N mutant form of human PrP^C. sFI prions were transmitted to Tg(MHu2M) mice and were found to produce an identical pattern of neuropathology to that in Tg mice infected with FFI prions, arguing that PrP^{Sc} isoforms associated with sporadic fatal insomnia and fatal familial insomnia have the same conformation. These findings argue that the conformation of PrP^{Sc}, not the amino acid sequence, determines the strain-specified disease phenotype.

Other studies have shown that different sporadic and iatrogenic CJD cases associated with specific codon 129 genotypes can be typed according to PrP^{Sc} fragment sizes following PK treatment and western blotting of brain extracts (Parchi *et al.*, 1996). A characteristic banding pattern of PrP^{Sc} glycoforms found in vCJD patients and BSE infected animals distinguishes vCJD PrP^{Sc} from the patterns observed in classical CJD (Collinge *et al.*, 1996; Hill *et al.*, 1997). Transgenic mice expressing mutations at one or both glycosylation consensus sites have been studied to investigate the role of the asparagine-linked oligosaccharides of PrP (DeArmond *et al.*, 1997). Mutation of the first site altered PrP^C trafficking and prevented infection with two prion strains; deletion of the second did not alter PrP^C trafficking, permitted infection with one prion strain, and altered the pattern of PrP^{Sc} deposition.

Initial transmission studies of human CJD cases in Tg mice were extended to include a larger number of additional sporadic cases. Transmission of these cases to Tg(MHu2M) and Tg(HuPrP) mice expressing either Valine (V) or Methionine (M) at codon 129 revealed that this polymorphism and the strain–specified conformation of PrP[Sc] profoundly influenced the length of the incubation time and patterns of PrP[Sc] deposition in recipient mice (Campbell *et al.*, 2000; Korth *et al.*, 2003). Also, the size of the protease-resistant PrP[Sc] fragment in human brains, was reproduced on primary and secondary passages of vCJD, sCJD, fCJD(E200K), and FFI prions in Tg(MHu2M) mice. The constancy of the strain specified tertiary structure on serial passage through Tg(MHu2M) mice contrasts with other parameters used to characterize prion strains, such as incubation periods and neuropathology profiles. Separate transgenic studies on the role of the codon 129 genotype in CJD transmission also showed that transgenic mice expressing human PrP M129 were more susceptible to sporadic CJD derived from M/M patients than CJD inocula from M/V or V/V patients (Asante *et al.*, 2002b). Transmission of BSE and vCJD to transgenic mice expressing human PrP 129M, resulted in the neuropathological and molecular phenotype of vCJD. In addition to producing a vCJD-like phenotype, BSE transmission also resulted in a molecular phenotype indistinguishable from that of sporadic CJD suggesting that more than one BSE-derived prion strain might infect humans (Asante *et al.*, 2002b).

Tg(MH2M) mice expressing chimeric hamster/mouse PrP and Syrian hamsters mice were used to study the emergence of new prion strains by primary and secondary of SHa strains Sc237 and DY. Prion strains were defined by incubation time, neuropathological lesion profiles, and PrP[Sc] conformation. Only Sc237 manifested a species barrier effect in Tg(MH2M) mice. Serial passage of brain homogenate from sick Tg(MH2M) mice inoculated with Sc237 into Tg(MH2M) mice resulted in a decreased incubation on second passage with a reproduction of prominent GSS amyloid pathology observed in the first passage. The original Sc237 and the MH2M(Sc237) inocula were analyzed using a conformational stability assay, which assigns profiles to different prion strains based on the rate of PrP[Sc] denaturation in the presence of guanidine hydrochloride (GdnHCl) prior to PK digestion. Interestingly, the $[GdnHCl]_{1/2}$ values of SHa (Sc237) and TgMH2M(Sc237) inocula were significantly different, while the SHa (DY) and MH2M(DY), which lack a species barrier, displayed very similar conformational states based on these values (Peretz *et al.*, 2002).

6. Transgenic Studies of Prion Pathogenesis

It appears that accumulation of PrP^{Sc} may not be sole cause of pathology
in prion diseases since certain examples of prion disease occur without
accumulation of protease-resistant PrP^{Sc} (Hsiao *et al.*, 1990; Telling
et al., 1996a); moreover the time course of neurodegeneration is not
equivalent to the time course of PrP^{Sc} accumulation in mice expressing
low levels of PrP^{C} (Büeler *et al.*, 1994). An alternative mechanism of
PrP-induced neurodegeneration was suggested by transgenic studies
of mutant forms of PrP that disrupt PrP biogenesis in the endoplasmic
reticulum (Hegde *et al.*, 1998). Transgenic mice expressing mutations
in the stop transfer effector region between residues Lys 104 - Met 112
and the hydrophobic TM1 region between residues Ala113 - Ser135,
spontaneously develop neurodegenerative disease and accumulate an
aberrant form of PrP termed ^{Ctm}PrP which appears to be different from
conventional protease-resistant PrP^{Sc}. Accumulation of ^{Ctm}PrP is also
associated with a form of GSS that segregates with the codon 117
mutation of *PRNP*.

The role of alternatively processed forms of PrP in prion
pathogenesis is currently a topic of intense debate. Recent studies
have focused on the role of a novel cytosolic form of PrP in
neurodegeneration. Proteasome inhibition resulted in retrograde
transport of PrP from the endoplasmic reticulum and cytoplasmic
accumulation of a PrP^{Sc}-like form that was apparently self-
perpetuating (Ma and Lindquist, 2002). Transgenic mice expressing
cytosolic PrP presented with ataxia as early as 7 weeks of age (Ma *et
al.*, 2002). The neuropathology of these mice consisted of cerebellar
atrophy, neuronal loss in the granular and molecular layers and severe
gliosis in the cerebellum. PK resistant PrP^{Sc} was not detected. The
authors proposed that circumstances where proteasomal function is
either diminished, such in normal aging, or overwhelmed, such as
in familial prion diseases where point mutations could initiate PrP
misfolding, result in propagation of cytosolic PrP and accumulation
into amorphous aggregates. It is currently not known if cytosolic
PrP is infectious. Subsequent studies in primary human and mouse
cerebellar granular neurons demonstrated that cytosolic PrP is not
necessarily neurotoxic (Drisaldi *et al.*, 2003; Roucou *et al.*, 2003).

7. Ectopic Expression Studies

Although the pathological consequences of prion infection occur in the CNS, PrP^C has a wide tissue distribution and the exact cell types responsible for agent propagation and pathogenesis are still uncertain. Next to brain, lung tissue has the highest level of expression and PrP^C is detectable on the surface of lymphocytes, in heart, skeletal muscle, intestinal tract, spleen, testis, ovary and some other organs. In the CNS, PrP is expressed in neurons throughout the life of the animal, with levels of PrP mRNA varying among different types of neurons (Locht *et al.*, 1986). PrP mRNA is also expressed in astrocytes and oligodendrocytes throughout the brain of postnatal hamsters and rats (Moser *et al.*, 1995). The level of glial PrP mRNA expression in neonatal animals is comparable to that of neurons and increases two-fold during postnatal development.

Using different gene control elements it has been possible to direct the expression of PrP to specific cell types. While previous reports found little or no prion infectivity in skeletal muscle, two types of transgenic mice in which expression of PrP^C is directed exclusively to muscle under the control of the muscle creatine kinase and chicken α-actin promoters demonstrated that this tissue is capable of propagating prion infectivity (Bosque *et al.*, 2002). Transgenic mice in which expression of SHaPrP was regulated by the neuron specific enolase promoter indicated that neuron-specific expression PrP^C was sufficient to mediate susceptibility to hamster scrapie (Race *et al.*, 1995). Interestingly, astrocytes have been found to be the earliest site of PrP^{Sc} accumulation in the brain (Diedrich *et al.*, 1991) suggesting that these cells may play an important role in scrapie propagation and/or pathogenesis or even that astrocytes themselves may be the cells in which prion propagation occurs. Transgenic mice expressing hamster PrP under the control of the astrocyte-specific glial fibrilary acidic protein (GFAP) accumulated infectivity and PrP^{Sc} to high levels and developed disease after ~220 days (Raeber *et al.*, 1997). The interferon regulatory factor-1 promoter/Eμ enhancer, Lck promoter and albumin promoter/enhancer have been used to direct PrP expression to the spleen, T lymphocytes and liver respectively (Raeber *et al.*, 1999). High prion titers were found in the spleens of inoculated transgenic mice expressing PrP under the control of the interferon regulatory factor-1 promoter/Eμ enhancer, while mice expressing PrP under the control of the Lck and albumin promoters failed to replicate prions.

References

Asante, E.A., I. Gowland, J.M. Linehan, S.P. Mahal, and J. Collinge. 2002a. Expression pattern of a mini human PrP gene promoter in transgenic mice. Neurobiol. Dis. 10: 1-7.

Asante, E.A., J.M. Linehan, M. Desbruslais, S. Joiner, I. Gowland, A.L. Wood, J. Welch, A.F. Hill, S.E. Lloyd, J.D. Wadsworth, and J. Collinge. 2002b. BSE prions propagate as either variant CJD-like or sporadic CJD-like prion strains in transgenic mice expressing human prion protein. EMBO J. 21: 6358-6366.

Atarashi, R., N. Nishida, K. Shigematsu, S. Goto, T. Kondo, S. Sakaguchi, and S. Katamine. 2003. Deletion of N-terminal residues 23-88 from prion protein (PrP) abrogates the potential to rescue PrP-deficient mice from PrP-like protein/doppel-induced neurodegeneration. J. Biol. Chem. 278: 28944-28949.

Bessen, R.A., and R.F. Marsh. 1992. Biochemical and physical properties of the prion protein from two strains of the transmissible mink encephalopathy agent. J. Virol. 66: 2096-2101.

Borchelt, D.R., J. Davis, M. Fischer, M.K. Lee, H.H. Slunt, T. Ratovitsky, J. Regard, N.G. Copeland, N.A. Jenkins, S.S. Sisodia, and D.L. Price. 1996. A vector for expressing foreign genes in the brains and hearts of transgenic mice. Genet. Anal. 13: 159-163.

Bosque, P.J., C. Ryou, G. Telling, D. Peretz, G. Legname, S.J. DeArmond, and S.B. Prusiner. 2002. Prions in skeletal muscle. Proc. Natl. Acad. Sci. USA. 99: 3812-3817.

Brown, D.R., W.J. Schulz-Schaeffer, B. Schmidt, and H.A. Kretzschmar. 1997a. Prion protein-deficient cells show altered response to oxidative stress due to decreased SOD-1 activity. Exp. Neurol. 146: 104-112.

Brown, D.R., K. Qin, J.W. Herms, A. Madlung, J. Manson, R. Strome, P.E. Fraser, T. Kruck, A. von Bohlen, W. Schulz-Schaeffer, A. Giese, D. Westaway, and H. Kretzschmar. 1997b. The cellular prion protein binds copper *in vivo*. Nature. 390: 684-687.

Büeler, H., A. Raeber, A. Sailer, M. Fischer, A. Aguzzi, and C. Weissmann. 1994. High prion and PrP^{Sc} levels but delayed onset of disease in scrapie-inoculated mice heterozygous for a disrupted PrP gene. Mol. Med. 1: 19-30.

Büeler, H., A. Aguzzi, A. Sailer, R.-A. Greiner, P. Autenried, M. Aguet, and C. Weissmann. 1993. Mice devoid of PrP are resistant to scrapie. Cell. 73: 1339-1347.

Büeler, H., M. Fischer, Y. Lang, H. Bluethmann, H.-P. Lipp, S.J. DeArmond, S.B. Prusiner, M. Aguet, and C. Weissmann. 1992. Normal development and behaviour of mice lacking the neuronal cell-surface PrP protein. Nature. 356: 577-582.

Buschmann, A., E. Pfaff, K. Reifenberg, H.M. Muller, and M.H. Groschup. 2000. Detection of cattle-derived BSE prions using transgenic mice overexpressing bovine PrP(C). Arch. Virol. Suppl.: 75-86.

Campbell, S., U. Dennehy, and G. Telling. 2000. Analyzing the influence of PrP primary structure on prion pathogenesis in transgenic mice. Arch. Virol. Suppl.: 87-94.

Castilla, J., A. Gutierrez Adan, A. Brun, B. Pintado, M.A. Ramirez, B. Parra, D. Doyle, M. Rogers, F.J. Salguero, C. Sanchez, J.M. Sanchez-Vizcaino, and J.M. Torres. 2003. Early detection of PrPres in BSE-infected bovine PrP transgenic mice. Arch. Virol. 148: 677-691.

Chiesa, R., P. Piccardo, B. Ghetti, and D.A. Harris. 1998. Neurological illness in transgenic mice expressing a prion protein with an insertional mutation. Neuron. 21: 1339-1351.

Chiesa, R., B. Drisaldi, E. Quaglio, A. Migheli, P. Piccardo, B. Ghetti, and D.A. Harris. 2000. Accumulation of protease-resistant prion protein (PrP) and apoptosis of cerebellar granule cells in transgenic mice expressing a PrP insertional mutation. Proc. Natl. Acad. Sci. USA. 97: 5574-5579.

Chiesa, R., A. Pestronk, R.E. Schmidt, W.G. Tourtellotte, B. Ghetti, P. Piccardo, and D.A. Harris. 2001. Primary myopathy and accumulation of PrPSc-like molecules in peripheral tissues of transgenic mice expressing a prion protein insertional mutation. Neurobiol. Dis. 8: 279-288.

Collinge, J., K.C.L. Sidle, J. Meads, J. Ironside, and A.F. Hill. 1996. Molecular analysis of prion strain variation and the aetiology of "new variant" CJD. Nature. 383: 685-690.

Collinge, J., M.A. Whittington, K.C. Sidle, C.J. Smith, M.S. Palmer, A.R. Clarke, and J.G.R. Jefferys. 1994. Prion protein is necessary for normal synaptic function. Nature. 370: 295-297.

Collinge, J., M.S. Palmer, K.C.L. Sidle, I. Gowland, R. Medori, J. Ironside, and P. Lantos. 1995. Transmission of fatal familial insomnia to laboratory animals (Lett.). Lancet. 346: 569-570.

Crozet, C., A. Bencsik, F. Flamant, S. Lezmi, J. Samarut, and T. Baron. 2001a. Florid plaques in ovine PrP transgenic mice infected with an experimental ovine BSE. EMBO Rep 2: 952-956.

Crozet, C., F. Flamant, A. Bencsik, D. Aubert, J. Samarut, and T. Baron. 2001b. Efficient transmission of two different sheep scrapie isolates in transgenic mice expressing the ovine PrP gene. J. Virol. 75: 5328-5334.

DeArmond, S.J., H. Sánchez, F. Yehiely, Y. Qiu, A. Ninchak-Casey, V. Daggett, A.P. Camerino, J. Cayetano, M. Rogers, D. Groth, M. Torchia, P. Tremblay, M.R. Scott, F.E. Cohen, and S.B. Prusiner. 1997. Selective neuronal targeting in prion disease. Neuron. 19: 1337-1348.

Diedrich, J.F., P.E. Bendheim, Y.S. Kim, R.I. Carp, and A.T. Haase. 1991. Scrapie-associated prion protein accumulates in astrocytes during scrapie infection. Proc. Natl. Acad. Sci. USA 88: 375-379.

Drisaldi, B., R.S. Stewart, C. Adles, L.R. Stewart, E. Quaglio, E. Biasini, L. Fioriti, R. Chiesa, and D.A. Harris. 2003. Mutant PrP is delayed in its exit from the endoplasmic reticulum, but neither wild-type nor mutant PrP undergoes retrotranslocation prior to proteasomal degradation. J. Biol. Chem. 278: 21732-21743.

Fischer, M., T. Rülicke, A. Raeber, A. Sailer, M. Moser, B. Oesch, S. Brandner, A. Aguzzi, and C. Weissmann. 1996. Prion protein (PrP) with amino-proximal deletions restoring susceptibility of PrP knockout mice to scrapie. EMBO J. 15: 1255-1264.

Flechsig, E., D. Shmerling, I. Hegyi, A.J. Raeber, M. Fischer, A. Cozzio, C. von Mering, A. Aguzzi, and C. Weissmann. 2000. Prion protein devoid of the octapeptide repeat region restores susceptibility to scrapie in PrP knockout mice. Neuron. 27: 399-408.

Flechsig, E., I. Hegyi, R. Leimeroth, A. Zuniga, D. Rossi, A. Cozzio, P. Schwarz, T. Rulicke, J. Gotz, A. Aguzzi, and C. Weissmann. 2003. Expression of truncated PrP targeted to Purkinje cells of PrP knockout mice causes Purkinje cell death and ataxia. EMBO J. 22: 3095-3101.

Fraser, H., M.E. Bruce, A. Chree, I. McConnell, and G.A.H. Wells. 1992. Transmission of bovine spongiform encephalopathy and scrapie to mice. J. Gen. Virol. 73: 1891-1897.

Gombojav, A., I. Shimauchi, M. Horiuchi, N. Ishiguro, M. Shinagawa, T. Kitamoto, I. Miyoshi, S. Mohri, and M. Takata. 2003. Susceptibility of transgenic mice expressing chimeric sheep, bovine and human PrP genes to sheep scrapie. J. Vet. Med. Sci. 65: 341-347.

Hegde, R.S., J.A. Mastrianni, M.R. Scott, K.A. DeFea, P. Tremblay, M. Torchia, S.J. DeArmond, S.B. Prusiner, and V.R. Lingappa. 1998.

A transmembrane form of the prion protein in neurodegenerative disease. Science. 279: 827-834.

Herms, J.W., H.A. Kretzschmar, S. Titz, and B.U. Keller. 1995. Patch-clamp analysis of synaptic transmission to cerebellar purkinje cells of prion protein knockout mice. Eur. J. Neurosci. 7: 2508-2512.

Hill, A.F., M. Desbruslais, S. Joiner, K.C.L. Sidle, I. Gowland, J. Collinge, L.J. Doey, and P. Lantos. 1997. The same prion strain causes vCJD and BSE. Nature. 389: 448-450.

Hsiao, K.K., M. Scott, D. Foster, D.F. Groth, S.J. DeArmond, and S.B. Prusiner. 1990. Spontaneous neurodegeneration in transgenic mice with mutant prion protein. Science. 250: 1587-1590.

Hsiao, K.K., D. Groth, M. Scott, S.-L. Yang, H. Serban, D. Rapp, D. Foster, M. Torchia, S.J. DeArmond, and S.B. Prusiner. 1994. Serial transmission in rodents of neurodegeneration from transgenic mice expressing mutant prion protein. Proc. Natl. Acad. Sci. USA 91: 9126-9130.

Kaneko K., B.H. Wille, H., Zhang, H., Groth, D., Torchia, M., Tremblay, P., Safar, J., Prusiner, S.B., DeArmond, S.J., Baldwin, M.A., Cohen, F.E. 2000. A Synthetic Peptide Initiates Gerstmann-Straussler-Scheinker (GSS) Disease in Transgenic Mice. J. Mol. Biol.295: 997-1007.

Kitamoto, T., S. Mohri, J.W. Ironside, I. Miyoshi, T. Tanaka, N. Kitamoto, S. Itohara, N. Kasai, M. Katsuki, J. Higuchi, T. Muramoto, and R.W. Shin. 2002. Follicular dendritic cell of the knock-in mouse provides a new bioassay for human prions. Biochem. Biophys. Res. Commun. 294: 280-286.

Korth, C., K. Kaneko, D. Groth, N. Heye, G. Telling, J. Mastrianni, P. Parchi, P. Gambetti, R. Will, J. Ironside, C. Heinrich, P. Tremblay, S.J. DeArmond, and S.B. Prusiner. 2003. Abbreviated incubation times for human prions in mice expressing a chimeric mouse-human prion protein transgene. Proc. Natl. Acad. Sci. USA. 100: 4784-4789.

Lledo, P.-M., P. Tremblay, S.J. DeArmond, S.B. Prusiner, and R.A. Nicoll. 1996. Mice deficient for prion protein exhibit normal neuronal excitability and synaptic transmission in the hippocampus. Proc. Natl. Acad. Sci. USA. 93: 2403-2407.

Locht, C., B. Chesebro, R. Race, and J.M. Keith. 1986. Molecular cloning and complete sequence of prion protein cDNA from mouse brain infected with the scrapie agent. Proc. Natl. Acad. Sci. USA. 83: 6372-6376.

Ma, J., and S. Lindquist. 2002. Conversion of PrP to a self-perpetuating PrPSc-like conformation in the cytosol. Science. 298: 1785-1788.

Ma, J., R. Wollmann, and S. Lindquist. 2002. Neurotoxicity and neurodegeneration when PrP accumulates in the cytosol. Science. 298: 1781-1785.

Mallucci, G.R., S. Ratte, E.A. Asante, J. Linehan, I. Gowland, J.G. Jefferys, and J. Collinge. 2002. Post-natal knockout of prion protein alters hippocampal CA1 properties, but does not result in neurodegeneration. EMBO J. 21: 202-210.

Manson, J.C., A.R. Clarke, P.A. McBride, I. McConnell, and J. Hope. 1994a. PrP gene dosage determines the timing but not the final intensity or distribution of lesions in scrapie pathology. Neurodegeneration. 3: 331-340.

Manson, J.C., A.R. Clarke, M.L. Hooper, L. Aitchison, I. McConnell, and J. Hope. 1994b. 129/Ola mice carrying a null mutation in PrP that abolishes mRNA production are developmentally normal. Mol. Neurobiol. 8: 121-127.

Mastrianni, J., F. Nixon, R. Layzer, S.J. DeArmond, and S.B. Prusiner. 1997. Fatal sporadic insomnia: fatal familial insomnia phenotype without a mutation of the prion protein gene. Neurology. 48 [Suppl.]: A296.

Mastrianni, J.A., S. Capellari, G.C. Telling, D. Han, P. Bosque, S.B. Prusiner, and S.J. DeArmond. 2001. Inherited prion disease caused by the V210I mutation: Transmission to transgenic mice. Neurology. 57: 2198-205.

Moore, R.C., P. Mastrangelo, E. Bouzamondo, C. Heinrich, G. Legname, S.B. Prusiner, L. Hood, D. Westaway, S.J. DeArmond, and P. Tremblay. 2001. Doppel-induced cerebellar degeneration in transgenic mice. Proc. Natl. Acad. Sci. USA. 98: 15288-15293.

Moore, R.C., I.Y. Lee, G.L. Silverman, P.M. Harrison, R. Strome, C. Heinrich, A. Karunaratne, S.H. Pasternak, M.A. Chishti, Y. Liang, P. Mastrangelo, K. Wang, A.F.A. Smit, S. Katamine, G.A. Carlson, F.E. Cohen, S.B. Prusiner, D.W. Melton, P. Tremblay, L.E. Hood, and D. Westaway. 1999. Ataxia in prion protein (PrP) deficient mice is associated with upregulation of the novel PrP-like protein doppel. J. Mol. Biol. 292: 797-817.

Moser, M., R.J. Colello, U. Pott, and B. Oesch. 1995. Developmental expression of the prion protein gene in glial cells. Neuron. 14: 509-517.

Mouillet-Richard, S., M. Ermonval, C. Chebassier, J.L. Laplanche, S. Lehmann, J.M. Launay, and O. Kellermann. 2000. Signal transduction through prion protein. Science. 289: 1925-1928.

Muramoto, T., S.J. DeArmond, M. Scott, G.C. Telling, F.E. Cohen, and S.B. Prusiner. 1997. Heritable disorder resembling neuronal storage disease in mice expressing prion protein with deletion of an alpha-helix. Nat. Med. 3: 750-755.

Parchi, P., R. Castellani, S. Capellari, B. Ghetti, K. Young, S.G. Chen, M. Farlow, D.W. Dickson, A.A.F. Sima, J.Q. Trojanowski, R.B. Petersen, and P. Gambetti. 1996. Molecular basis of phenotypic variability in sporadic Creutzfeldt-Jakob disease. Ann. Neurol. 39: 767-778.

Peretz, D., R.A. Williamson, G. Legname, Y. Matsunaga, J. Vergara, D.R. Burton, S.J. DeArmond, S.B. Prusiner, and M.R. Scott. 2002. A change in the conformation of prions accompanies the emergence of a new prion strain. Neuron. 34: 921-932.

Prusiner, S.B., D. Groth, A. Serban, R. Koehler, D. Foster, M. Torchia, D. Burton, S.-L. Yang, and S.J. DeArmond. 1993. Ablation of the prion protein (PrP) gene in mice prevents scrapie and facilitates production of anti-PrP antibodies. Proc. Natl. Acad. Sci. USA. 90: 10608-10612.

Prusiner, S.B., M. Scott, D. Foster, K.-M. Pan, D. Groth, C. Mirenda, M. Torchia, S.-L. Yang, D. Serban, G.A. Carlson, P.C. Hoppe, D. Westaway, and S.J. DeArmond. 1990. Transgenetic studies implicate interactions between homologous PrP isoforms in scrapie prion replication. Cell. 63: 673-686.

Race, R.E., S.A. Priola, R.A. Bessen, D. Ernst, J. Dockter, G.F. Rall, L. Mucke, B. Chesebro, and M.B.A. Oldstone. 1995. Neuron-specific expression of a hamster prion protein minigene in transgenic mice induces susceptibility to hamster scrapie agent. Neuron. 15: 1183-1191.

Raeber, A.J., A. Sailer, I. Hegyi, M.A. Klein, T. Rulike, M. Fischer, S. Brandner, A. Aguzzi, and C. Weissmann. 1999. Ectopic expression of prion protein (PrP) in T lymphocytes or hepatocytes of PrP knockout mice is insufficient to sustain prion replication. Proc. Natl. Acad. Sci. USA. 96: 3987-3992.

Raeber, A.J., R.E. Race, S. Brandner, S.A. Priola, A. Sailer, R.A. Bessen, L. Mucke, J. Manson, A. Aguzzi, M.B. Oldstone, C. Weissmann, and B. Chesebro. 1997. Astrocyte-specific expression of hamster prion protein (PrP) renders PrP knockout mice susceptible to hamster scrapie. EMBO J. 16: 6057-6065.

Rossi, D., A. Cozzio, E. Flechsig, M.A. Klein, T. Rulicke, A. Aguzzi, and C. Weissmann. 2001. Onset of ataxia and Purkinje cell loss in PrP null mice inversely correlated with Dpl level in brain. EMBO J. 20: 694-702.

Roucou, X., Q. Guo, Y. Zhang, C.G. Goodyer, and A.C. LeBlanc. 2003. Cytosolic prion protein is not toxic and protects against Bax-mediated cell death in human primary neurons. J. Biol. Chem.278: 40877-40881.

Sakaguchi, S., S. Katamine, K. Shigematsu, A. Nakatani, R. Moriuchi, N. Nishida, K. Kurokawa, R. Nakaoke, H. Sato, K. Jishage, J. Kuno, T. Noda, and T. Miyamoto. 1995. Accumulation of proteinase K-resistant prion protein (PrP) is restricted by the expression level of normal PrP in mice inoculated with a mouse-adapted strain of the Creutzfeldt-Jakob disease agent. J. Virol. 69: 7586-7592.

Sakaguchi, S., S. Katamine, N. Nishida, R. Moriuchi, K. Shigematsu, T. Sugimoto, A. Nakatani, Y. Kataoka, T. Houtani, S. Shirabe, H. Okada, S. Hasegawa, T. Miyamoto, and T. Noda. 1996. Loss of cerebellar Purkinje cells in aged mice homozygous for a disrupted PrP gene. Nature. 380: 528-531.

Scott, M., D. Groth, D. Foster, M. Torchia, S.-L. Yang, S.J. DeArmond, and S.B. Prusiner. 1993. Propagation of prions with artificial properties in transgenic mice expressing chimeric PrP genes. Cell. 73: 979-988.

Scott, M., D. Foster, C. Mirenda, D. Serban, F. Coufal, M. Wälchli, M. Torchia, D. Groth, G. Carlson, S.J. DeArmond, D. Westaway, and S.B. Prusiner. 1989. Transgenic mice expressing hamster prion protein produce species-specific scrapie infectivity and amyloid plaques. Cell. 59: 847-857.

Scott, M.R., R. Köhler, D. Foster, and S.B. Prusiner. 1992. Chimeric prion protein expression in cultured cells and transgenic mice. Protein Sci. 1: 986-997.

Scott, M.R., J. Safar, G. Telling, O. Nguyen, D. Groth, M. Torchia, R. Koehler, P. Tremblay, D. Walther, F.E. Cohen, S.J. DeArmond, and S.B. Prusiner. 1997. Identification of a prion protein epitope modulating transmission of bovine spongiform encephalopathy prions to transgenic mice. Proc. Natl. Acad. Sci. USA. 94: 14279-14284.

Shmerling, D., M. Fischer, T.T. Blättler, I. Hegy, S. Brandner, A. Aguzzi, and C. Weissmann. 1997. Expression in mice of amino-proximally truncated but not of full-length PrP causes a cerebellar syndrome (Abstr.). 29th Annual Meeting of the Union of the Swiss Societies for Experimental Biology in Geneva, March 20-21: A46.

Supattapone, S., P. Bosque, T. Muramoto, H. Wille, C. Aagaard, D. Peretz, H.-O.B. Nguyen, C. Heinrich, M. Torchia, J. Safar, F.E.

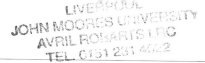
Cohen, S.J. DeArmond, S.B. Prusiner, and M. Scott. 1999. Prion protein of 106 residues creates an artificial transmission barrier for prion replication in transgenic mice. Cell. 96: 869-878.

Telling, G.C., T. Haga, M. Torchia, P. Tremblay, S.J. DeArmond, and S.B. Prusiner. 1996a. Interactions between wild-type and mutant prion proteins modulate neurodegeneration in transgenic mice. Genes & Dev. 10: 1736-1750.

Telling, G.C., M. Scott, J. Mastrianni, R. Gabizon, M. Torchia, F.E. Cohen, S.J. DeArmond, and S.B. Prusiner. 1995. Prion propagation in mice expressing human and chimeric PrP transgenes implicates the interaction of cellular PrP with another protein. Cell. 83: 79-90.

Telling, G.C., M. Scott, K.K. Hsiao, D. Foster, S.L. Yang, M. Torchia, K.C. Sidle, J. Collinge, S.J. DeArmond, and S.B. Prusiner. 1994. Transmission of Creutzfeldt-Jakob disease from humans to transgenic mice expressing chimeric human-mouse prion protein. Proc. Natl. Acad. Sci. USA. 91: 9936-9940.

Telling, G.C., P. Parchi, S.J. DeArmond, P. Cortelli, P. Montagna, R. Gabizon, J. Mastrianni, E. Lugaresi, P. Gambetti, and S.B. Prusiner. 1996b. Evidence for the conformation of the pathologic isoform of the prion protein enciphering and propagating prion diversity. Science. 274: 2079-2082.

Tobler, I., S.E. Gaus, T. Deboer, P. Achermann, M. Fischer, T. Rülicke, M. Moser, B. Oesch, P.A. McBride, and J.C. Manson. 1996. Altered circadian activity rhythms and sleep in mice devoid of prion protein. Nature. 380: 639-642.

Tremblay, P., Z. Meiner, M. Galou, C. Heinrich, C. Petromilli, T. Lisse, J. Cayetano, M. Torchia, W. Mobley, H. Bujard, S.J. DeArmond, and S.B. Prusiner. 1998. Doxycyline control of prion protein transgene expression modulates prion disease in mice. Proc. Natl. Acad. Sci. USA. 95: 12580-12585.

Vilotte, J.L., S. Soulier, R. Essalmani, M.G. Stinnakre, D. Vaiman, L. Lepourry, J.C. Da Silva, N. Besnard, M. Dawson, A. Buschmann, M. Groschup, S. Petit, M.F. Madelaine, S. Rakatobe, A. Le Dur, D. Vilette, and H. Laude. 2001. Markedly increased susceptibility to natural sheep scrapie of transgenic mice expressing ovine prp. J. Virol. 75: 5977-5984.

Westaway, D., S.J. DeArmond, J. Cayetano-Canlas, D. Groth, D. Foster, S.-L. Yang, M. Torchia, G.A. Carlson, and S.B. Prusiner. 1994. Degeneration of skeletal muscle, peripheral nerves, and the central nervous system in transgenic mice overexpressing wild-type prion proteins. Cell. 76: 117-129.

Westaway, D., C.A. Mirenda, D. Foster, Y. Zebarjadian, M. Scott, M. Torchia, S.-L. Yang, H. Serban, S.J. DeArmond, C. Ebeling, S.B. Prusiner, and G.A. Carlson. 1991. Paradoxical shortening of scrapie incubation times by expression of prion protein transgenes derived from long incubation period mice. Neuron. 7: 59-68.

Whittington, M.A., K.C.L. Sidle, I. Gowland, J. Meads, A.F. Hill, M.S. Palmer, J.G.R. Jefferys, and J. Collinge. 1995. Rescue of neurophysiological phenotype seen in PrP null mice by transgene encoding human prion protein. Nat. Genet. 9: 197-201.

From: Prions and Prion Diseases: Current Perspectives
Edited by: Glenn C. Telling

Chapter 7

Peripheral Pathogenesis
of Prion Diseases

Adriano Aguzzi

Abstract

Prion neuroinvasion consists of an ordered sequence of events resulting in infection of the central nervous system (CNS). Successful oral challenge requires transepithelial migration of prions, which may be accomplished by M-cells. Depletion of lymphocytes from the intestinal mucosa by ablation of $\alpha_4\beta_7$ integrins does not prevent pathogenesis, yet mice exhibiting reduced number of Peyer's patches are virtually uninfectible orally. After gaining access to the body from peripheral sites, prions colonize lymphoid organs of mice, humans, and sheep: the failure of peripherally administered prions to elicit disease in immune deficient mice indicates that this is crucial for pathogenesis. B-lymphocytes are required for neuroinvasion upon intraperitoneal administration, probably (but not necessarily only) because they provide lymphotoxins to secondary lymphoid organs, thereby maintaining follicular dendritic cells (FDCs): genetic or pharmacological interference with lymphotoxin signaling effectively impairs pathogenesis. The sympathetic nervous system appears to be involved in prion transfer to brain, since sympathectomy delays or prevents pathogenesis, whereas sympathetic hyperinnervation accelerates it. Various components of the complement system are modifiers of neuroinvasion efficiency, and their pharmacological or genetic ablation interferes with neuroinvasion. Although isogenic prions are immunologically inert, expression of anti-PrP^C antibodies in transgenic mice has uncovered that autoreactivity to PrP^C does not necessarily tolerize B-cells, and that sustained anti-PrP^C IgM titers can prevent peripheral prion pathogenesis.

1. Prion Biology: Some Basic Facts

Prion diseases are inevitably fatal neurodegenerative conditions which affect humans and a wide variety of animals (Aguzzi *et al.*, 2001c). Although prion diseases may present with certain morphological and pathophysiological parallels to other progressive encephalopathies, such as Alzheimer's and Parkinson's disease (Aguzzi and Raeber, 1998), they are unique in that they are transmissible. Homogenization of brain tissue from affected individuals and intracerebral inoculation into another individual of the same species will typically reproduce the disease. This important fact was recognized more than half a century ago in the case of scrapie (Cuille and Chelle, 1939), a prototypic prion disease that affects sheep and goats.

In the 1960's and '70s, the work of Gajdusek on Kuru and Creutzfeldt-Jakob disease (Gajdusek *et al.*, 1966; Gibbs *et al.*, 1968), partly inspired by a suggestion of Hadlow (Hadlow, 1959) on the similarity of scrapie and Kuru, established that these diseases are also transmissible to primates, to mice, and in some unfortunate iatrogenic instances, also to other humans (Brown *et al.*, 2000). Therefore, prion diseases are also called transmissible spongiform encephalopathies (TSEs), a term that emphasizes their infectious character.

While only less than 1% of all reported cases of Creutzfeldt-Jakob disease (CJD) can be traced to a defined infectious source, the identification of bovine spongiform encephalopathy (BSE) (Wells *et al.*, 1987) and its subsequent epizootic spread has highlighted prion-contaminated meat-and-bone meal as a tremendously efficient vector for bovine prion diseases (Weissmann and Aguzzi, 1997), which does not completely lose its infectious potential even after extensive autoclaving (Taylor, 2000). BSE is most likely transmissible to humans, too, and strong circumstantial evidence (Aguzzi, 1996; Aguzzi and Weissmann, 1996b; Collinge *et al.*, 1996; Bruce *et al.*, 1997; Hill *et al.*, 1997a) suggests that BSE is the cause of variant Creutzfeldt-Jakob disease (vCJD) which has claimed more than 130 lives in the United Kingdom (Will *et al.*, 1996; http://www.doh.gov.uk/

cjd/stats/aug02.htm, 2002) and, to a much smaller extent, in some other countries (Chazot *et al.*, 1996). When transmitted to primates, BSE produces a pathology strikingly similar to that of vCJD (Aguzzi and Weissmann, 1996b; Lasmezas *et al.*, 1996b). While there is no direct indication yet that variant Creutzfeldt-Jakob disease has been transmitted from one human to another, this is certainly a very worrying scenario (Aguzzi, 2000). Not only surgical instruments may represent potential vectors of transmission, but possibly also transfusions. Significantly, sheep infected with BSE can efficiently transmit the agent to other sheep via blood transfusion (Houston *et al.*, 2000; Hunter *et al.*, 2002).

Several aspects of CJD epidemiology continue to be enigmatic. For example, CJD incidence in Switzerland has risen two-fold in 2001, and appears to be increasing even further in the year 2002 (Glatzel *et al.*, 2002). A screen for recognized or hypothetical risk factors for CJD has, to date, not exposed any causal factors. Several scenarios may account for the increase in incidence, including improved reporting, iatrogenic transmission, and transmission of a prion zoonosis.

Prion diseases typically exhibit a very long latency period between the time of infection and the clinical manifestation: this is the reason why these diseases were originally thought to be caused by "slow viruses". From the viewpoint of interventional approaches, this peculiarity may be exploitable, since it opens a possible window of intervention after infection has occurred, but before brain damage is being initiated. Prions spend much of this latency time executing neuroinvasion, which is the process of reaching the CNS after entering the body from peripheral sites (Aguzzi, 1997a; Nicotera, 2001). During this process, little or no damage occurs to brain, and one might hope that its interruption may prevent neurodegeneration.

1.1. Stanley Prusiner's Protein-only Hypothesis

The most widely accepted hypothesis on the nature of the infectious agent causing TSEs (which was termed prion by Stanley B. Prusiner) (Prusiner, 1982) predicates that it consists essentially of PrP^{Sc}, an abnormally folded, protease-resistant, beta-sheet rich isoform of a normal cellular protein termed PrP^{C}. According to this fascinating theory, the prion does not contain any informational nucleic acids, and its infectivity propagates simply by recruitment and "autocatalytic" conformational conversion of cellular prion protein into disease-associated PrP^{Sc} (Aguzzi and Weissmann, 1997).

A large body of experimental and epidemiological evidence is compatible with the protein-only hypothesis, and very stringently designed experiments have failed to disprove it. It would go well beyond the scope of this article to review all efforts that have been undertaken to this effect. Perhaps most impressively, knockout mice carrying a homozygous deletion of the *Prnp* gene that encodes PrP^{C}, fail to develop disease upon inoculation with infectious brain homogenate (Büeler *et al.*, 1993), nor does their brain carry prion infectivity (Sailer *et al.*, 1994). Reintroduction of *Prnp* by transgenesis – even in a shortened, redacted form – restores infectibility and prion replication in *Prnp*$^{o/o}$ mice (Fischer *et al.*, 1996; Shmerling *et al.*, 1998; Supattapone *et al.*, 1999; Flechsig *et al.*, 2000). In addition, all familial cases of human TSEs are characterized by *Prnp* mutations (Aguzzi and Weissmann, 1996a; Prusiner *et al.*, 1998).

Informational nucleic acids of >50 nucleotides in length do not participate in prion infectivity (Kellings *et al.*, 1993; Riesner *et al.*, 1993), but shorter non-coding oligonucleotides have not been formally excluded – a fact that may have some relevance in view of the surprising discoveries related to RNA-mediated gene silencing. *Prnp* exhibits a long reading frame on its non-coding strand (Moser *et al.*, 1993) which is conserved among mammals (Aguzzi, 1997b; Rother *et al.*, 1997), suggesting that it may be transcribed – maybe only in specific pathological situations. One might wonder, therefore, whether this peculiar property of *Prnp* might result in the production of double-stranded transcripts with silencing properties. However, bona fide antisense transcription of the *Prnp* locus has never been demonstrated *in vivo*, nor in cultured cells.

1.2. Some Major Open Questions in Prion Biology

In recent years, prion research has progressed at a faster pace than many of us would have thought possible. As a consequence, many enigmas surrounding prion diseases have now been solved. However, the areas that are still obscure do not relate only to the details: some of these concern the core of the prion concept (Chesebro, 1998). In my opinion, there are four large groups of questions regarding the basic science of prion replication and of development of transmissible spongiform encephalopathies diseases that deserve to be addressed with a vigorous research effort:

- Which are the molecular mechanisms of prion replication? How does the disease-associated prion protein, PrP^{Sc}, achieve the conversion of its cellular sibling, PrP^C, into a likeness of itself? Which other proteins assist this process? Can we inhibit this process? If so, how?

- What is the essence of prion strains, which are operationally defined as variants of the infectious agent capable of retaining stable phenotypic traits upon serial passage in syngeneic hosts? The existence of strains is very well known in virology, but it was not predicted to exist in the case of an agent that propagates epigenetically.

- How do prions reach the brain after having entered the body? Which molecules and which cell types are involved in this process of neuroinvasion? Which inhibitory strategies are likely to succeed?

- The mechanisms of neurodegeneration in spongiform encephalopathies are not understood. Which are the pathogenetic cascades that are activated upon accumulation of disease-associated prion protein, and ultimately lead to brain damage?

- What is the physiological function of the highly conserved, normal prion protein, PrP^C? The *Prnp* gene encoding PrP^C was identified in 1985 (Oesch *et al.*, 1985; Basler *et al.*, 1986), *Prnp* knockout mice were described in 1992 (Büeler *et al.*, 1992), and some PrP^C-interacting proteins have been identified (Oesch *et al.*, 1990; Rieger *et al.*, 1997; Yehiely *et al.*, 2002; Zanata *et al.*, 2002). Yet the function of PrP^C remains unknown.

The present review article will discuss some of the progress recently achieved in some of the areas delineated above, with special reference to the topics that have directly interested my laboratory.

1.3. Brain Damage in Prion Diseases

An interesting question regards the molecular mechanism underlying neuropathological changes, in particular cell death, resulting from prion disease. Depletion of PrP^C is an unlikely cause, in view of the finding that abrogation of PrP does not cause scrapie-like neuropathological changes (Büeler *et al.*, 1992), even when elicited postnatally (Mallucci *et al.*, 2002). More likely, toxicity of PrP^{Sc} or some PrP^C-dependent process is responsible.

To address the question of neurotoxicity, brain tissue of *Prnp$^{o/o}$* mice was exposed to a continuous source of PrP^{Sc}. To this purpose, telencephalic tissue from transgenic mice overexpressing PrP (Fischer *et al.*, 1996) was transplanted into the forebrain of *Prnp$^{o/o}$* mice and the "pseudochimeric" brains were inoculated with scrapie prions. All grafted and scrapie-inoculated mice remained free of scrapie symptoms for at least 70 weeks; this exceeded at least sevenfold the survival time of scrapie-infected donor mice (Brandner *et al.*, 1996a). Therefore, the presence of a continuous source of PrP^{Sc} and of scrapie prions does not exert any clinically detectable adverse effects on a mouse devoid of PrP^C. On the other hand, the grafts developed characteristic histopathological features of scrapie after inoculation. The course of the disease in the graft was very similar to that observed in the brain of scrapie-inoculated wild-type mice (Brandner *et al.*, 1998). Importantly, grafts had extensive contact with the recipient brain, and prions could navigate between the two compartments, as shown by the fact that inoculation of wild-type animals engrafted with PrP-expressing neuroectodermal tissue resulted in scrapie pathology in both graft and host tissue. Nonetheless, histopathological changes never extended into host tissue, even at the latest stages (>450 days), although PrP^{Sc} was detected in both grafts and recipient brain, and immunohistochemistry revealed PrP deposits in the hippocampus, and occasionally in the parietal cortex, of all animals (Brandner *et al.*, 1996a). Thus, prions moved from the grafts to some regions of the PrP-deficient host brain without causing pathological changes or clinical disease. The distribution of PrP^{Sc} in the white matter tracts of the host brain suggests diffusion within the extra-cellular space (Jeffrey *et al.*, 1994) rather than axonal transport.

These findings suggest that the expression of PrP^C by an infected cell, rather than the extracellular deposition of PrP^{Sc}, is the critical prerequisite for the development of scrapie pathology. Perhaps

PrP^{Sc} is inherently non-toxic and PrP^{Sc} plaques found in spongiform encephalopathies are an epiphenomenon rather than a cause of neuronal damage. This hypothesis appears to be supported by the recent data (Ma and Lindquist, 2002; Ma *et al.*, 2002) indicating that exaggerated retrograde transport of the prion protein from the endoplasmatic reticulum into the cytosol, or functionally equivalent inhibition of proteasome function, might induce a self-propagating, extremely cytotoxic cellular form, which may ultimately be responsible for neuronal damage. These results are very exciting, and it will be important to validate them by investigating whether the aggregated self-propagating material can also be transmitted between individual animals in a classic transmission experiment.

One may therefore propose that availability of PrP^{C} for some intracellular process elicited by the infectious agent, perhaps the formation of a toxic form of PrP (PrP*; (Weissmann, 1991)) other than PrP^{Sc} is responsible for spongiosis, gliosis, and neuronal death. This would be in agreement with the fact that in several instances, and especially in fatal familial insomnia, spongiform pathology is detectable although very little PrP^{Sc} is present (Aguzzi and Weissmann, 1997).

2. Peripheral Entry Sites of Prions

The fastest and most efficient method for inducing spongiform encephalopathy in the laboratory is intracerebral inoculation of brain homogenate. Inoculation of 1'000'000 infectious units (defined as the amount of infectivity that will induce TSE with 50% likelihood in a given host) will yield disease in approximately half year; a remarkably strict inverse relationship can be observed between the logarithm of the inoculated dose and the incubation time (Prusiner *et al.*, 1982) (Figure 1).

However, the above situation does not correspond to what typically happens in the field. There, acquisition of prion infectivity through any of several peripheral routes is the rule. However, prion diseases can also be initiated by feeding (Wells *et al.*, 1987; Kimberlin and Wilesmith, 1994; Anderson *et al.*, 1996), by intravenous and intraperitoneal injection (Kimberlin and Walker, 1978) as well as from the eye by conjunctival instillation (Scott *et al.*, 1993), corneal grafts (Duffy *et al.*, 1974) and intraocular injection (Fraser, 1982).

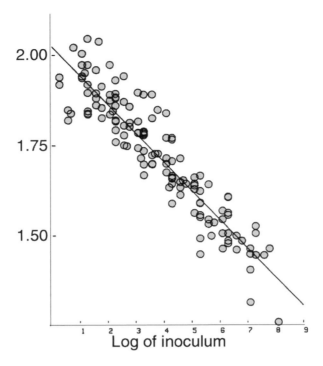

Log of inoculum

Figure 1. Prion bioassay by the incubation time method. This figure is reproduced from an important study by Prusiner and colleagues that demonstrates an inverse logarithmic relationship between size of infectious inoculum and latency period of scrapie in experimental animals (Prusiner *et al.*, 1982). The relationship is so precise that incubation time can actually be used to back-calculate the prion titer of a test article, e.g. for inactivation studies. Measurement of prion titers by the incubation time method has become a standard procedure which can be used for both wild-type and transgenic animals.

2.1. The Pathway of Orally Administered Prions

Upon oral challenge, an early rise in prion infectivity can be observed in the distal ileum of infected organisms: this applies to several species but was most extensively investigated in the sheep (Wells *et al.*, 1994; Vankeulen *et al.*, 1996). There, Peyer's patches acquire strong immunopositivity for the prion protein. Immunohistochemical stains with antibodies to the prion protein typically reveal a robust signal in primary B-cell follicles and germinal centers, which roughly colocalizes with the complement receptor, CD35, in a wide variety of secondary lymphoid organs including appendix and tonsils (Hill *et al.*, 1997b). Although conventional light microscopy does not allow differentiating between PrP^C and PrP^{Sc}, Western blot analysis has not left any doubt about the fact that Peyer's patches do accumulate the disease-associated form of the prion protein.

The latter is true also in the mouse model of scrapie, which is being used as a convenient experimental paradigm by many laboratories including ours. Administration of mouse-adapted scrapie prions (Rocky Mountain Laboratory or RML strain, originally derived from the Chandler sheep scrapie isolate) induces a surge in intestinal prion infectivity as early as a few days after inoculation (Marco Prinz, Gerhard Huber and AA, unpublished results).

All of the above evidence conjures the suggestion that Peyer's patches may represent a portal of entry for orally administered prions on their journey from the luminal aspect of the gastroenteric tube to the CNS. However, the question as to whether the same applies for BSE-affected cattle has been answered less unambiguously.

In a monumental study of BSE pathogenesis in cattle carried out at the UK Veterinary Laboratory Agency, cows of various ages were fed with 100g, 10g, 1g, or 100mg of brain homogenate derived from BSE-sick cows (Bradley, 2000). A large variety of tissues was taken at various points in time, homogenized, and transmitted intracerebrally to indicator organisms in order to assess their prion content. This study was designed to be performed over a time frame of more than a decade and is still underway at the time of writing: it has uncovered a transient surge in infectivity in the distal ileum of cows at approximately 6 months post infection. Infectivity then subsides, but it appears to return to the terminal ileum at the end stages of disease, maybe by means of some sort of retrograde transport (Wells *et al.*, 1998). Although this was not formally confirmed, it appears likely that Peyer's patches are the sites of prion accumulation in the gastrointestinal tract of cattle challenged orally with prions.

2.2. Oral Prion Susceptibility Correlates With Number But Not Structure of Peyer's Patches

Reduced mucosal lymphocyte trafficking does not impair, as expected, the susceptibility to orally initiated prion disease. In contrast, mice deficient in both tumor necrosis factor and lymphotoxin-α (TNF$\alpha^{-/-}$ x LT$\alpha^{-/-}$) or in lymphocytes (RAG-1$^{-/-}$, μMT), in which numbers of Peyer's patches are reduced in number, are highly resistant to oral challenge, and their intestines were virtually devoid of prion infectivity at all times after challenge. Therefore, lymphoreticular requirements for enteric and for intraperitoneal uptake of prions differ from each other and that susceptibility to prion infection following oral challenge correlates with the number of Peyer's patches, but is independent of the number of intestinal mucosa-associated lymphocytes (Prinz *et al.*, 2003a).

2.3. Diagnosis of Prion Disease: Identification of the Infectious Agent

As explained in the glossary, the term "prion" is used here in an operational meaning: it is not meant to be necessarily identical with PrP^{Sc}, but it simply denotes the infectious agent, whatever this agent exactly consists of (Aguzzi and Weissmann, 1997). If one takes this viewpoint, PrP^{Sc} defined originally as the protease-resistant moiety associated with prion disease would be a mere surrogate marker for the prion. Even if this position may seem all too negativistic in view of the wealth of compelling evidence in favor of the protein-only hypothesis, it is still advisable to maintain that we do not understand any of the molecular details of prion replication. In other words, we still need to be open to the possibility that infectivity might be brought about by a conformer of PrP^C that is not identical with what was originally defined as PrP^{Sc} (see glossary). This position appears to gain increasing acceptance, as documented by the recent description of "protease sensitive PrP^{Sc}" by the group of Stanley Prusiner (Safar *et al.*, 1998; Safar *et al.*, 2000).

Because protease resistance and infectivity of PrP^{Sc} may not necessarily be congruent, it is desirable to expand the range of reagents that will interact specifically with disease-associated prion protein, but not (or to a lesser degree) with its counterpart, PrP^C. One IgM monoclonal antibody termed 15B3 has been described a few years ago to fulfill the above conditions (Korth *et al.*, 1997), but this antibody has not become generally available, and no follow-up studies have been published. Disappointingly, expression of the complementarity-determining regions of 15B3 in the context of a phage, as well as expression of its heavy chain variable region in transgenic mice, failed to reproduce the specificity of PrP^{Sc} binding (Heppner *et al.*, 2001b).

We have described the puzzling phenomenon that a prevalent constituent of blood serum, plasminogen, captures efficiently PrP[Sc] but not PrP[C] when immobilized onto the surface of magnetic beads (Fischer *et al.*, 2000; Maissen *et al.*, 2001). The significance of this phenomenon *in vivo* has yet to be addressed, and it has been claimed on the basis of experiments that binding occurs only in the presence of specific detergents (Shaked *et al.*, 2002). While the latter study was performed only *in vitro*, and related only to the binding of a mixture of serum proteins rather than purified plasminogen, this does not detract from the usefulness of the binding phenomenon for diagnostic purposes: plasminogen binds to PrP[Sc] from a variety of species and genotypes (Maissen *et al.*, 2001), and the captured PrP[Sc] retains the glycotype profile characteristic of the prion strain from which it was isolated (M. Maissen and A. Aguzzi, unpublished data).

2.4. Transepithelial Enteric Passage of Prions: A Role for M Cells?

We have set out to investigate some of the preconditions of transepithelial passage of prions. Membranous epithelial cells (M cells) are key sites of antigen sampling for the mucosal-associated lymphoid system (MALT) and have been recognized as major ports of entry for enteric pathogens in the gut via transepithelial transport (Neutra *et al.*, 1996). Interestingly, maturation of M cells is dependent on signals transmitted by intraepithelial B cells. The group of Jean-Pierre Kraehenbuhl (Lausanne) has developed efficient *in vitro* systems, in which epithelial cells can be instructed to undergo differentiation to cells that resemble M cells by morphological and functional-physiological criteria. Therefore, we investigated whether M cells are a plausible site of prion entry in a co-culture model (Kerneis *et al.*, 1997) in which colon carcinoma cells are seeded with B-lymphoblastoid cells, inducing differentiation of M cells. After 24h, scrapie infectivity was determined within the basolateral compartment by bioassay with *tga20* mice, which overexpress a *Prnp* transgene and develop scrapie rapidly after infection (Fischer *et al.*, 1996). Even at low prion doses (3 logLD$_{50}$), we found infectivity in at least one M cell-containing co-culture. In contrast, there was hardly any prion transport in Caco-2 cultures without M cells (Heppner *et al.*, 2001a).

Table 1. Susceptibility of various strains of immunodeficient mice to intracerebrally or intraperitoneally administered prions (Klein et al., 1997). We measured the latency of scrapie (days from inoculation to terminal disease) upon delivery of a standard prion inoculum. All mice developed spongiform encephalopathy after i.c. inoculation. In contrast, all mice that carried a defect in B-cell differentiation stayed healthy after i.p. inoculation of RML scrapie prions. Intriguingly, in this and in several subsequent studies, TNF receptor 1 deficient mice developed disease with the same kinetics as wild-type mice, although there are no morphologically detectable FDCs in the spleens of TNFRI-/- mice.

Defect	Genotype	Intracerebral route		Intraperitoneal route	
		Scrapie	Time to terminal disease	Scrapie	Time to terminal disease
T	CD-4$^{o/o}$ $	7/7	159±11	8/8	191±1
T	CD-8$^{o/o}$ $	6/6	157±15	6/6	202±5
T	βσ$_2$-μ$^{o/o}$ $	8/8	162±11	7/7	211±6
T	Perforin$^{-/-}$ $	3/4*	171±2	4/4	204±3
T and B	SCID $	7/8*	160±11	6/8**	226±15
T and B	RAG 2$^{o/o}$ $	7/7	167±2	0/7	healthy (> 504)
T and B	RAG 1$^{o/o}$ $	3/3	175±2	0/5	healthy (>258)
T and B	AGR$^{o/o}$ @	6/6	184±10	0/7	healthy (> 425)
B	IgM$^{o/o}$ $	8/8	181±6	0/8	healthy (> 510)
IgG	t11umt $	5/5	170±3	4/4	223±2
FDC	TNFR1$^{o/o}$ £	7/7	165±3	9/9	216±4
Controls	129Sv	4/4	167±9	4/4	193±3
	C57BL/6	4/4	166±2	4/4	206±2

*: One perforin-deficient and one SCID mouse suffered from intercurrent death 135 and 141 days after inoculation, respectively; $: genetic background was C57BL/6; £: genetic background was 129sv; @: genetic background was C57BL/6 x 129sv; ** two SCID mice remained healthy and were sacrificed 303 and 323 days after inoculation

These findings indicate that M cell differentiation is necessary and sufficient for active transepithelial prion transport *in vitro*. M cell-dependent uptake of foreign antigens or particles is known to be followed by rapid transcytosis directly to the intraepithelial pocket, where key players of the immune system, e.g. macrophages, dendritic cells and lymphocytes (Neutra *et al.*, 1996), are located. Therefore, prions may exploit M cell-dependent transcytosis to gain access to the immune system.

While these findings suggest that M cells are a plausible candidate for the mucosal portal of prion infection, it still remains to be established whether the pathway delineated above does indeed represent the first portal of entry of orally administered prions into the body. This will necessitate *in vivo* experimentation, for example by ablation of M-cells through suicide transgenetic strategies, or by M-cell specific expression of *Prnp* transgenes.

3. Lymphocytes and Prion Pathogenesis

Innate or acquired deficiency of lymphocytes impairs peripheral prion pathogenesis, whereas no aspects of pathogenesis are affected by the presence or absence of lymphocytes upon direct transmission of prions to the CNS (Kitamoto *et al.*, 1991; Lasmezas *et al.*, 1996a). Klein and colleagues were then able to pinpoint the lymphocyte requirement to B-cells (Klein *et al.*, 1997) (Aguzzi, 1997a): at first blush this was very surprising, since there had been no suggestions that any aspect of humoral immunity would be involved in prion diseases. In the same study, it was shown that T-cells deficiency brought about by ablation of the T-cell receptor α (TCRα) chain did not affect prion pathogenesis (Table 1).

TCRα deficient mice, however, still contain TCR$\gamma\delta$ T-lymphocytes. Although the latter represent a subpopulation of T-cells, the experiments described did not allow to exclude a role for TCR$\gamma\delta$ T-lymphocytes. Therefore, we challenged also TCR TCRβ/δ deficient mice with prions. Incubation times after intracerebral and intraperitoneal inoculation of limiting or saturating doses of prions, however, elicited disease in these mice with the same kinetics as in wild-type mice (Michael Klein and AA, unpublished data). Also, accumulation of PrP^{Sc} and development of histopathological changes in the brain were indistinguishable in these two strains of mice. We therefore conclude that the complete absence of T-cells has no measurable impact on prion diseases. Therefore, it is unlikely that the T-cell infiltrates which have been reported to occur in the CNS during the course of prion infections (Betmouni *et al.*, 1996; Betmouni and Perry, 1999; Perry *et al.*, 2002) represent more than an epiphenomenon.

In 1996-7, when the results of the studies described above were being collected, we had no precise idea of what the mechanistic role of B-cells might be in prion pathogenesis. It had become rather clear, on the other hand, that lymphocytes alone could not account for the entirety of prion pathogenesis, and an additional sessile compartment had to be involved: adoptive transfer of $Prnp^{+/+}$ bone marrow to $Prnp^{o/o}$ recipient mice did not suffice to restore infectibility of $Prnp$-expressing brain grafts, indicating that neuroinvasion was still defective (Blättler *et al.*, 1997).

It then emerged that peripheral prion pathogenesis required the physical presence of B-cells, yet intraperitoneal infection occurred efficiently even in B-cell deficient hosts that had been transferred with B-cell from $Prnp$ knockout mice (Klein *et al.*, 1998). Therefore, presence of B-cells – but not expression of the cellular prion protein by these cells – is indispensable for pathogenesis upon intraperitoneal infection in the mouse scrapie model (Aguzzi *et al.*, 2001a).

The above results have been reproduced and confirmed several times over the years by many laboratories in various experimental paradigms: the requirement for B-cells in particular appears to be very stringent in most instances investigated. However, it has emerged that not all strains of prions induce identical patterns of peripheral pathogenesis, even when propagated in the same, isogenic strain of host organism.

3.1. Straining the Lymphocytes

Another interesting discrepancy that remains to be addressed concerns the actual nature of the cells that replicate and accumulate prions in lymphoid organs. Four series of rigorously controlled experiments over five years (Aguzzi *et al.*, 2000; Kaeser *et al.*, 2001; Klein *et al.*, 2001; Prinz *et al.*, 2002) unambiguously reproduced the original observation by Thomas Blättler and colleagues that transfer of wild-type bone marrow cells (or fetal liver cells) to $Prnp$ deficient mice restored accumulation and replication of prions in spleen (Blättler *et al.*, 1997). By contrast, Karen Brown and colleagues reported a diametrically opposite outcome of similar experiments when mice were inoculated with prions of the ME7 strain (Brown *et al.*, 1999). Maybe this discrepancy may identify yet another significant difference in the cellular tropism of different prion strains.

Bone marrow transplantation may (1) transfer an ill-defined population with the capability to replenish splenic stroma and to replicate prions, or – less probably – (2) donor-derived PrP^C expressing hematopoietic cells may confer prion replication capability to recipient stroma by virtue of "GPI painting", i.e. the posttranslational cell-to-cell transfer of glycophosphoinositol linked extracellular membrane proteins (Kooyman *et al.*, 1998). Some evidence might be accrued for either possibility: stromal splenic FDCs have been described by some authors to possibly derive from hematopoietic precursors, particularly when donors and recipients were young (Szakal *et al.*, 1995; Kapasi *et al.*, 1998). Conversely, instances have been described in which transfer of GPI-linked proteins occurs *in vivo* with suprisingly high efficiency (Kooyman *et al.*, 1995). Most recently, GPI painting has been described specifically for the cellular prion protein (Liu *et al.*, 2002).

The paradox described above may uncover an analogous phenomenon in peripheral prion pathogenesis. The latter question may be important, since the molecular and cellular basis of peripheral tropism of prion strains is likely to be directly linked to the potential danger of BSE in sheep (Glatzel and Aguzzi, 2001; Bruce *et al.*, 2002; Kao *et al.*, 2002), as well the potential presence of vCJD prions in human blood (Aguzzi, 2000).

4. Prion Hideouts in Lymphoid Organs

Prion infectivity rises very in a matter of days in the spleen of intraperitoneally infected mice. Although B-lymphocytes are crucial for neuroinvasion, the bulk of infectivity is not contained in lymphocytes. Instead, most splenic prion infectivity resides in a "stromal" fraction. Lymphocytes may be important for trafficking prions within lymphoid organs, but FDCs are the prime candidate prion reservoir. Recently, elegant immuno electron microscopic studies have evidenced that prion immunoreactivity is situated in the immediate neighborhood of iccosomes (Jeffrey *et al.*, 2000).

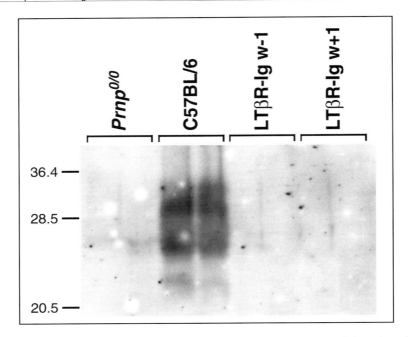

Figure 2. Depletion of FDCs by pharmacological inhibition of lymphotoxin signalling. Time course of FDC depletion in spleen of LTβR-Ig treated mice. Frozen sections of treated (left) and control mice (right) immunostained with FDC specific antibody FDC-M1 at different times points after injection of LTβR-Ig (original magnifications: upper row 8x25; lower row 8x63). Germinal centers FDCs networks were depleted already one week after treatment as described (Mackay and Browning, 1998; Montrasio *et al.*, 2000). Some FDC-M1 positive cells, which may represent residual FDCs or tingible body macrophages, were still detectable in the spleens of treated mice.

A definitive assessment of the contribution of FDCs to prion pathogenesis continues to be problematic since the histogenesis and the molecular characteristics of these cells are ill-defined. FDCs express S-100 proteins, as well as the complement receptors 2 (CD35) and 4 (identical to the marker FDC-M2). All of these markers, however, are also expressed by additional cell types, even within lymphoid organs (Bofill *et al.*, 2000). The FDC-M1 marker recognized by hybridoma clone 4C11 appears somewhat more specific.

Gene deletion experiments in mice have shown that signaling by both TNF and lymphotoxins is required for FDC development (Fu *et al.*, 1997; Koni *et al.*, 1997; Endres *et al.*, 1999). Membrane-bound lymphotoxin-α/β (LT-α/β) heterotrimers signal through the LT-β receptor (LT-βR) (Ware *et al.*, 1995) thereby activating a signaling pathway required for the development and maintenance of secondary lymphoid organs (Mackay *et al.*, 1997; Matsumoto *et al.*, 1997). Membrane LT-α/β heterotrimers are mainly expressed by activated lymphocytes (Browning *et al.*, 1993; 1995). Maintenance of pre-existing FDCs in a differentiated state requires continuous interaction with B lymphocytes expressing surface LT-α/β (Gonzalez *et al.*, 1998). Inhibition of the LT-α/β pathway in mice by treatment with LTβR-immunoglobulin fusion protein (LTβR-Ig) (Crowe *et al.*, 1994) leads to the disappearance of mature, functional FDCs (defined as cells that express markers such as FDC-M1, FDC-M2 or CD35) within one day, both in spleen and in lymph nodes (Mackay *et al.*, 1997; Mackay and Browning, 1998). Prolonged administration of the LTβR-Ig protein leads to disruption of B cell follicles.

All of the above prompted us to study the effect of selective ablation of functional FDCs on the pathogenesis of scrapie in mice. FDC-depletion was maintained by weekly administration of the LTβR-Ig fusion protein for eight weeks. Histological examination of spleen sections revealed that FDC networks had disappeared one week after treatment (Figure 2), as expected. In mice, following peripheral inoculation, infectivity in the spleen rises within days and reaches a plateau after a few weeks (Bruce, 1985; Rubenstein *et al.*, 1991; Büeler *et al.*, 1993). Even after intracerebral inoculation, spleens of C57BL/6 animals contain infectivity already 4 days post infection (p.i.), (Büeler *et al.*, 1993). However, Western blot analysis (Figure 3) revealed that eight weeks after inoculation spleens of control mice showed strong bands of protease-resistant PrP, whereas mice injected weekly with LTβR-Ig, starting either one week before or one week after inoculation, showed no detectable signal (<1/50th of the controls).

Prion infectivity in three spleens for each time point was assayed by intracerebral inoculation into indicator mice (Fischer *et al.*, 1996). In spleens of mice treated with LTβR-Ig 1 week before intraperitoneal inoculation no infectivity could be detected after 3 or 8 weeks (<1.0 log ID50 units/ml 10% homogenate). Traces of infectivity, possibly representing residual inoculum, were present in the one-week samples. In mice treated with LTβR-Ig 1 week after inoculation the titers were about 2.2 and 4.1 logID50 units/ml 10% homogenate at 3 weeks and at borderline detectability at 8 weeks after infection, suggesting that some prion accumulation took place in the first weeks

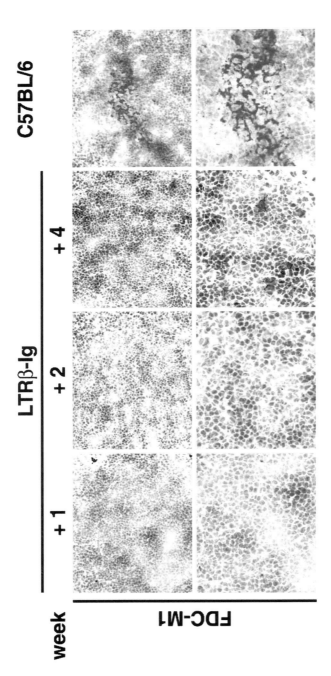

Figure 3. Inhibition of lymphotoxin signalling blocks prion replication in lymphoid organs. Accumulation of PrP^{Sc} in spleens of LTβR-Ig treated mice eight weeks after inoculation. Immunoblot analysis of spleen extracts (200μg total protein) from LTβR-Ig-treated and control mice sacrificed 8 weeks after inoculation. All spleen samples were treated with proteinase K. Mice analyzed (left to right): $Prnp^{0/0}$, untreated, LTβR-Ig treated C57BL/6 mice either 1 week before (LTβR-Ig w-1) or after (LTβR-Ig w+1) prion inoculation. Spleens of two mice for each group were analyzed. Immunoreactive PrP was detected using a rabbit antiserum against mouse-derived PrP (1B3) (Farquhar *et al.*, 1994) and enhanced chemiluminescence. The position of the molecular weight standards (in kDa) is indicated on the left side of the fluorogram. The three diagnostic bands of PrP^{Sc} are only recognized in untreated mice. LTβR-Ig treatment led to complete disappearance of the signal (reduction >50 fold).

after inoculation but was reversed under treatment with LTβR-Ig by 8 weeks. Spleens of scrapie-inoculated control mice treated with unspecific pooled human immunoglobulins (huIgG) 1 week after inoculation had 4.5 and 5.5 $logID_{50}$ units/ml 10% homogenate 8 weeks after intraperitoneal inoculation.

To assess whether prolonged depletion of mature FDCs would perturb the progression of the disease, C57BL/6 mice were subjected to weekly administration of the fusion protein up to eight weeks post inoculation and observed for more than 340 days. Mice receiving LTβR-Ig starting one week after inoculation developed the disease with about 25 days delay as compared to control mice. Whenever FDC depletion was initiated 1 week before inoculation, the effect on incubation time was even more pronounced. Immunofluorescence analysis of spleen sections of terminally sick animals revealed that the FDC networks were reconstituted after interruption of LTβR-Ig administration at 8 weeks and that PrP^C and/or PrP^{Sc} co-localized with FDCs. Immunoblot analysis showed that PrP^{Sc} accumulation was restored upon reappearance of FDCs in the spleens of terminal sick LTβR-Ig treated mice.

Therefore, FDCs are essential for the deposition of PrP^{Sc} and generation of infectivity in the spleen and suggest that they participate in the process of neuroinvasion. PrP knockout mice expressing PrP transgenes only in B cells do not sustain prion replication in the spleen or elsewhere, suggesting that prions associated with splenic B cells (Raeber *et al.*, 1999a) may be acquired from FDCs. Finally, these results suggest that strategies aimed at depleting FDCs might be envisaged for post-exposure prophylaxis (Aguzzi and Collinge, 1997) of prion infections initiated at extracerebral sites.

Additional experimentation appears to indicate that PrP^C-expressing hematopoietic cells are required in addition to FDCs for efficient lymphoreticular prion propagation (Blättler *et al.*, 1997; Kaeser *et al.*, 2001). This apparent discrepancy called for additional studies of the molecular requirements for prion replication competence in lymphoid stroma. We therefore set out to study peripheral prion pathogenesis in mice lacking TNFα, LTα/β, or their receptors. After intracerebral inoculation, all treated mice developed clinical symptoms of scrapie with incubation times, attack rates, and histopathological characteristics similar to those of wild-type mice, indicating that TNF/LT signaling is not relevant to cerebral prion pathogenesis. Upon intraperitoneal prion challenge, mice defective in LT signalling (LT$\alpha^{-/-}$, LT$\beta^{-/-}$, LTβR$^{-/-}$, or LTαTNF$\alpha^{-/-}$) proved virtually non-infectible with $\leq$5 logLD$_{50}$ scrapie infectivity, and establishment of subclinical disease (Frigg *et al.*, 1999) was prevented. In contrast, TNFR1$^{-/-}$ mice were almost fully susceptible to all inoculum sizes, and TNF$\alpha^{-/-}$ mice showed dose-dependent susceptibility. TNFR2$^{-/-}$ mice had intact FDCs and germinal centers, and were fully susceptible to scrapie. Unexpectedly, all examined lymph nodes of TNFR1$^{-/-}$ and TNF$\alpha^{-/-}$ mice had consistently high infectivity titers. Even inguinal lymph nodes, which are distant from the injection site and do not drain the peritoneum, contained infectivity titers equal to all other lymph nodes. Therefore, TNF deficiency prevents lymphoreticular prion accumulation in spleen but not in lymph nodes.

Why is susceptibility to peripheral prion challenge preserved in the absence of TNFR1 or TNFα, whereas deletion of LT signaling components confers high resistance to peripheral prion infection? After all, each of these defects (except TNFR2$^{-/-}$) abolishes FDCs. For one thing, prion pathogenesis in the lymphoreticular system appears to be compartmentalized, with lymph nodes (rather than spleen) being important reservoirs of prion infectivity during disease. Second, prion replication appears to take place in lymph nodes even in the absence of mature FDCs.

In *Prnp*$^{o/o}$ mice grafted with TNFR1$^{-/-}$ hematopoietic cells, high infectivity loads were detectable in lymph nodes but not spleens, indicating that TNFR1 deficient, PrP^C expressing hematopoietic cells may support prion propagation within lymph nodes. The PrP signal colocalized with a subset of macrophages in TNFR1$^{-/-}$ lymph nodes. Since marginal zone macrophages are in close contact to FDCs and also interact with marginal zone B cells, this cell type is certainly a candidate supporter in prion uptake and replication. On the other hand, it has been reported that in short-time infection experiments depletion of macrophages appears to enhance the amount of recoverable infectivity, implying that macrophages may degrade prions rather than transport them (Beringue *et al.*, 2000). The finding that a cell type other than mature FDCs is involved in prion replication and accumulation within lymph nodes may be relevant to the development of post-exposure prophylaxis strategies.

5. Neuroinvasion Proper: The Role of Sympathetic Nerves

In the last several years, a model has emerged that predicts prion neuroinvasion to consist of two distinct phases (Aguzzi *et al.*, 2001c). The details of the first phase are discussed above: widespread colonization of lymphoreticular organs is achieved by mechanisms that depend on B-lymphocytes (Klein *et al.*, 1997; 1998), FDCs (Montrasio *et al.*, 2000), and complement factors (Klein *et al.*, 2001). The second phase has long been suspected to involve peripheral nerves, and possibly the autonomic nervous system, and may depend on expression of PrP^C by these nerves (Glatzel and Aguzzi, 2000). We have been attempting to test the requirement for expression of PrP^C in peripheral nerves for several years, and have developed a gene transfer protocol to spinal ganglia aimed at resolving this question (Glatzel *et al.*, 2000); however, we were never able to recover infectivity in spinal cords of *Prnp*$^{o/o}$ mice whose spinal nerves had been transduced by Prnp-expressing adenoviruses (Glatzel and Aguzzi, unpublished results). Also, fast axonal transport does not appear to be involved in prion neuroinvasion, since mice that are severely impaired in this transport mechanism experience prion pathogenesis with kinetics similar to that of wild-type mice (Künzi *et al.*, 2002).

There is substantial evidence suggesting that prion transfer from the lymphoid system to the CNS occurs along peripheral nerves in a PrP^C dependent fashion (Blättler *et al.*, 1997; Glatzel and Aguzzi, 2000; Race *et al.*, 2000). Studies focusing on the temporal and spatial dynamics of neuroinvasion have suggested that the autonomic nervous system might be responsible for transport from lymphoid organs to the CNS (Clarke and Kimberlin, 1984; Cole and Kimberlin, 1985; Beekes *et al.*, 1998; McBride and Beekes, 1999).

The innervation pattern of lymphoid organs is mainly sympathetic (Felten and Felten, 1988). Markus Glatzel and colleagues have shown that denervation by injection of the drug 6-hydroxydopamine (6-OHDA), as well as "immunosympathectomy" by injection of antibodies against nerve growth factor (NFG) leads to a rather dramatic decrease in the density of sympathetic innervation of lymphoid organs, and significantly delayed the development of scrapie (Glatzel *et al.*, 2001). Sympathectomy appears to delay the transport of prions from lymphatic organs to the thoracic spinal cord, which is the entry site of

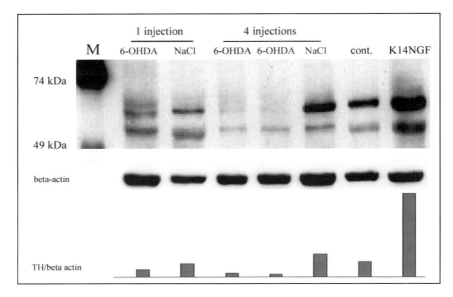

Figure 4. The sympathetic innervation of the spleen can be ablated by 6-hydroxydopamine. The density of sympathetic innervation correlates with the concentration of tyrosine hydroxylase, which was measured by Western blot analysis in spleens of adult 6-OHDA treated mice, K14NGF transgenic mice and controls. The two main tyrosine hydroxylase bands are visible at ca. 55 kDa. While, there is a certain increase in sympathetic innervation with age, it is obvious that 6-OHDA treatment depletes, and the K14NGF transgene increases sympathetic innervation. Lower panel: quantification of tyrosine hydroxylase content by chemiluminescence scanning. The bars represent the ratio between signal intensity of the tyrosine hydroxylase band and that of the corresponding beta-actin bands (arbitrary units).

sympathetic nerves to the CNS. Transgenic mice overexpressing NGF under control of the K14 promoter, whose spleens are hyperinnervated, developed scrapie significantly earlier than non-transgenic mice. No alteration in lymphocyte subpopulations was detected in spleens at any time point investigated. In particular, we did not detect any significant differences in the content of FDCs of treated and untreated mice, which negates the possibility that the observed protection may be due to modulation of FDC microanatomy.

While the sympathetic nervous system may represent a major component of the second phase of prion neuroinvasion, many details remain to be elucidated. It is not known whether prions can be transferred directly from FDCs to sympathetic endings, or whether additional cell types are involved. The latter possibility is particularly enticing, since FDCs have not been shown to entertain physical contact with sympathetic nervous system terminals. Moreover, it is unclear how prions are actually transported within peripheral nerves. Axonal and non-axonal transport mechanisms may be involved, and non-neuronal cells (such as Schwann cells) may play a role. Within the framework of the protein-only hypothesis, one may hypothesize a "domino" mechanism, by which incoming PrP^{Sc} converts resident PrP^C on the axolemmal surface, thereby propagating spatially the infection. While speculative, this model is attractive since it may accommodate the finding that the velocity of neural prion spread is extremely slow (Kimberlin *et al.*, 1983) and may not follow the canonical mechanisms of fast axonal transport. Indeed, recent studies may favor a non-axonal transport mechanism that results in periaxonal deposition of PrP^{Sc} (Hainfellner and Budka, 1999; Glatzel and Aguzzi, 2000). The fact that denervated mice eventually developed scrapie may be due to an alternative, low-efficiency route of entry that may become uncovered by the absence of sympathetic fibers (Figure 4). Entry through the vagal nerve has been proposed in studies of the dynamics of vacuolation following oral and intraperitoneal challenge with prions (Baldauf *et al.*, 1997; Beekes *et al.*, 1998). The surprising finding that infectious titers in hyperinnervated spleens are at least two logs higher and show enhanced PrP^{Sc} accumulations compared to control mice suggests that sympathetic nerves, besides being involved in the transport of prions, may also accumulate and replicate prions in lymphatic organs (Clarke and Kimberlin, 1984). Obviously this finding has implications related to the permanence and possibly eradication of prions in subclinically infected hosts.

5.1. Spread of Prions Within the CNS

Ocular administration of prions has proved particularly useful to study neural spread of the agent, since the retina is a part of the CNS and intraocular injection does not produce direct physical trauma to the brain, which may disrupt the blood-brain barrier and impair other aspects of brain physiology. The assumption that spread of prions occurs axonally rests mainly on the demonstration of diachronic spongiform changes along the retinal pathway following intraocular infection (Fraser, 1982). To investigate whether spread of prions within the CNS is dependent on PrP^C expression in the visual pathway, PrP-producing neural grafts were used as sensitive indicators of the presence of prion infectivity in the brain of an otherwise PrP-less host. Following inoculation with prions into the eye of grafted *Prnp$^{o/o}$*

mice, none of the grafts showed signs of scrapie. Therefore, it was concluded that infectivity administered to the eye of PrP-deficient hosts cannot induce scrapie in a PrP-expressing brain graft (Brandner *et al.*, 1996b).

Engraftment of *Prnp^{o/o}* mice with PrP^C-producing tissue might lead to an immune response to PrP which, in turn, was shown to be in principle capable of neutralizing infectivity (Heppner *et al.*, 2001b). In order to definitively rule out the possibility that prion transport was disabled by a neutralizing immune response, *Prnp^{o/o}* mice were rendered tolerant by expressing PrP^C under the control of the *lck* promoter. These mice overexpress PrP on T-lymphocytes, but are resistant to scrapie and do not replicate prions in brain, spleen and thymus after intraperitoneal inoculation with scrapie prions (Raeber *et al.*, 1999b). Engraftment of these mice with PrP-overexpressing neuroectoderm did not lead to the development of antibodies to PrP after intracerebral or intraocular inoculation, presumably due to clonal deletion of PrP-immunoreactive T-lymphocytes. As before, intraocular inoculation with prions did not provoke scrapie in the graft, supporting the conclusion that lack of PrP^C, rather than immune response to PrP, prevented prion spread (Brandner *et al.*, 1996b). Therefore, PrP^C appears to be necessary for the spread of prions along the retinal projections and within the intact CNS.

These results indicate that intracerebral spread of prions is based on a PrP^C -paved chain of cells, perhaps because they are capable of supporting prion replication. When such a chain is interrupted by interposed cells that lack PrP^C, as in the case described here, no propagation of prions to the target tissue can occur. Perhaps prions require PrP^C for propagation across synapses: PrP^C is present in the synaptic region (Fournier *et al.*, 1995) and certain synaptic properties are altered in *Prnp^{o/o}* mice (Collinge *et al.*, 1994; Whittington *et al.*, 1995). Perhaps transport of prions within, or on the surface of neuronal processes, is PrP^C-dependent. Within the framework of the protein-only hypothesis (Griffith, 1967; Prusiner, 1989), these findings may be accommodated by a "domino-stone" model in which spreading of scrapie prions in the CNS occurs per continuitatem through conversion of PrP^C by adjacent PrP^{Sc} (Aguzzi, 1997a).

6. Innate Immunity and Antiprion Defense

6.1. Macrophages and Toll-like Receptors

Cells of the monocyte/macrophage lineage typically represent the first line of defense against an extremely broad variety of pathogens. In the case of prions, it might be conceivable that macrophages protect against prions. However, it would be equally conceivable that macrophages, by virtue of their phagocytic properties and of their intrinsic mobility, may function as Trojan horses that transport prion infectivity between sites of replication within the body. This interesting question has not yet been fully resolved.

In a short-term prion infection paradigm, Beringue and colleagues administered dichloromethylene disphosphonate encapsulated into liposomes to mice: this eliminates for a short period of time all spleen macrophages. Accumulation of newly synthesized PrP^{Sc} was accelerated, suggesting that macrophages participate in the clearance of prions, rather than being involved in PrP^{Sc} synthesis. On the basis of the results presented above, Beringue and colleagues have suggested that activation or targeting of macrophages may represent a therapeutic pathway to explore in TSE infection. This suggestion was taken up by Sethi and Kretzschmar, who recently reported that activation of Toll like receptors (TLRs), which function as general stimulators of innate immunity by driving expression of various sets of the immune regulatory molecules, can effect postexposure prophylaxis in an experimental model of intraperitoneal scrapie infection (Sethi *et al.*, 2002). In this experimental paradigm, administration of prions intraperitoneally elicited disease after approximately 180 days, whereas the administration of CpG oligodeoxynucleotides 7 hours after prion inoculation and daily for 20 days led to disease-free intervals of "more than 330 days" – although it appears that all inoculated mice died of scrapie shortly thereafter (communicated by H. Kretzschmar at the TSE conference in Edinburgh, September 2002).

This finding is very surprising, since most available evidence indicates that general activation of the immune system would typically sensitize mice to prions, rather than protect them. The mechanism by which activation of toll-like receptor can result in post-exposure prophylaxis are wholly unclear at present, particularly in view of the fact that mice lacking Myd88 (Adachi *et al.*, 1998), which is an essential mediator of TLR signaling, develop prion disease with exactly the same sensitivity and kinetics as wild-type mice (Prinz *et al.* 2003b).

6.2. The Role of the Complement System

Another prominent component at the crossroad between innate and adaptive immunity is represented by the complement system. Opsonization by complement system components also appears to be relevant to prion pathogenesis: mice genetically engineered to lack complement factors (Klein *et al.*, 2001), or mice depleted of the C3 complement component by administration of cobra venom (Mabbott *et al.*, 2001), exhibit a remarkable resistance to peripheral prion inoculation. This phenomenon may, once again, be related to the pathophysiology of FDCs, which typically function as antigen traps. Trapping mechanisms essentially consist of capture of immune complexes by Fcγ receptors, and binding of opsonized antigens (linked covalently to C3d and C4b complement adducts) to the CD21/CD35 complement receptors.

Capture mediated by Fcγ receptors does not appear to be very important in prion disease: for one thing, knockout mice lacking Fcγ receptors (Takai *et al.*, 1994; Hazenbos *et al.*, 1996; Takai *et al.*, 1996; Park *et al.*, 1998) are just as susceptible to intraperitoneally administered scrapie as wild-type mice. Further, introduction into μMT mice of a generic immunoglobulin μ chain fully restored prion neuroinvasion irrespective of whether this heavy chain allowed for secretion of immunoglobulins, or only for production of membrane-bound immunoglobulins. We therefore conclude that circulating immunoglobulins are certainly not crucial to prion replication in lymphoid organs and to neuroinvasion.

A second mechanism exploited by FDCs for antigen trapping involves covalent linking of proteolytic fragments of the complement components C3 and C4 (Szakal and Hanna, 1968; Carroll, 1998). The CD21/CD35 complement receptors on FDCs bind C3b, iC3b, C3d, and C4b through short consensus repeats in their extracellular domain. Ablation of C3, or of its receptor CD21/CD35, as well as C1q (alone or combined with BF/C2$^{-/-}$), delayed neuroinvasion significantly after *intraperitoneal* inoculation when a limiting dose of prions was administered. These effects suggest that opsonization of the infectious agent may enhance its accessibility to germinal centers by facilitating docking to FDCs.

Very large prion inocula ($>10^6$ infectious units) appear to override the requirement for a functional complement receptor in prion pathogenesis. This is similar to systemic viral infections and coreceptor-dependent retention within the follicular compartment, whose necessity can be overridden by very high affinity antigens (Fischer *et al.*, 1998) or adjuvants (Wu *et al.*, 2000). Additional retention mechanisms for prions may therefore exist in FDCs, which are not complement-dependent, or depend on hitherto unidentified complement receptors.

7. Adaptive Immunity and Pre-Exposure Prophylaxis Against Prions

For many conventional viral agents, vaccination is the most effective method of infection control. But is it at all possible to induce protective immunity *in vivo* against prions? Prions are extremely sturdy and their resistance against sterilization is proverbial. Pre-incubation with anti-PrP antisera was reported to reduce the prion titer of infectious hamster brain homogenates by up to 2 log units (Gabizon *et al.*, 1988) and an anti-PrP antibody was found to inhibit formation of PrP^{Sc} in a cell-free system (Horiuchi and Caughey, 1999). Also, antibodies (Klein *et al.*, 2001) and F(ab) fragments raised against certain domains of PrP (Peretz *et al.*, 2001) can suppress prion replication in cultured cells. However, it is difficult to induce humoral immune responses against PrP^C and PrP^{Sc}. This is most likely due to tolerance of the mammalian immune system to PrP^C, which is an endogenous protein expressed rather ubiquitously. Ablation of the *Prnp* gene (Büeler *et al.*, 1992), which encodes PrP^C, renders mice highly susceptible to immunization with prions (Brandner *et al.*, 1996b), and many of the best available monoclonal antibodies to the prion protein have been generated in *Prnp*$^{o/o}$ mice (Prusiner *et al.*, 1993). However, *Prnp*$^{o/o}$ mice are unsuitable for testing vaccination regimens since they do not support prion pathogenesis (Büeler *et al.*, 1993).

We have therefore asked whether genes encoding high-affinity anti-PrP antibodies (originally generated in *Prnp*$^{o/o}$ mice) may be utilized to reprogram B cell responses of prion-susceptible mice that express

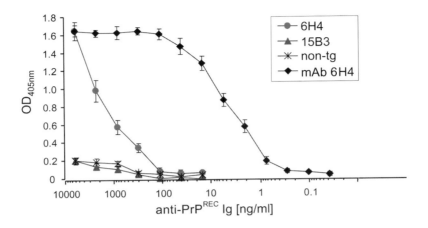

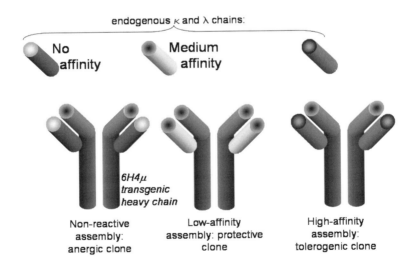

Figure 5. Affinity and avidity of antibodies against the prion protein. While the heavy chain is kept constant in the transgene, it may pair with a large repertoire of endogenous light chains. Some of the pairs will yield very high affinity antibodies, others will have low affinity, but the majority may have no affinity at all for the prion protein. It is possible that some degree of clonal deletion occurs, and that combinations with the highest affinity are eliminated.

PrP^C. Indeed, introduction of the epitope-interacting region the heavy chain of 6H4, a high-affinity anti-PrP monoclonal antibody (Korth *et al.*, 1997) into the germ line of mice sufficed to produce high-titer anti-PrP^C immunity. The 6H4μ heavy chain transgene induced similar anti-PrP^C titers in *Prnp*$^{o/o}$, *Prnp*$^{o/+}$ and *Prnp*$^{+/+}$ mice, indicating that deletion of autoreactive B-cells does not prevent anti-PrP immunity. The buildup of anti-PrP^C titers, however, was more sluggish in the presence of endogenous PrP^C, suggesting that some clonal deletion is actually occurring.

How can these observations be interpreted? The total anti-PrP^C titer results from pairing of one transgenic μ heavy chain with a large repertoire of endogenous κ and λ chains: some pairings may lead to reactive moieties, while others may be anergic (Figure 5). Maybe the B cell clones with the highest affinity to PrP^C are being eliminated by the immune tolerization machinery, and only clones with medium affinity are retained. This would explain the delay in titer buildup in the presence of PrP^C, and would be in agreement with our affinity measurements – which indicate that the total molar avidity of 6H4m serum is approximately 100-fold lower that that of the original 6H4 antibody from which the transgene was derived. But most intriguingly, expression of the 6H4μ heavy chain sufficed to block peripheral prion pathogenesis upon intraperitoneal inoculation of the prion agent (Heppner *et al.*, 2001a). PrP^C is a normal protein expressed by most tissues of the body. Therefore an anti-PrP immune response may conceivably induce an autoimmune disease, and defeat any realistic prospect for prion vaccination. We did not observe any blatant autoimmune disease as a consequence of anti-prion immunization – unless PrP^C was artificially transgenically expressed at non-physiological, extremely high levels.

The strategy outlined above delivers proof-of-principle that a protective humoral response against prions can be mounted by the mammalian immune system, and suggests that B cells are not intrinsically tolerant to PrP^C. If the latter is generally true, lack of immunity to prions may be due to T-helper tolerance. The latter problem is not trivial, but may perhaps be overcome by presenting PrP^C to the immune system along with highly active adjuvants. These findings, therefore, encourage a reassessment of the possible value of active and passive immunization (Westaway and Carlson, 2002), and perhaps of reprogramming B cell repertoires by μ chain transfer, in prophylaxis or in therapy of prion diseases.

7.1. Prion Immunization and its Reduction to Practice

Approximately one year later, a first example of a reduction to practice of the approach proposed by Heppner was demonstrated by the laboratory of Thomas Wisnieswki (Sigurdsson *et al.*, 2002). The authors have explored whether immunization with recombinant prion protein might be protective against prion diseases. Two paradigms where chosen: prophylactic immunization, and rescue after infection (post-exposure prophylaxis). The first remarkable surprise of the study was that it was indeed possible to induce antibody responses in wild-type mice – although the actual titers were not determined. Immunization was simply achieved by injecting 50 μg of recombinant prion protein emulsified in Freund's adjuvant. This procedure had been utilized extensively in the Zurich laboratory (F.L. Heppner, M. Polymenidou, E. Pellicioli, M. Bachmann, and A. Aguzzi, unpublished observations), but had never produced any reasonable titers in wild-type mice. Upon inoculation with a high dose or with a lower dose of prion inoculum, vaccinated mice exhibited a modest delay in development of prion disease. Although the success of the study, from the viewpoint of survival of the mice, might be regarded as limited, the study is in my view very promising in that it shows that vaccination approaches can be translated into models that are closer to the real-life situation than immunoglobulin-transgenic mice. The possibility of producing protective immunity against prions has captured the imagination of a considerable number of scientists, and additional reports are appearing of original ways to break tolerance and induce an immune response in animals that express the normal prion protein, including mixing of the immunogenic moiety with bacterial chaperones, among others (Koller *et al.*, 2002).

8. Perspectives in Prion Therapy

In the long run, one would hope that this accrued knowledge will be put into useful practice. To that effect, it is encouraging to note that a sizeable number of the steps in prion transport which have been discussed above appear to be rate-limiting. Because of that, these steps lend themselves as a target for interventions, which may be therapeutic or prophylactic (Aguzzi *et al.*, 2001b). A number of substances appear capable of influencing the outcome of a contact of mammalian organisms with prions: a non-exhaustive list includes compounds as diverse as Congo red (Caughey and Race, 1992), amphotericin B (Pocchiari *et al.*, 1987), anthracyclin derivatives (Tagliavini *et al.*, 1997), sulfated polyanions (Caughey and Raymond, 1993), pentosan polysulphate (Farquhar *et al.*, 1999), soluble lymphotoxin-β receptors (Montrasio *et al.*, 2000), porphyrins (Priola *et al.*, 2000), branched polyamines (Supattapone *et al.*, 2001), and beta-sheet breaker peptides (Soto *et al.*, 2000). However, it is sobering that none of the substances have yet made it to any validated clinical use: quinacrine appears to represent the most recent unfulfilled promise (Collins *et al.*, 2002). On the other hand, the tremendous interest in this field has attracted researchers from various neighboring disciplines, including immunology, genetics, and pharmacology, and therefore it is to hope that rational and efficient methods for managing prion infections will be developed in the future.

9. Acknowledgments

The work of my lab is supported by the Canton of Zurich, the Faculty of Medicine at the University of Zurich, the Swiss Federal Offices of Education and Science, of Health, and of Animal Health, the Swiss National Foundation, the National Center for Competence in Research on neural plasticity and repair, the Migros foundation, the Coop foundation, the U.K. Department for Environment, Food, and Rural Affairs, and the Stammbach foundation.

References

Adachi, O., Kawai, T., Takeda, K., Matsumoto, M., Tsutsui, H., Sakagami, M., Nakanishi, K., and Akira, S. 1998. Targeted disruption of the *MyD88* gene results in loss of IL-1- and IL- 18-mediated function. Immunity. 9: 143-50.

Aguzzi, A. 1996. Between cows and monkeys. Nature. 381: 734.

Aguzzi, A. 1997a. Neuro-immune connection in spread of prions in the body? Lancet. 349: 742-743.

Aguzzi, A. 1997b. Prions and antiprions. Biol. Chem. 378: 1393-1395.

Aguzzi, A. 2000. Prion diseases, blood and the immune system: concerns and reality. Haematologica. 85: 3-10.

Aguzzi, A., Brandner, S., Fischer, M.B., Furukawa, H., Glatzel, M., Hawkins, C., Heppner, F.L., Montrasio, F., Navarro, B., Parizek, P., Pekarik, V., Prinz, M., Raeber, A.J., Rockl, C., and Klein, M.A. 2001a. Spongiform encephalopathies: insights from transgenic models. Adv. Virus. Res. 56: 313-352.

Aguzzi, A., and Collinge, J. 1997. Post-exposure prophylaxis after accidental prion inoculation. Lancet. 350: 1519-1520.

Aguzzi, A., Glatzel, M., Montrasio, F., Prinz, M., and Heppner, F.L. 2001b. Interventional strategies against prion diseases. Nat. Rev. Neurosci. 2: 745-749.

Aguzzi, A., Klein, M.A., Montrasio, F., Pekarik, V., Brandner, S., Furukawa, H., Kaser, P., Rockl, C., and Glatzel, M. 2000. Prions: pathogenesis and reverse genetics. Ann. N. Y. Acad. Sci. 920: 140-157.

Aguzzi, A., Montrasio, F., and Kaeser, P.S. 2001c. Prions: health scare and biological challenge. Nat. Rev. Mol. Cell. Biol. 2: 118-126.

Aguzzi, A., and Raeber, A.J. 1998. Transgenic models of neurodegeneration. Neurodegeneration: of transgenic. mice and men. Brain Pathol. 8: 695-697.

Aguzzi, A., and Weissmann, C. 1996a. Sleepless in Bologna: transmission of fatal familial insomnia. Trends Microbiol. 4: 129-131.

Aguzzi, A., and Weissmann, C. 1996b. Spongiform encephalopathies: a suspicious signature. Nature. 383: 666-667.

Aguzzi, A., and Weissmann, C. 1997. Prion research: the next frontiers. Nature. 389: 795-798.

Anderson, R.M., Donnelly, C.A., Ferguson, N.M., Woolhouse, M.E., Watt, C.J., Udy, H.J., MaWhinney, S., Dunstan, S.P., Southwood, T.R., Wilesmith, J.W., Ryan, J.B., Hoinville, L.J., Hillerton, J.E., Austin, A.R., and Wells, G.A. 1996. Transmission dynamics and epidemiology of BSE in British cattle. Nature. 382: 779-788.

Baldauf, E., Beekes, M., and Diringer, H. 1997. Evidence for an alternative direct route of access for the scrapie agent to the brain bypassing the spinal cord. J. Gen. Virol. 78: 1187-1197.

Basler, K., Oesch, B., Scott, M., Westaway, D., Walchli, M., Groth, D.F., McKinley, M.P., Prusiner, S.B., and Weissmann, C. 1986. Scrapie and cellular PrP isoforms are encoded by the same chromosomal gene. Cell. 46: 417-428.

Beekes, M., McBride, P.A., and Baldauf, E. 1998. Cerebral targeting indicates vagal spread of infection in hamsters fed with scrapie. J. Gen. Virol. 79(3): 601-607.

Beringue, V., Demoy, M., Lasmezas, C.I., Gouritin, B., Weingarten, C., Deslys, J.P., Andreux, J.P., Couvreur, P., and Dormont, D. 2000. Role of spleen macrophages in the clearance of scrapie agent early in pathogenesis. J. Pathol. 190: 495-502.

Betmouni, S., and Perry, V.H. 1999. The acute inflammatory response in CNS following injection of prion brain homogenate or normal brain homogenate [In Process Citation]. Neuropathol. Appl. Neurobiol. 25: 20-28.

Betmouni, S., Perry, V.H., and Gordon, J.L. 1996. Evidence for an early inflammatory response in the central nervous system of mice with scrapie. Neuroscience. 74: 1-5.

Blättler, T., Brandner, S., Raeber, A.J., Klein, M.A., Voigtländer, T., Weissmann, C., and Aguzzi, A. 1997. PrP-expressing tissue required for transfer of scrapie infectivity from spleen to brain. Nature. 389: 69-73.

Bofill, M., Akbar, A.N., and Amlot, P.L. 2000. Follicular dendritic cells share a membrane-bound protein with fibroblasts. J. Pathol. 191: 217-226.

Bradley, R. 2000. Veterinary research at the Central Veterinary Laboratory, Weybridge, with special reference to scrapie and bovine spongiform encephalopathy. Rev. Sci. Tech. 19: 819-830.

Brandner, S., Isenmann, S., Kuhne, G., and Aguzzi, A. 1998. Identification of the end stage of scrapie using infected neural grafts. Brain Pathol. 8: 19-27.

Brandner, S., Isenmann, S., Raeber, A., Fischer, M., Sailer, A., Kobayashi, Y., Marino, S., Weissmann, C., and Aguzzi, A. 1996a. Normal host prion protein necessary for scrapie-induced neurotoxicity. Nature. 379: 339-343.

Brandner, S., Raeber, A., Sailer, A., Blattler, T., Fischer, M., Weissmann, C., and Aguzzi, A. 1996b. Normal host prion protein PrPC. is required for scrapie spread within the central nervous system. Proc. Natl. Acad. Sci. USA. 93: 13148-13151.

Brown, K.L., Stewart, K., Ritchie, D.L., Mabbott, N.A., Williams, A., Fraser, H., Morrison, W.I., and Bruce, M.E. 1999. Scrapie replication in lymphoid tissues depends on prion protein-expressing follicular dendritic cells. Nat. Med. 5: 1308-1312.

Brown P., Preece, M., Brandel, J.P., Sato, T., McShane, L., Zerr, I., Fletcher, A., Will, R.G., Pocchiari, M., Cashman, N.R., d'Aignaux, J.H., Cervenakova, L., Fradkin, J., Schonberger, L.B., and Collins, S.J. 2000. Iatrogenic Creutzfeldt-Jakob disease at the millennium. Neurology. 55: 1075-1081.

Browning, J.L., Dougas, I., Ngam-ek, A., Bourdon, P.R., Ehrenfels, B.N., Miatkowski, K., Zafari, M., Yampaglia, A.M., Lawton, P., Meier, W. *et al.* 1995. Characterization of surface lymphotoxin forms. Use of specific monoclonal antibodies and soluble receptors. J. Immunol. 154: 33-46.

Browning, J.L., Ngam-ek, A., Lawton, P., DeMarinis, J., Tizard, R., Chow, E.P., Hession, C., O'Brine-Greco, B., Foley, S.F., and Ware, C.F. 1993. Lymphotoxin beta, a novel member of the TNF family that forms a heteromeric complex with lymphotoxin on the cell surface. Cell. 72: 847-856.

Bruce, M.E. 1985. Agent replication dynamics in a long incubation period model of mouse scrapie. J. Gen. Virol. 66: 2517-2522.

Bruce, M.E., Boyle, A., Cousens, S., McConnell, I., Foster, J., Goldmann, W., and Fraser, H. 2002. Strain characterization of natural sheep scrapie and comparison with BSE. J. Gen. Virol. 83: 695-704.

Bruce, M.E., Will, R.G., Ironside, J.W., McConnell, I., Drummond, D., Suttie, A., McCardle, L., Chree, A., Hope, J., Birkett, C., Cousens, S., Fraser, H., and Bostock, C.J. 1997. Transmissions to mice indicate that 'new variant' CJD is caused by the BSE agent [see comments]. Nature. 389: 498-501.

Büeler, H.R., Aguzzi, A., Sailer, A., Greiner, R.A., Autenried, P., Aguet, M., and Weissmann, C. 1993. Mice devoid of PrP are resistant to scrapie. Cell. 73: 1339-4137.

Büeler, H.R., Fischer, M., Lang, Y., Bluethmann, H., Lipp, H.P., DeArmond, S.J., Prusiner, S.B., Aguet, M., and Weissmann, C. 1992. Normal development and behaviour of mice lacking the neuronal cell-surface PrP protein. Nature. 356: 577-582.

Carroll, M.C. 1998. CD21/CD35 in B cell activation. Semin. Immunol. 10: 279-286.

Caughey, B., and Race, R.E. 1992. Potent inhibition of scrapie-associated PrP accumulation by congo red. J. Neurochem. 59: 768-771.

Caughey, B., and Raymond, G.J. 1993. Sulfated polyanion inhibition of scrapie-associated PrP accumulation in cultured cells. J. Virol. 67: 643-650.

Chazot, G., Broussolle, E., Lapras, C., Blättler, T., Aguzzi, A., and Kopp, N. 1996. New variant of Creutzfeldt-Jakob disease in a 26-year-old French man [letter]. Lancet. 347: 1181.

Chesebro, B. 1998. BSE and prions: uncertainties about the agent. Science. 279: 42-43.

Clarke, M.C., and Kimberlin, R.H. 1984. Pathogenesis of mouse scrapie: distribution of agent in the pulp and stroma of infected spleens. Vet. Microbiol. 9: 215-225.

Cole, S., and Kimberlin, R.H. 1985. Pathogenesis of mouse scrapie: dynamics of vacuolation in brain and spinal cord after intraperitoneal infection. Neuropathol. Appl. Neurobiol. 11: 213-227.

Collinge, J., Sidle, K.C., Meads, J., Ironside, J., and Hill, A.F. 1996. Molecular analysis of prion strain variation and the aetiology of 'new variant' CJD. Nature. 383: 685-690.

Collinge, J., Whittington, M.A., Sidle, K.C., Smith, C.J., Palmer, M.S., Clarke, A.R., and Jefferys, J.G. 1994. Prion protein is necessary for normal synaptic function. Nature. 370: 295-297.

Collins, S.J., Lewis, V., Brazier, M., Hill, A.F., Fletcher, A., and Masters, C.L. 2002. Quinacrine does not prolong survival in a murine Creutzfeldt-Jakob disease model. Ann. Neurol. 52: 503-506.

Crowe, P.D., VanArsdale, T.L., Walter, B.N., Ware, C.F., Hession, C., Ehrenfels, B., Browning, J.L., Din, W.S., Goodwin, R.G., and Smith, C.A. 1994. A lymphotoxin-beta-specific receptor. Science. 264: 707-710.

Cuille, J., and Chelle, P.L. 1939. Experimental transmission of trembling to the goat. C R Seances Acad. Sci. 208: 1058-1160.

Duffy, P., Wolf, J., Collins, G., DeVoe, A.G., Streeten, B., and Cowen, D. 1974. Possible person-to-person transmission of Creutzfeldt-Jakob disease. N. Engl. J. Med. 290: 692-693.

Endres, R., Alimzhanov, M.B., Plitz, T., Futterer, A., Kosco-Vilbois, M.H., Nedospasov, S.A., Rajewsky, K., and Pfeffer, K. 1999. Mature follicular dendritic cell networks depend on expression of lymphotoxin beta receptor by radioresistant stromal cells and of lymphotoxin beta and tumor necrosis factor by B cells. J. Exp. Med. 189: 159-168.

Farquhar, C., Dickinson, A., and Bruce, M. 1999. Prophylactic potential of pentosan polysulphate in transmissible spongiform encephalopathies [letter]. Lancet. 353: 117.

Farquhar, C.F., Dornan, J., Somerville, R.A., Tunstall, A.M., and Hope, J. 1994. Effect of Sinc genotype, agent isolate and route of infection on the accumulation of protease-resistant PrP in non-central nervous system tissues during the development of murine scrapie. J. Gen. Virol. 75: 495-504.

Felten, D.L., and Felten, S.Y. 1988. Sympathetic noradrenergic innervation of immune organs. Brain Behav. Immun. 2: 293-300.

Fischer, M., Rülicke, T., Raeber, A., Sailer, A., Moser, M., Oesch, B., Brandner, S., Aguzzi, A., and Weissmann, C. 1996. Prion protein PrP. with amino-proximal deletions restoring susceptibility of PrP knockout mice to scrapie. EMBO J. 15: 1255-1264.

Fischer, M.B., Goerg, S., Shen, L., Prodeus, A.P., Goodnow, C.C., Kelsoe, G., and Carroll, M.C. 1998. Dependence of germinal center B cells on expression of CD21/CD35 for survival. Science. 280: 582-585.

Fischer, M.B., Roeckl, C., Parizek, P., Schwarz, H.P., and Aguzzi, A. 2000. Binding of disease-associated prion protein to plasminogen. Nature. 408: 479-483.

Flechsig, E., Shmerling, D., Hegyi, I., Raeber, A.J., Fischer, M., Cozzio, A., von Mering, C., Aguzzi, A., and Weissmann, C. 2000. Prion protein devoid of the octapeptide repeat region restores susceptibility to scrapie in PrP knockout mice. Neuron. 27: 399-408.

Fournier, J.G., Escaig Haye, F., Billette de Villemeur, T., and Robain, O. 1995. Ultrastructural localization of cellular prion protein PrPc. in synaptic boutons of normal hamster hippocampus. C. R. Acad. Sci. III. 318: 339-344.

Fraser, H. 1982. Neuronal spread of scrapie agent and targeting of lesions within the retino-tectal pathway. Nature. 295: 149-150.

Frigg, R., Klein, M.A., Hegyi, I., Zinkernagel, R.M., and Aguzzi, A. 1999. Scrapie pathogenesis in subclinically infected B-cell-deficient mice. J. Virol. 73: 9584-9588.

Fu, Y.X., Huang, G., Matsumoto, M., Molina, H., and Chaplin, D.D. 1997. Independent signals regulate development of primary and secondary follicle structure in spleen and mesenteric lymph node. Proc. Natl. Acad. Sci. USA. 94: 5739-5743.

Gabizon, R., McKinley, M.P., Groth, D., and Prusiner, S.B. 1988. Immunoaffinity purification and neutralization of scrapie prion infectivity. Proc. Natl. Acad. Sci. USA. 85: 6617-6621.

Gajdusek, D.C., Gibbs, C.J., and Alpers, M. 1966. Experimental transmission of a Kuru-like syndrome to chimpanzees. Nature. 209: 794-796.

Gibbs, C.J., Jr., Gajdusek, D.C., Asher, D.M., Alpers, M.P., Beck, E., Daniel, P.M., and Matthews, W.B. 1968. Creutzfeldt-Jakob disease spongiform encephalopathy.: transmission to the chimpanzee. Science. 161: 388-389.

Glatzel, M., and Aguzzi, A. 2000. PrP C. expression in the peripheral nervous system is a determinant of prion neuroinvasion. J. Gen. Virol. 81: 2813-2821.

Glatzel, M., and Aguzzi, A. 2001. The shifting biology of prions. Brain Res Brain Res Rev 36: 241-8.

Glatzel, M., Flechsig, E., Navarro, B., Klein, M.A., Paterna, J.C., Bueler, H., and Aguzzi, A. 2000. Adenoviral and adeno-associated viral transfer of genes to the peripheral nervous system. Proc. Natl. Acad. Sci. USA. 97: 442-447.

Glatzel, M., Heppner, F.L., Albers, K.M., and Aguzzi, A. 2001. Sympathetic innervation of lymphoreticular organs is rate limiting for prion neuroinvasion. Neuron. 31: 25-34.

Glatzel, M., Rogivue, C., Ghani, A., Streffer, J., Amsler, L., and Aguzzi, A. 2002. Incidence of Creutzfeldt-Jakob disease in Switzerland. Lancet. 360: 139-141.

Gonzalez, M., Mackay, F., Browning, J.L., Kosco-Vilbois, M.H., and Noelle, R.J. 1998. The sequential role of lymphotoxin and B cells in the development of splenic follicles. J. Exp. Med. 187: 997-1007.

Griffith, J.S. 1967. Self-replication and scrapie. Nature. 215: 1043-1044.

Hadlow, W.J. 1959. Scrapie and kuru. Lancet. 2: 289-290.

Hainfellner, J.A., and Budka, H. 1999. Disease associated prion protein may deposit in the peripheral nervous system in human transmissible spongiform encephalopathies. Acta. Neuropathol. Berl. 98: 458-60.

Hazenbos, W.L., Gessner, J.E., Hofhuis, F.M., Kuipers, H., Meyer, D., Heijnen, I.A., Schmidt, R.E., Sandor, M., Capel, P.J., Daeron, M., van de Winkel, J.G., and Verbeek, J.S. 1996. Impaired IgG-dependent anaphylaxis and Arthus reaction in Fc gamma RIII CD16. deficient mice. Immunity. 5: 181-8.

Heppner, F.L., Christ, A.D., Klein, M.A., Prinz, M., Fried, M., Kraehenbuhl, J.P., and Aguzzi, A. 2001a. Transepithelial prion transport by M cells. Nat. Med. 7: 976-977.

Heppner, F.L., Musahl, C., Arrighi, I., Klein, M.A., Rulicke, T., Oesch, B., Zinkernagel, R.M., Kalinke, U., and Aguzzi, A. 2001b. Prevention of scrapie pathogenesis by transgenic expression of anti-prion protein antibodies. Science. 294: 178-182.

Hill, A.F., Desbruslais, M., Joiner, S., Sidle, K.C., Gowland, I., Collinge, J., Doey, L.J., and Lantos, P. 1997a. The same prion strain causes vCJD and BSE [letter] [see comments]. Nature. 389: 448-450.

Hill, A.F., Zeidler, M., Ironside, J., and Collinge, J. 1997b. Diagnosis of new variant Creutzfeldt-Jakob disease by tonsil biopsy. Lancet. 349: 399.

Horiuchi, M., and Caughey, B. 1999. Specific binding of normal prion protein to the scrapie form via a localized domain initiates its conversion to the protease-resistant state [In Process Citation]. EMBO J. 18: 3193-3203.

Houston, F., Foster, J.D., Chong, A., Hunter, N., and Bostock, C.J. 2000. Transmission of BSE by blood transfusion in sheep. Lancet. 356: 999-1000.

http://www.doh.gov.uk/cjd/stats/aug02.htm. 2002. Monthly Creutzfeldt-Jakob disease statistics. In. Department of Health.

Hunter, N., Foster, J., Chong, A., McCutcheon, S., Parnham, D., Eaton, S., MacKenzie, C., and Houston, F. 2002. Transmission of prion diseases by blood transfusion. J. Gen. Virol. 83: 2897-2905.

Jeffrey, M., Goodsir, C.M., Bruce, M., McBride, P.A., Scott, J.R., and Halliday, W.G. 1994. Correlative light and electron microscopy studies of PrP localisation in 87V scrapie. Brain Res. 656: 329-343.

Jeffrey, M., McGovern, G., Goodsir, C.M., K, L.B., and Bruce, M.E. 2000. Sites of prion protein accumulation in scrapie-infected mouse spleen revealed by immuno-electron microscopy [In Process Citation]. J. Pathol. 191: 323-332.

Kaeser, P.S., Klein, M.A., Schwarz, P., and Aguzzi, A. 2001. Efficient lymphoreticular prion propagation requires PrP (c) in stromal and hematopoietic cells. J. Virol. 75: 7097-7106.

Kao, R.R., Gravenor, M.B., Baylis, M., Bostock, C.J., Chihota, C.M., Evans, J.C., Goldmann, W., Smith, A.J., and McLean, A.R. 2002. The potential size and duration of an epidemic of bovine spongiform encephalopathy in British sheep. Science. 295: 332-335.

Kapasi, Z.F., Qin, D., Kerr, W.G., Kosco-Vilbois, M.H., Shultz, L.D., Tew, J.G., and Szakal, A.K. 1998. Follicular dendritic cell FDC. precursors in primary lymphoid tissues. J. Immunol. 160: 1078-1084.

Kellings, K., Meyer, N., Mirenda, C., Prusiner, S.B., and Riesner, D. 1993. Analysis of nucleic acids in purified scrapie prion preparations. Arch. Virol. Suppl. 7: 215-225.

Kerneis, S., Bogdanova, A., Kraehenbuhl, J.P., and Pringault, E. 1997. Conversion by Peyer's patch lymphocytes of human enterocytes into M cells that transport bacteria. Science. 277: 949-952.

Kimberlin, R.H., Hall, S.M., and Walker, C.A. 1983. Pathogenesis of mouse scrapie. Evidence for direct neural spread of infection to the CNS after injection of sciatic nerve. J. Neurol. Sci. 61: 315-325.

Kimberlin, R.H., and Walker, C.A. 1978. Pathogenesis of mouse scrapie: effect of route of inoculation on infectivity titres and dose-response curves. J. Comp. Pathol. 88: 39-47.

Kimberlin, R.H., and Wilesmith, J.W. 1994. Bovine spongiform encephalopathy. Epidemiology, low dose exposure and risks. Ann. N. Y. Acad. Sci. 724: 210-220.

Kitamoto, T., Muramoto, T., Mohri, S., Dohura, K., and Tateishi, J. 1991. Abnormal isoform of prion protein accumulates in follicular dendritic cells in mice with Creutzfeldt-Jakob disease. J.Virol. 65: 6292-6295.

Klein, M.A., Frigg, R., Flechsig, E., Raeber, A.J., Kalinke, U., Bluethmann, H., Bootz, F., Suter, M., Zinkernagel, R.M., and Aguzzi, A. 1997. A crucial role for B cells in neuroinvasive scrapie. Nature. 390: 687-690.

Klein, M.A., Frigg, R., Raeber, A.J., Flechsig, E., Hegyi, I., Zinkernagel, R.M., Weissmann, C., and Aguzzi, A. 1998. PrP expression in B lymphocytes is not required for prion neuroinvasion. Nat. Med. 4: 1429-1433.

Klein, M.A., Kaeser, P.S., Schwarz, P., Weyd, H., Xenarios, I., Zinkernagel, R.M., Carroll, M.C., Verbeek, J.S., Botto, M., Walport, M.J., Molina, H., Kalinke, U., Acha-Orbea, H., and Aguzzi, A. 2001. Complement facilitates early prion pathogenesis. Nat. Med. 7: 488-492.

Koller, M.F., Grau, T., and Christen, P. 2002. Induction of antibodies against murine full-length prion protein in wild-type mice. J. Neuroimmunol. 132: 113-116.

Koni, P.A., Sacca, R., Lawton, P., Browning, J.L., Ruddle, N.H., and Flavell, R.A. 1997. Distinct roles in lymphoid organogenesis for lymphotoxins alpha and beta revealed in lymphotoxin beta-deficient mice. Immunity. 6: 491-500.

Kooyman, D.L., Byrne, G.W., and Logan, J.S. 1998. Glycosyl phosphatidylinositol anchor. Exp. Nephrol. 6: 148-151.

Kooyman, D.L., Byrne, G.W., McClellan, S., Nielsen, D., Tone, M., Waldmann, H., Coffman, T.M., McCurry, K.R., Platt, J.L., and Logan, J.S. 1995. *In vivo* transfer of GPI-linked complement restriction factors from erythrocytes to the endothelium. Science 269: 89-92.

Korth, C., Stierli, B., Streit, P., Moser, M., Schaller, O., Fischer, R., Schulz-Schaeffer, W., Kretzschmar, H., Raeber, A., Braun, U., Ehrensperger, F., Hornemann, S., Glockshuber, R., Riek, R., Billeter, M., Wuthrich, K., and Oesch, B. 1997. Prion PrPSc.-specific epitope defined by a monoclonal antibody. Nature. 390: 74-77.

Künzi, V., Glatzel, M., Nakano, M.Y., Greber, U.F., Van Leuven, F., and Aguzzi, A. 2002. Unhampered prion neuroinvasion despite impaired fast axonal transport in transgenic mice overexpressing four-repeat tau. J. Neurosci. 22: 7471-7477

Lasmezas, C.I., Cesbron, J.Y., Deslys, J.P., Demaimay, R., Adjou, K.T., Rioux, R., Lemaire, C., Locht, C., and Dormont, D. 1996a. Immune system-dependent and -independent replication of the scrapie agent. J. Virol. 70: 1292-1295.

Lasmezas, C.I., Deslys, J.P., Demaimay, R., Adjou, K.T., Lamoury, F., Dormont, D., Robain, O., Ironside, J., and Hauw, J.J. 1996b. Bse Transmission to Macaques. Nature. 381: 743-744.

Liu, T., Li, R., Pan, T., Liu, D., Petersen, R.B., Wong, B.S., Gambetti, P., and Sy, M.S. 2002. Intercellular transfer of the cellular prion protein. J. Biol. Chem. 277: 47671-47678.

Ma, J., and Lindquist, S. 2002. Conversion of PrP to a self-perpetuating PrPSc-like conformation in the cytosol. Science. 298: 1785-1788.

Ma, J., Wollmann, R., and Lindquist, S. 2002. Neurotoxicity and Neurodegeneration When PrP Accumulates in the Cytosol. Science. 298: 1781-1785.

Mabbott, N.A., Bruce, M.E., Botto, M., Walport, M.J., and Pepys, M.B. 2001. Temporary depletion of complement component C3 or genetic deficiency of C1q significantly delays onset of scrapie. Nat. Med. 7: 485-487.

Mackay, F., and Browning, J.L. 1998. Turning off follicular dendritic cells. Nature. 395: 26-27.

Mackay, F., Majeau, G.R., Lawton, P., Hochman, P.S., and Browning, J.L. 1997. Lymphotoxin but not tumor necrosis factor functions to maintain splenic architecture and humoral responsiveness in adult mice. Eur. J. Immunol.27: 2033-2042.

Maissen, M., Roeckl, C., Glatzel, M., Goldmann, W., and Aguzzi, A. 2001. Plasminogen binds to disease-associated prion protein of multiple species. Lancet. 357: 2026-2028.

Mallucci, G.R., Ratte, S., Asante, E.A., Linehan, J., Gowland, I., Jefferys, J.G., and Collinge, J. 2002. Post-natal knockout of prion protein alters hippocampal CA1 properties, but does not result in neurodegeneration. EMBO J. 21: 202-210.

Matsumoto, M., Fu, Y.X., Molina, H., Huang, G., Kim, J., Thomas, D.A., Nahm, M.H., and Chaplin, D.D. 1997. Distinct roles of lymphotoxin alpha and the type I tumor necrosis factor TNF. receptor in the establishment of follicular dendritic cells from non-bone marrow-derived cells. J. Exp. Med. 186: 1997-2004.

McBride, P.A., and Beekes, M. 1999. Pathological PrP is abundant in sympathetic and sensory ganglia of hamsters fed with scrapie. Neurosci. Lett. 265: 135-138.

Montrasio, F., Frigg, R., Glatzel, M., Klein, M.A., Mackay, F., Aguzzi, A., and Weissmann, C. 2000. Impaired prion replication in spleens of mice lacking functional follicular dendritic cells. Science. 288: 1257-1259.

Moser, M., Oesch, B., and Bueler, H. 1993. An anti-prion protein? Nature. 362: 213-214.

Neutra, M.R., Frey, A., and Kraehenbuhl, J.P. 1996. Epithelial M cells: gateways for mucosal infection and immunization. Cell. 86: 345-348.

Nicotera, P. 2001. A route for prion neuroinvasion. Neuron. 31: 345-348.

Oesch, B., Teplow, D.B., Stahl, N., Serban, D., Hood, L.E., and Prusiner, S.B. 1990. Identification of cellular proteins binding to the scrapie prion protein. Biochem. 29: 5848-5855.

Oesch, B., Westaway, D., Walchli, M., McKinley, M.P., Kent, S.B., Aebersold, R., Barry, R.A., Tempst, P., Teplow, D.B., Hood, L.E., and Weissmann, C. 1985. A cellular gene encodes scrapie PrP 27-30 protein. Cell. 40: 735-746.

Park, S.Y., Ueda, S., Ohno, H., Hamano, Y., Tanaka, M., Shiratori, T., Yamazaki, T., Arase, H., Arase, N., Karasawa, A., Sato, S., Ledermann, B., Kondo, Y., Okumura, K., Ra, C., and Saito, T. 1998. Resistance of Fc receptor- deficient mice to fatal glomerulonephritis. J. Clin. Invest. 102: 1229-1238.

Peretz, D., Williamson, R.A., Kaneko, K., Vergara, J., Leclerc, E., Schmitt-Ulms, G., Mehlhorn, I.R., Legname, G., Wormald, M.R., Rudd, P.M., Dwek, R.A., Burton, D.R., and Prusiner, S.B. 2001. Antibodies inhibit prion propagation and clear Cell. cultures of prion infectivity. Nature. 412: 739-743.

Perry, V.H., Cunningham, C., and Boche, D. 2002. Atypical inflammation in the central nervous system in prion disease. Curr. Opin. Neurol. 15: 349-354.

Pocchiari, M., Schmittinger, S., and Masullo, C. 1987. Amphotericin B delays the incubation period of scrapie in intracerebrally inoculated hamsters. J. Gen. Virol. 68: 219-223.

Prinz, M., Huber, G., Macpherson, A.J.S., Heppner, F.L., Glatzel, M., Eugster, H., Wagner, N., and Aguzzi, A. 2003a. Oral prion infection requires normal numbers of Peyer's patches but not of enteric lymphocytes. submitted. Am. J. Pathol. 162(4): 1103-1111

Prinz, M., Montrasio, F., Klein, M.A., Schwarz, P., Priller, J., Odermatt, B., Pfeffer, K., and Aguzzi, A. 2002. Lymph nodal prion replication and neuroinvasion in mice devoid of follicular dendritic cells. Proc. Natl. Acad. Sci. USA. 99: 919-924.

Prinz, M., Heikenwalder, M., Schwarz, P., Takeda, K., Akira, S., and Aguzzi, A. 2003b. Prion pathogenesis in the absence of Toll-like receptor signalling. EMBO Rep. 4:195-199.

Priola, S.A., Raines, A., and Caughey, W.S. 2000. Porphyrin and phthalocyanine antiscrapie compounds [see comments]. Science. 287: 1503-1506.

Prusiner, S.B. 1982. Novel proteinaceous infectious particles cause scrapie. Science. 216, 136-44.

Prusiner, S.B. 1989. Scrapie prions. Annu. Rev. Microbiol. 43: 345-374.

Prusiner, S.B., Cochran, S.P., Groth, D.F., Downey, D.E., Bowman, K.A., and Martinez, H.M. 1982. Measurement of the scrapie agent using an incubation time interval assay. Ann Neurol 11: 353-358.

Prusiner, S.B., Groth, D., Serban, A., Koehler, R., Foster, D., Torchia, M., Burton, D., Yang, S.L., and DeArmond, S.J. 1993. Ablation of the prion protein PrP. gene in mice prevents scrapie and facilitates production of anti-PrP antibodies. Proc. Natl. Acad. Sci. USA. 90: 10608-10612.

Prusiner, S.B., Scott, M.R., DeArmond, S.J., and Cohen, F.E. 1998. Prion protein biology. Cell. 93: 337-348.

Race, R., Oldstone, M., and Chesebro, B. 2000. Entry versus blockade of brain infection following oral or intraperitoneal scrapie administration: role of prion protein expression in peripheral nerves and spleen. J. Virol. 74: 828-833.

Raeber, A.J., Klein, M.A., Frigg, R., Flechsig, E., Aguzzi, A., and Weissmann, C. 1999a. PrP-dependent association of prions with splenic but not circulating lymphocytes of scrapie-infected mice. EMBO J. 18: 2702-2706.

Raeber, A.J., Sailer, A., Hegyi, I., Klein, M.A., T, R., Fischer, M., Brandner, S., Aguzzi, A., and Weissmann, C. 1999b. Ectopic expression of prion protein PrP in T lymphocytes or hepatocytes of PrP knockout mice is insufficient to sustain prion replication. Proc. Natl. Acad. Sci. USA. 96: 3987-3992.

Rieger, R., Edenhofer, F., Lasmezas, C.I., and Weiss, S. 1997. The human 37-kDa laminin receptor precursor interacts with the prion protein in eukaryotic cells [see comments]. Nat. Med. 3: 1383-1388.

Riesner, D., Kellings, K., Wiese, U., Wulfert, M., Mirenda, C., and Prusiner, S.B. 1993. Prions and nucleic acids: search for "residual" nucleic acids and screening for mutations in the PrP-gene. Dev. Biol. Stand. 80: 173-181.

Rother, K.I., Clay, O.K., Bourquin, J.P., Silke, J., and Schaffner, W. 1997. Long non-stop reading frames on the antisense strand of heat shock protein 70 genes and prion protein PrP genes are conserved between species [see comments]. Biol. Chem. 378: 1521-1530.

Rubenstein, R., Merz, P.A., Kascsak, R.J., Scalici, C.L., Papini, M.C., Carp, R.I., and Kimberlin, R.H. 1991. Scrapie-infected spleens: analysis of infectivity, scrapie-associated fibrils, and protease-resistant proteins. J. Infect. Dis. 164: 29-35.

Safar, J., Cohen, F.E., and Prusiner, S.B. 2000. Quantitative traits of prion strains are enciphered in the conformation of the prion protein. Arch. Virol. Suppl. 16: 227-235.

Safar, J., Wille, H., Itri, V., Groth, D., Serban, H., Torchia, M., Cohen, F.E., and Prusiner, S.B. 1998. Eight prion strains have PrP Sc. molecules with different conformations [see comments]. Nat. Med. 4: 1157-1165.

Sailer, A., Büeler, H., Fischer, M., Aguzzi, A., and Weissmann, C. 1994. No propagation of prions in mice devoid of PrP. Cell. 77: 967-968.

Scott, J.R., Foster, J.D., and Fraser, H. 1993. Conjunctival instillation of scrapie in mice can produce disease. Vet. Microbiol. 34: 305-309.

Sethi, S., Lipford, G., Wagner, H., and Kretzschmar, H. 2002. Postexposure prophylaxis against prion disease with a stimulator of innate immunity. Lancet. 360: 229-230.

Shaked, Y., Engelstein, R., and Gabizon, R. 2002. The binding of prion proteins to serum components is affected by detergent extraction conditions. J. Neurochem. 82: 1-5.

Shmerling, D., Hegyi, I., Fischer, M., Blattler, T., Brandner, S., Gotz, J., Rulicke, T., Flechsig, E., Cozzio, A., von Mering, C., Hangartner, C., Aguzzi, A., and Weissmann, C. 1998. Expression of amino-terminally truncated PrP in the mouse leading to ataxia and specific cerebellar lesions. Cell. 93: 203-214.

Sigurdsson, E.M., Brown, D.R., Daniels, M., Kascsak, R.J., Kascsak, R., Carp, R., Meeker, H.C., Frangione, B., and Wisniewski, T. 2002. Immunization delays the onset of prion disease in mice. Am. J. Pathol. 161: 13-17.

Soto, C., Kascsak, R.J., Saborio, G.P., Aucouturier, P., Wisniewski, T., Prelli, F., Kascsak, R., Mendez, E., Harris, D.A., Ironside, J.,

Tagliavini, F., Carp, R.I., and Frangione, B. 2000. Reversion of prion protein conformational changes by synthetic beta- sheet breaker peptides. Lancet. 355: 192-197.

Supattapone, S., Bosque, P., Muramoto, T., Wille, H., Aagaard, C., Peretz, D., Nguyen, H.O., Heinrich, C., Torchia, M., Safar, J., Cohen, F.E., DeArmond, S.J., Prusiner, S.B., and Scott, M. 1999. Prion protein of 106 residues creates an artifical transmission barrier for prion replication in transgenic mice. Cell. 96: 869-878.

Supattapone, S., Wille, H., Uyechi, L., Safar, J., Tremblay, P., Szoka, F.C., Cohen, F.E., Prusiner, S.B., and Scott, M.R. 2001. Branched polyamines cure prion-infected neuroblastoma cells. J. Virol. 75: 3453-3461.

Szakal, A.K., and Hanna, M.G., Jr. 1968. The ultrastructure of antigen localization and viruslike particles in mouse spleen germinal centers. Exp. Mol. Pathol. 8: 75-89.

Szakal, A.K., Kapasi, Z.F., Haley, S.T., and Tew, J.G. 1995. Multiple lines of evidence favoring a bone marrow derivation of follicular dendritic cells FDCs. Adv. Exp. Med. Biol. 378: 267-272.

Tagliavini, F., McArthur, R.A., Canciani, B., Giaccone, G., Porro, M., Bugiani, M., Lievens, P.M., Bugiani, O., Peri, E., Dall'Ara, P., Rocchi, M., Poli, G., Forloni, G., Bandiera, T., Varasi, M., Suarato, A., Cassutti, P., Cervini, M.A., Lansen, J., Salmona, M., and Post, C. 1997. Effectiveness of anthracycline against experimental prion disease in Syrian hamsters. Science. 276: 1119-1122.

Takai, T., Li, M., Sylvestre, D., Clynes, R., and Ravetch, J.V. 1994. FcR gamma chain deletion results in pleiotrophic effector Cell. defects. Cell. 76: 519-529.

Takai, T., Ono, M., Hikida, M., Ohmori, H., and Ravetch, J.V. 1996. Augmented humoral and anaphylactic responses in Fc gamma RII-deficient mice. Nature. 379: 346-349.

Taylor, D.M. 2000. Inactivation of transmissible degenerative encephalopathy agents: A review [see comments]. Vet. J. 159: 10-17.

Vankeulen, L.J.M., Schreuder, B.E.C., Meloen, R.H., Mooijharkes, G., Vromans, M.E.W., and Langeveld, J.P.M. 1996. Immunohistochemical Detection of Prion Protein in Lymphoid Tissues of Sheep With Natural Scrapie. J. Clin. Microbiol. 34: 1228-1231.

Ware, C.F., VanArsdale, T.L., Crowe, P.D., and Browning, J.L. 1995. The ligands and receptors of the lymphotoxin system. Curr. Top. Microbiol. Immunol. 198: 175-218.

Weissmann, C. 1991. Spongiform encephalopathies. The prion's progress. Nature. 349: 569-571.

Weissmann, C., and Aguzzi, A. 1997. Bovine spongiform encephalopathy and early onset variant Creutzfeldt- Jakob disease. Curr. Opin. Neurobiol. 7: 695-700.

Wells, G.A., Dawson, M., Hawkins, S.A., Green, R.B., Dexter, I., Francis, M.E., Simmons, M.M., Austin, A.R., and Horigan, M.W. 1994. Infectivity in the ileum of cattle challenged orally with bovine spongiform encephalopathy. Vet. Rec. 135: 40-41.

Wells, G.A., Hawkins, S.A., Green, R.B., Austin, A.R., Dexter, I., Spencer, Y.I., Chaplin, M.J., Stack, M.J., and Dawson, M. 1998. Preliminary observations on the pathogenesis of experimental bovine spongiform encephalopathy BSE.: an update. Vet. Rec. 142: 103-106.

Wells, G.A., Scott, A.C., Johnson, C.T., Gunning, R.F., Hancock, R.D., Jeffrey, M., Dawson, M., and Bradley, R. 1987. A novel progressive spongiform encephalopathy in cattle. Vet. Rec. 121: 419-420.

Westaway, D., and Carlson, G.A. 2002. Mammalian prion proteins: enigma, variation and vaccination. Trends Biochem. Sci. 27: 301-307.

Whittington, M.A., Sidle, K.C., Gowland, I., Meads, J., Hill, A.F., Palmer, M.S., Jefferys, J.G., and Collinge, J. 1995. Rescue of neurophysiological phenotype seen in PrP null mice by transgene encoding human prion protein. Nat. Genet. 9: 197-201.

Will R.G., Ironside J.W., Zeidler M., Cousens S.N., Estibeiro K., Alperovitch A., Poser S., Pocchiari M., Hofman A., and Smith. 1996. A new variant of Creutzfeldt-Jakob disease in the UK. Lancet. 347: 921-925.

Wu, X., Jiang, N., Fang, Y.F., Xu, C., Mao, D., Singh, J., Fu, Y.X., and Molina, H. 2000. Impaired affinity maturation in Cr2-/- mice is rescued by adjuvants without improvement in germinal center development. J. Immunol. 165: 3119-27.

Yehiely, F., Bamborough, P., Costa, M.D., Perry, B.J., Thinakaran, G., Cohen, F.E., Carlson, G.A., and Prusiner, S.B. 2002. Identification of candidate proteins binding to prion protein. Neurobiol Dis. 1997; 3(4): 339-355.

Zanata, S.M., Lopes, M.H., Mercadante, A.F., Hajj, G.N., Chiarini, L.B., Nomizo, R., Freitas, A.R., Cabral, A.L., Lee, K.S., Juliano, M.A., De Oliveira, E., Jachieri, S.G., Burlingame, A., Huang, L., Linden, R., Brentani, R.R., and Martins, V.R. 2002. Stress-inducible protein 1 is a cell surface ligand for cellular prion that triggers neuroprotection. EMBO J. 21: 3307-3316.

From: Prions and Prion Diseases: Current Perspectives
Edited by: Glenn C. Telling

Chapter 8

Immunological Advances in Prion Diseases

R. Anthony Williamson

Abstract

The major component of prions, the infectious agents causing transmissible spongiform encephalopathies (TSEs), is an abnormally folded conformer (PrP^{Sc}) of the cellular prion protein, PrP^C. During prion propagation, PrP^{Sc} appears to act as a molecular template, sequestering endogenous PrP^C and somehow triggering its conformational conversion, thereby yielding additional molecules of PrP^{Sc}. Presently, there are no effective approaches to either prevent or treat prion infections. However, antibodies specifically binding to certain regions of PrP have emerged as a class of potent inhibitors of prion replication, seemingly capable of resolving infection both *in vitro* and *in vivo*. Mechanistically, these molecules likely operate either by hindering PrP^C-PrP^{Sc} interactions, or by stabilizing PrP^C in its native conformation, thereby preventing its structural reconfiguration. These findings indicate that the immune system, which remains largely impassive throughout the course of a natural prion infection, may if stimulated correctly, yield meaningful protection from exposure to prions. Significantly, however, antibodies are generally excluded from central nervous system (CNS), and are therefore likely to be limited in their ability to readily reverse established prion infections that have spread to, or originated within, these tissues. Looking to the future, antibodies that efficiently inhibit prion propagation are now being harnessed as tools with which to rationally identify small molecules possessing equivalent anti-prion properties, but that may more readily gain access to the CNS.

1. Transmissible Spongiform Encephalopathies (TSEs)

TSEs or prion diseases are a group of neurodegenerative disorders afflicting animals (bovine spongiform encephalopathy (BSE), scrapie, chronic wasting disease (CWD)) and humans (Creutzfeldt Jakob disease (CJD), Gerstmann-Straussler-Scheinker disease, fatal familial insomnia) that may manifest via infectious, inherited and apparently sporadic mechanisms. In each case, the central molecular event is the aberrant metabolism of the cellular prion protein, PrP^C, a glycosyl-phosphatidylinositol (GPI) anchored protein of 209 amino acids, expressed on many different tissue types, but particularly within the CNS.

Neuropathologically, neuronal loss, spongiosis, gliosis, and the accumulation of PrP^{Sc}, a misassembled conformationally altered β-sheet rich form of PrP^C typically characterize TSEs. This abnormal disease-associated PrP isoform, also termed PrP^{res} because it is frequently partially resistant to proteolytic degradation, is widely accepted to be the major component of infectious prion particles (Prusiner, 1998b; Caughey, 2001; Collinge, 2001; Dormont, 2002; Weissmann *et al.*, 2002; Aguzzi and Weissmann, 1998). The presence of PrP^{Sc} in the host continues to be the only consistent correlate for prion infectivity and disease.

Investigations of prion diseases have taken on added importance with the emergence of variant Creutzfeldt-Jakob disease (vCJD), which has epidemiological, clinical and neuropathological characteristics that distinguish it from other prion diseases in humans. vCJD bears closest resemblance to BSE, and it is highly probable that the appearance of vCJD results directly from exposure to prion contaminated foodstuffs following the BSE epidemic which has afflicted over 180, 000 British cattle (Collinge *et al.*, 1996; Anderson *et al.*, 1996; Bruce, 2000; Bruce *et al.*, 1997; Hill *et al.*, 1997; Knight and Stewart, 1998). The number of individuals exposed to BSE prions is likely to have been very large, and, as a result of the prolonged asymptomatic incubation times associated with TSEs, considerable concern continues to manifest over the eventual magnitude of the vCJD cohort. Although statistical analyses now indicate that the probable eventual number of vCJD cases within the UK will be limited to several hundreds, there remains a significant degree of uncertainty in these predictions (Boelle *et al.*, 2003; Ghani *et al.*, 2003; Hilton *et al.*, 2002; Andrews *et al.*, 2003). For example, of the 132 confirmed cases of vCJD, genotyping has been performed in 115, with every case exhibiting methionine homozygosity at codon 129 of the PrP gene. Thus, the possibility remains that genotypes other than methionine homozygotes, if infected with BSE prions, may possess longer incubation periods and/or present with clinical manifestations other than vCJD. This uncertainty is compounded by evidence in experimental animal models arguing that prions from one species can persist for extended periods in a heterologous host at subclinical levels, and even adapt to their new environment over several serial passages, as evidenced by sequential reduction in incubation times to symptomatic disease (Race and Chesebro, 1998; Race *et al.*, 2001; 2002). These findings have potentially grave implications for the future, in that uninfected individuals may have unknowingly received prion-infected blood from donors silently propagating prions. Indeed, exhaustive measures have been taken by the blood product industry to exclude individuals most likely to have been exposed to BSE prions from the donor pool. Nonetheless, the need to develop effective prion diagnostics and therapeutics remains urgent (Ironside and Head, 2003).

Within the United States, no single case of BSE has yet been reported. However there is increasing concern over the spread of chronic wasting disease (CWD), a prion disease afflicting both free-ranging and domestic deer and elk. In endemic regions of the states of Colorado and Wyoming, as many as 15% of mule deer and elk are infected (Miller *et al.*, 2000; Sigurdson *et al.*, 2002). The origin of CWD and its efficient transmission between herd members is not understood. Importantly, the potential for CWD prion transmission to humans and other animals remains unknown, although limited *in vitro* studies suggest that non-cervid species may have limited susceptibility to CWD prions (Raymond *et al.*, 2000). More alarmingly, CWD prions have transmitted to cattle experimentally, with 5 of 13 bovines receiving intracerebral (i.c.) inoculations of CWD-infected tissues, having thus far developed disease.

1.1. Prion Propagation and Neurotoxicity

Prion propagation, that is the events leading to PrP^{Sc} formation and the acquisition of prion infectivity, is poorly understood. Evidence from structural studies (Pan *et al.*, 1993; Cohen *et al.*, 1994; Zhang *et al.*, 1997; Peretz *et al.*, 1997) argues that PrP^{Sc} formation involves an extensive conformational rearrangement of PrP^C in which a large amount of β-sheet structure is acquired in a misassembly process. Three-dimensional NMR structures for α-helical conformations of recombinant Syrian hamster (SHa), mouse, bovine and human recombinant PrP molecules thought to resemble PrP^C have been solved (Riek *et al.*, 1996; Donne *et al.*, 1997; Wüthrich and Riek, 2001). The principal conclusion is that the C-terminal part of the molecule (residues 125-231) forms a three helix-bundle (helices A, B and C containing residues 144-156, 172-193 and 220-227, respectively), with a small amount of β-structure, whereas the N-terminal portion (residues 23-124) lacks structure in the absence of Cu binding, but possibly acquires limited organization in the metallated form (Garnett and Viles, 2003; Burns *et al.*, 2003). In contrast, other than its comparatively high β-sheet content, very little is known about the conformation of PrP^{Sc}, which is generally found in an associated form that does not lend itself to structural study because it is neither crystalline nor readily soluble.

In the favored model of prion replication, PrP^{Sc} acts as a conformational template that promotes, perhaps with the aid of an additional cellular factor(s), the conversion of PrP^C to PrP^{Sc} (Prusiner, 1998a; Caughey, 2001). Direct interaction between the normal and aberrant forms of PrP is therefore believed to be a crucial step in the pathway toward formation of additional PrP^{Sc} from PrP^C (Caughey, 2001). In the absence of PrP^{Sc}, the PrP^C to PrP^{Sc} transition would be expected to be an extremely rare event, unlikely to occur within the lifetime of the host unless mutations in the PrP gene unfavorably destabilized PrP^C conformation, e.g. in inherited forms of prion disease. PrP^C is largely located within cell-surface cholesterol and sphingolipid-rich membrane domains but has been shown to constitutively cycle between the cell surface and the endocytic compartment (Harris, 2001). Initial interaction between PrP^C and PrP^{Sc} and the subsequent generation of nascent PrP^{Sc} is therefore likely to occur either at the cell surface or within this pathway.

Although the appearance of PrP^{Sc} is intimately associated with the generation of nascent prion infectivity and with progression of prion neuropathology and disease, it is important to note that PrP^{Sc} itself does not appear to be intrinsically toxic. To demonstrate this point, Bradner *et al.* transplanted prion-infected grafts of wild-type brain into the brains of Prnp$^{o/o}$ mice (Brandner *et al.*, 1996). PrP^{Sc} was produced efficiently in the grafted tissue, which displayed typical histological hallmarks of prion-associated pathology. In contrast, although PrP^{Sc} was deposited into the tissues surrounding the graft, these cells remained healthy. Thus any neurotoxic properties associated with PrP^{Sc} only became manifest in the presence of PrP^C.

Overall, the dearth of information about prion neuropathology and PrP^{Sc}-PrP^C interactions at the molecular level has hindered the development of disease models that comfortably accommodate some key aspects of prion biology - such as the observed species barriers to prion transmission, or the prion strains phenomenon - and has undermined the drive toward effective prion therapeutics.

2. Interventional Strategies for TSEs

There is a pressing demand for clinical strategies to successfully treat or prevent prion disease. Unfortunately, by the time clinical symptoms are evident, significant and probably irreversible injury to the brain will likely already have been sustained (Aguzzi *et al.*, 2001). Under these circumstances, therapeutic interventions may only slow or halt further progression of the disease. However, should efficient *ante mortem* diagnosis of prion diseases become available in the future, individuals infected with prions can expect to be identified and treated much earlier in the disease course, having sustained lower levels of CNS damage. Conceptually, the most obvious strategy to combat prion disease is to prevent accumulation of PrP^{Sc}, the appearance of which is so closely associated with generation of prion infectivity and profound damage of CNS tissues. In principle, this could be achieved by increasing clearance of the disease-associated PrP conformers, or by hindering their formation. A host of chemically diverse molecules have been identified that modulate the accumulation of PrP^{Sc} and prion infectivity *in vitro*. These include pentosan polysulphate (Farquhar *et al.*, 1999), Congo red (Caughey and Race, 1992), amphoterecin B (Pocchiari *et al.*, 1987), anthracycline derivatives (Tagliavini *et al.*, 1997), sulphated polyanions (Caughey and Raymond, 1993), porphyrins (Priola *et al.*, 2000), soluble lymphotoxin-β receptors (Montrasio *et al.*, 2000), branched polyamines (Supattapone *et al.*, 2001), bis-acridines (May *et al.*, 2003), curcumin (Caughey *et al.*, 2003), CpG oligodeoxynucleotides (Sethi *et al.*, 2002) and β-sheet breaker peptides (Soto *et al.*, 2000). Dispiritingly, none of these compounds has yielded convincingly beneficial effects in prion infected animals. More recently, quinacrine was shown to inhibit prion replication in infected cell cultures (Korth *et al.*, 2001). Since quinacrine is a well-characterized drug administered in the treatment of malaria, clinical trials were initiated in the UK and the US in which this drug was given to patients suffering from sporadic CJD or vCJD. After these trials had begun, however, quinacrine was determined not to prolong survival in a murine CJD model (Collins *et al.*, 2002; Barret *et al.*, 2003).

Despite the lack of progress to date, optimism for the future is not unwarranted. As our understanding of many different facets of both prion biology and other diseases of protein conformation progressively improves, exciting new opportunities are presenting themselves that offer considerable promise for meaningful future therapeutic intervention in TSEs. Most notably, PrP-specific monoclonal antibodies (MAbs) have proven highly potent inhibitors of prion replication.

2.1. Antibodies Recognizing PrP

A puzzling facet of prion infection is the failure to induce a response in either the humoral or cellular arms of the host immune system. Mechanisms of immune tolerance readily explain the absence of naturally occurring autologous antibodies recognizing PrP^C, however, it is unclear why PrP^{Sc} which is clearly distinguishable from PrP^C on the basis of its conformation and physico-chemical properties, remains undetected by the immune system of the infected host.

In the laboratory, immune tolerance to PrP has historically hindered the production of specifically reactive polyclonal and monoclonal antibodies. The first PrP-specific antisera were produced in rabbits immunized with various synthetic PrP peptides or large quantities of PrP 27-30, the proteinase K resistant core of PrP^{Sc}, purified from the brains of scrapie prion-infected rodents (Bendheim *et al.*, 1984; Barry and Prusiner, 1986; Bode *et al.*, 1985; Takahashi *et al.*, 1986; Barry *et al.*, 1986; Wiley *et al.*, 1987). Certain of these antibodies were shown to react specifically with PrP in immunoblotting assays, providing direct evidence that PrP27-30 was a structural component of purified prion rods (Barry and Prusiner, 1986), and also to strain amyloid plaques in the brains of prion-infected hamster brains - provided the antigen was first denatured in guanidinium thiocyanate (GdnSCN) (Barry *et al.*, 1986; Wiley *et al.*, 1987).

Prior to the introduction of PrP-knockout mice (*Prnp$^{o/o}$*), production MAbs against PrP was hindered by immune tolerance. However, two MAbs, designated 3F4 and 13A5, were raised by immunizing normal (Prnp$^{+/+}$) mice with Syrian hamster PrP antigen. Each of these antibodies recognized hamster, but not mouse PrP^C and bound to epitopes defined by differences in primary sequence between the immunogen and host PrPs (Barry and Prusiner, 1986; Kascsak *et al.*, 1987; Rubenstein *et al.*, 1987; Bolton *et al.*, 1991). Robust serum antibody titers against PrP^C from mouse and other species were much more readily elicited in *Prnp$^{o/o}$* mice (Büeler *et al.*, 1992; Prusiner *et al.*, 1993). The first PrP-specific MAbs were recovered from mice immunized with purified mouse prion rods, composed of PrP 27-30. Surprisingly, analysis of serum IgG in the immunized animals revealed specific reactivity with prion rods only following GdnSCN denaturation. Antibody Fab libraries displayed on the surface of M13 phage were prepared from spleen, lymph node and bone marrow tissues of the immunized mice, and selected against GdnSCN-treated and untreated prion rods coated onto ELISA wells (Williamson *et al.*, 1996). In good agreement with the serum reactivity of the donor mice, PrP-specific recombinant antibody Fabs were recovered when library selection was performed against denatured PrP 27-30 antigen or recombinant PrP (90-231) refolded into an α-helical conformation, but not when selection was performed against non-denatured PrP 27-30. All of the recombinant MAbs rescued from mice immunized with PrP 27-30 rods recognized denatured PrP 27-30 and PrP^C only. It was reasoned that the lack of antibody response to antigenic determinants associated with PrP^{Sc} may have resulted from the loss of specific epitopes following oligomerization of purified PrP 27-30 into rods.

In a series of later experiments, to generate additional antibody diversity, *Prnp$^{o/o}$* mice were immunized with a host of different PrP antigen preparations, including biotinylated PrP 27-30 dispersed in detergent and incorporated into liposomes, recombinant PrP preparations refolded into both α-helical and β-sheet-rich

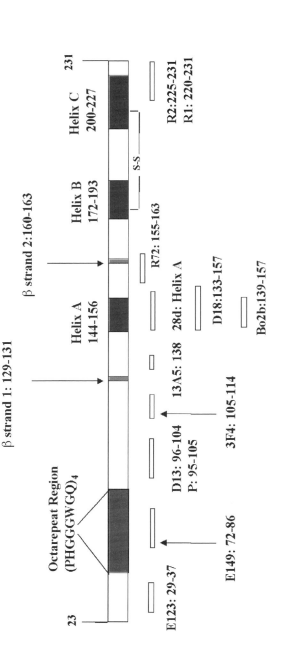

Table 1. Location of selected specific monoclonal antibody epitopes on the sequence of PrP 23–231. The octarepeat, α-helical regions, and β-strand regions of PrP are also shown.

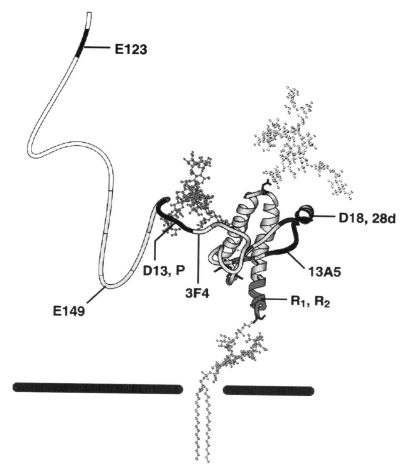

Figure 1. Selected PrP antibody epitopes superimposed on the solution structure of recombinant PrP 23-231 as refolded into an α-helical conformation thought to closely resemble that of PrP^C. Carbohydrate moieties linked to Asn 180 and Asn 196 are shown.

conformations, and a variety of PrP synthetic peptides. Phage-antibody libraries prepared from these animals were each selected against a plethora of PrP antigens immobilized onto ELISA wells. Dispersed forms of biotinylated PrP 27-30, which retain their infectivity as determined in bioassay (Gabizon and Prusiner, 1990), were captured onto streptavadin-coated ELISA wells. A large panel of distinct novel MAbs were recovered from these experiments (Peretz *et al.*, 1997; Williamson *et al.*, 1998; Leclerc *et al.*, 2003). The binding epitopes of each antibody were carefully mapped using several complementary strategies. Antibody reactivity was first measured against synthetic

PrP peptides of 15 residues in length, overlapping by 5 residues at the amino terminus, and collectively spanning PrP residues 90-231. In addition, each antibody was selected against random PrP fragment libraries, displayed on the phage surface (Petersen *et al.*, 1995; Williamson *et al.*, 1998). The linear and discontinuous epitopes identified using these approaches are shown in Table 1 and Figure 1.

All of the antibodies recovered from the phage libraries bound efficiently to PrP^C and to recombinant PrP preparations refolded into predominantly α-helical, PrP^C-like conformations. The majority of the antibodies also reacted well with cell surface PrP^C, and have proven useful tools with which to study PrP in its native environment (Leclerc *et al.*, 2003). Exceptions to this rule are antibody R72, that binds to PrP residues 155-163 and recognizes PrP only when coated onto ELISA wells, and antibody Bo2b, that binds to an unstructured 139-157 peptide, which contains sequence that corresponds to helix A in PrP^C. A small subset of antibodies, including MAbs D18 (recognizing an epitope between residues 134-158) and R1/R2 (recognizing epitopes between residues 220-231), bound non-denatured PrP 27-30 in an ELISA format, indicating that these epitopes may be present in both PrP^C and disease-associated infectious PrP conformers (Peretz *et al.*, 1997). No MAb, however, recognized an epitope exclusive to PrP^{Sc}.

Over recent years a number of different laboratories have now also successfully generated diverse panels of PrP-specific MAbs recognizing PrP sequence from many different species by immunizing *Prnp*$^{o/o}$ mice and then using conventional hybridoma methodologies to immortalize antibody-producing cells (Krasemann *et al.*, 1996; Grassi *et al.*, 2001; 2000; Demart *et al.*, 1999; Beringue *et al.*, 2003; Li *et al.*, 2000). Disappointingly, however, in each case, immunization has failed to elicit an antibody specifically recognizing PrP^{Sc} or PrP 27-30. One such molecule, 15B3, a low-affinity pentameric antibody of the IgM class, was reported to specifically recognize PrP^{Sc} (Korth *et al.*, 1997) but later reports have questioned these findings (Heppner *et al.*, 2001). Ultimately, as described in detail below, two very different approaches were required to finally generate antibodies reacting specifically with PrP^{Sc}.

MAbs specific for PrP antigens have been harnessed as research tools in many studies of prion biology. They have been used to study the exposure of various PrP epitopes in PrP^C and PrP^{Sc} (Li *et al.*,

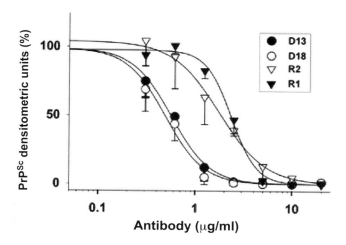

Figure 2. Dose-dependent inhibition of PrP^{Sc} formation in ScN2a cells by PrP-specific recombinant monoclonal antibody Fabs D18, D13, R2 and R1. Cells were independently cultured for a period of 7 days in the presence of 0-20 µg/ml of each antibody and PrP^{Sc} levels measured by immunoblotting. Under equivalent conditions, PrP-reactive antibodies R72, E123 and E149 did not alter PrP^{Sc} levels in ScN2a cells.

2000; Peretz *et al.*, 2002; 2001a; 1997; Demart *et al.*, 1999; Leclerc *et al.*, 2001), to examine the structure of PrP peptide in complex with antibody (Kanyo *et al.*, 1999), and for immunohistochemical (Wadsworth *et al.*, 2001; Moya *et al.*, 2000; Gonzalez *et al.*, 2003; Barclay *et al.*, 2002; Spraker *et al.*, 2002; Liu *et al.*, 2001b; 2001a; Laffling *et al.*, 2001; Esiri *et al.*, 2000; Beringue *et al.*, 2003) and ultrastructural studies related to the cell biology of PrP and neuropathology of prion disease (Peters *et al.*, 2003; Mironov *et al.*, 2003; Ivanova *et al.*, 2001; Keshet *et al.*, 2000; Jeffrey *et al.*, 2000; Kanu *et al.*, 2002). In addition, antibodies now lie at the forefront of efforts to develop prototypic diagnostic systems to detect early-stage prion infections in humans and livestock, and in the pursuit of clinically useful approaches for the treatment and prevention of prion disease. These latter two applications will now be discussed in greater detail.

2.2. Specific Antibodies Inhibit Prion Propagation

A body of experimental evidence indicates that PrP^C-PrP^{Sc} interactions lie at the heart of prion replication (reviewed in Caughey, 2001). It was therefore reasonable to speculate that reagents, such as antibodies, that bind specifically to the appropriate regions of either PrP conformer may interrupt prion production by inhibiting this interaction. To test this hypothesis, several recombinant antibody Fab fragments, D13, D18, R1, R2, E123, E149 and R72 (Table 1), recognizing different regions of PrP were added to culture medium of PrP^{Sc}-infected mouse neuroblastoma cells (ScN2a). Initially, a range of concentrations of each antibody was added to ScN2a cultures for a period of seven days. After this time, cells were harvested and the level of PrP^{Sc} in the culture analyzed by immunoblotting. As shown in Figure 2, densitometric analysis of the blotted PrP bands, representative of the level of PrP27-30 in the culture, indicated that as compared with non-treated cells, PrP^{Sc} was dramatically reduced in a dose-dependent manner. By this analysis, Fabs D13 and D18 appear to be approximately equally effective, having IC_{50} values of 0.45 μg/ml (9 nM) and 0.6 μg/ml (12 nM), respectively. In contrast, 7-day treatment with Fabs E123, E149 or R72, or 5-day treatment with polyclonal IgG recognizing both transmembrane and GPI-anchored forms of mouse N-CAM (Chuong *et al.*, 1982), did not reduce the level of PrP^{Sc} in ScN2a cultures even when these antibodies were used at high concentrations. During these experiments, the levels of PrP^C and glyceraldehyde-3-phosphate dehydrogenase in antibody-treated and untreated cells were found to be invariant, indicating that the PrP-specific antibodies used produced no overt cytotoxic effects that may have indirectly compromised the production of PrP^{Sc}.

To determine whether PrP^{Sc} remained undetectable after removal of PrP-specific antibody, ScN2a cells were independently passaged for 7, 14 or 21 days in the presence of 10 μg/ml of each of the recombinant Fabs. Antibody was then removed from the culture and the cells passaged for an additional period in Fab-free medium, after which time the level of PrP^{Sc} was re-measured. If cells were cultured for a 14 day period in the presence of Fab D18 then PrP^{Sc} remained at undetectable levels after 9 additional weeks of culture in antibody-free medium. As a second measure of prion titer, CD-1 Swiss mice were inoculated with antibody-treated (10 μg/ml) and untreated ScN2a cells. Mice inoculated intracerebrally (i.c.) with

D18, D13, or R2-treated cells were disease free after a period of 265 days, whereas mice inoculated with untreated or R72-treated cells had a mean incubation time to disease of 169 and 165 days, respectively. The prolonged incubation times indicate a reduction of over 3 logs in the infectious prion titer in treated cells (Butler *et al.*, 1988).

The above experiments did not determine to what degree PrP^{Sc} preexisting in the ScN2a cultures at the onset of antibody treatment was subsequently eliminated from the cells. The above data could be taken to indicate that PrP^{Sc} levels are rapidly diminished. Indeed, similar studies describing the inhibition of prion propagation with other molecules have been interpreted in this fashion (Caughey and Raymond, 1993; Caughey *et al.*, 1998; Chabry *et al.*, 1999; Perrier *et al.*, 2000; Supattapone *et al.*, 1999). However, this interpretation fails to account for the rapid expansion of ScN2a cell populations in culture. Thus, if an antibody inhibits formation of nascent PrP^{Sc} molecules, then each successive round of cell division may serve to dilute the effective concentration, but not necessarily the total amount, of residual PrP^{Sc} within the culture, thus creating a potentially misleading impression of PrP^{Sc} clearance. An accurate calibration of the rate with which PrP^{Sc} is purged from ScN2a cultures must therefore be normalized to account for any increase in cell population and commensurate reduction in PrP^{Sc} concentration that has taken place over the course of the experiment.

In a more thorough analysis of the kinetics of prion clearance, ScN2a cells were independently grown in the presence of 10 μg/ml of PrP-specific Fab. Cells were harvested after a period of 1, 2, 3, and 4 days of antibody treatment and the total mass of cell protein determined in each case as a measure of cell number. The PrP^{Sc} concentrations in Fab-treated and untreated cells at these time points were determined by immunoblotting. Total PrP^{Sc} in the culture at each time point was then calculated by factoring in the total cell mass in each case. When these data were plotted against the duration of antibody treatment, clear and more meaningful differences in the efficacy with which individual Fabs resolved prion infection became apparent (Figure 3).

By this analysis, Fab D18 was again determined to be the most effective antibody. The time taken from the initial treatment with D18 to eliminate 50% of PrP^{Sc} from the cells ($t_{1/2}$) was 28 h. This figure is

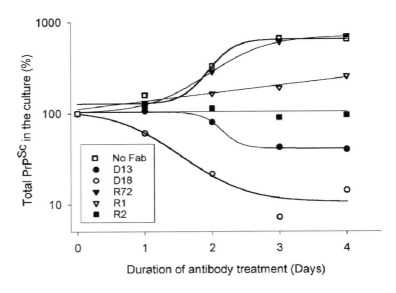

Figure 3. A time course of antibody-mediated PrP^{Sc} clearance from ScN2a cell cultures. Total amounts of PrP^{Sc} in ScN2a cells was determined by immunoblotting following culture for 1, 2, 3 or 4 days in the presence of 10 µg/ml PrP-specific antibodies D18, D13, R1, R2 and R72.

in good agreement with previous studies in which the $t_{1/2}$ of PrP^{Sc} in ScN2a cells was determined to lie in the region of 24 h (Borchelt *et al.*, 1990). The data suggest that at a concentration of 10 μg/ml, Fab D18 is able to completely abolish prion propagation and that preexisting PrP^{Sc} is subsequently eliminated from the cells. This finding indicates that a certain amount of PrP^{Sc} is continuously expunged from ScN2a cultures through degradation pathways and suggests for the first time that prion propagation is actually a highly dynamic process. Importantly, these data introduce the principle that if the formation of nascent PrP^{Sc} can be prevented, then the pre-existing PrP^{Sc} load can be efficiently degraded, with resulting clearance of prion infectivity.

What then were the factors dictating which antibodies efficiently inhibited prion replication, and which did not? Mechanistically, MAb inhibition of prion replication is most readily explained by antibody binding specifically to PrP^{C} molecules on the cell surface and thereby hindering docking of PrP^{Sc} template or other co-factor critical for the conversion of PrP^{C} to PrP^{Sc}. In agreement with this notion, Fab D18, which was the most potent antibody we evaluated, bound to a significantly greater number of cell surface PrP^{C} molecules than Fabs D13, R1 or R2. In contrast, Fab R72, which had no effect on prion propagation, and Fabs E123 and E149 that did not diminish PrP^{Sc} levels, either failed to recognize cell surface PrP^C or reacted comparatively weakly with MoPrP^C. Thus, the fraction of total cell surface PrP^C occupied by a given antibody would appear to be critically important in determining the inhibitory potency of that antibody.

The next question to be considered was whether or not the region of PrP bound by each antibody was of intrinsic importance to its inhibitory potency. To address this point, inhibition of prion replication was compared in conditions in which equal amounts of two different Fabs were bound to the ScN2a cell surface. At concentrations of 0.6 μg/ml although equivalent amounts of Fabs D18 and D13 were bound to the cell surface, D18 inhibited prion replication much more efficiently (Figure 2). Similarly, at concentrations of 0.6 μg/ml and 2.5 μg/ml respectively, Fabs D18 and R1 bound equivalently to ScN2a cells, but D18 was clearly more effective in reducing the level of PrP^{Sc} in the culture. Finally, at a concentration of 2.5 μg/ml Fabs D13 and R1 bound equivalently to the cell surface, but D13 more actively reduced PrP^{Sc} synthesis. This analysis illustrated that the region of PrP^C bound by a given antibody is one of the key determinants of its inhibitory capacity. A reasonable extension of this argument is that PrP sequence composing the D18 and D13 antibody epitopes is specifically bound by PrP^{Sc}. We shall return to this point later, as it suggested a novel approach for generating long-sought reagents reacting specifically with disease-associated conformers of PrP.

3. Prospects for Antibody Prophylaxis and Therapy of Prion Disease

The antibody studies described above (Peretz *et al.*, 2001b), together with similar findings produced in another laboratory (Enari *et al.*, 2001), indicated that certain MAbs are highly potent inhibitors of prion replication in infected cells. Other groups extended this line of investigation *in vivo* using a murine scrapie model. In these studies, experiments were performed to determine whether or not antibody could i) protect from challenge with infectious prions, and ii) resolve or delay established prion infections in the CNS and peripheral tissues, such as the spleen. Heppner *et al.* (2001) were able to demonstrate that antibody 6H4, binding a PrP epitope very similar to that recognized by the D18 antibody, when expressed constitutively in transgenic

(Prnp$^{+/o}$-6H4μ) mice was able to partly protect those mice from peripheral prion challenge. Equivalent mice expressing the putative PrP^{Sc}–specific 15B3 antibody were not protected from prion infection. This study introduced the tantalizing possibility that vaccine strategies capable of eliciting antibody responses against PrP may be useful in protecting the host from exposure to infectious prions. Of course, in pursuing this goal there are significant challenges to be faced, not the least of which is that PrP^C is a self-antigen, and poorly immunogenic under normal circumstances because of mechanisms of immune tolerance. In addition, should antibody be successfully elicited by vaccination, there is the possibility of unwanted autoimmune effects that would contraindicate a vaccine approach. Promisingly, no such effects were detected in the transgenic animals expressing good levels of 6H4.

Encouraged by immunization studies in transgenic mouse models of Alzheimer's disease in which vaccination with the βA4 peptide has produced clear pathological improvements (Wisniewski *et al.*, 2002), several laboratories have begun to explore how auto-immune tolerance to PrP immunogen may be overcome. In an attempt to identify immunogenic prion peptides, Lewis rats were immunized with two rat PrP peptides (composed of PrP residues 118-137, and 182-202) that fitted the MHC class II RT1.B 1 motif (Souan *et al.*, 2001). Both humoral and cellular immunity specific to the prion peptides were reported following peptide immunization, with no harmful effects evident in young animals. However, a cautionary note was sounded when severe skin inflammation featuring mononuclear cell infiltration, together with concomitant hair loss, was observed in about 20% of peptide-immunized 8-month-old rats. It is noteworthy that these same investigators found immunization of prion-infected mice with complete Freund's adjuvant prolonged survival, compared to unimunized control animals, whether the route of prion infection was intraperitoneal (i.p.) or i.c. (Tal *et al.*, 2003). Intriguingly, post-exposure prophylaxis with CpG oligodeoxynucleotides, known to stimulate innate immunity, was also reported to increase survival times of prion-infected mice, when compared to mice that received equivalent treatment with saline (Sethi *et al.*, 2002). Taking a slightly different approach, Gilch et al., immunized mice with a dimeric form of mouse PrP consisting of two PrP 23-231 moieties fused together via a seven-residue linker sequence (Gilch *et al.*, 2003). The resulting protein, when refolded following expression in *E. coli*, contained equivalent α-helical conformation to that of monomeric recombinant PrP 23-231. Using 'PrP dimer' in combination with various adjuvant formulations, serum anti-PrP titers of as high as 1: 20,000 were reported. These authors went on to demonstrate that the serum antibodies elicited by 'PrP dimer', bound to cell surface PrP and effectively inhibited prion replication in ScN2a cells, whereas antibodies raised against equivalently purified PrP 23-231 monomer did not.

Once it became clear that antibodies could help to protect the host from at least a peripheral prion inoculum (Heppner *et al.*, 2001), the next step was to determine if antibody prophylaxis could act therapeutically and resolve existing prion infections. To address this issue, two monoclonal antibodies, ICSM35 and ICSM 18, binding to PrP epitopes very similar to those recognized by the potently inhibitory D13 and D18 recombinant MAbs, respectively, were administered i.p. twice weekly to prion infected mice (White *et al.*, 2003). Under normal circumstances, in this model, i.p. inoculation of prions will lead to detectable PrP^{Sc} in the spleen of the infected animal 7 days post infection. PrP^{Sc} levels will continue to increase in the spleen until around day 30-post infection, when a maximum plateau is reached. If ICSM 18 antibody treatment of the infected mice was begun from either day 7 or day 30 post infection, PrP^{Sc} was reduced to almost undetectable levels. Equivalent treatment with ICSM 35 antibody led to a less complete inhibition of PrP^{Sc} replication, with levels of disease-associated PrP in spleen reduced by 75%. Survival of untreated i.p. inoculated mice was to an average of 197 days, whereas mice treated with either ICSM 18 or ICSM 35 from 7 or 30 days post infection remained healthy to over 500 days post infection. Whether or not prion infection in these mice was fully eradicated, or merely suppressed, remains to be determined. Significantly, no PrP^{Sc} was detected in the brain or spleen of ICSM 18 treated mice 250 days post infection. Although these findings are undoubtedly impressive, it must be noted that very large amounts (2 mg) of antibody were given on each occasion to the mice, producing a serum IgG concentration of approximately 300-400 µg/ml. Furthermore, treatment was ineffective if antibody was first delivered at the onset of clinical scrapie, or if mice were challenged with prions intracerebrally. As was the case with the Prnp$^{+/o}$-6H4µ mice, no gross unwanted autoimmune effects, as evidenced by monitoring of T- and B-cell or splenic follicular dendritic cell populations, were detected in the prolonged presence of high concentrations of PrP^C-specific antibody. It is important to note, however, that we do not yet know if PrP^C-specific antibody can induce adverse effects if delivered directly within the CNS compartment.

In summary, then, antibodies recognizing particular regions of PrP^C are potent inhibitors of prion replication *in vitro* and *in vivo*. In peripheral tissues, antibody can serve to protect from exposure to a prion inoculum, or severely deplete existing infections, even when prion titers are at their highest. However, once prion infections have accessed the CNS from peripheral sites, or are initiated within the CNS (such as may be encountered in prion disease of genetic origin), antibody appears incapable of effectively inhibiting the disease course. This shortcoming is probably reflective of the inability of antibody to traverse the blood brain barrier. Nonetheless, specific antibodies that potently inhibit prion replication identify regions of PrP as logical targets for the development of anti-prion small molecule drugs. Antibody-based competitive screens of large chemical libraries can provide a rational platform with which to identify small molecule drugs that both efficiently inhibit prion replication and more readily access the CNS compartment (Peretz *et al.*, 2001b).

4. Insights Into PrP^C-PrP^{Sc} Interactions in the Prion Replicative Complex

Antibody inhibition of prion propagation can most readily be explained through two distinct mechanisms. In the first of these, antibody binding directly interferes with the formation of PrP^C-PrP^{Sc} interactions and assembly of the prion replicative complex. In a second possibility, antibody binding stabilizes PrP^C native conformation, thereby increasing the magnitude of the activation energy barrier associated with the PrP^C to PrP^{Sc} conformational and quaternary structural changes, effectively depleting the concentration of substrate available to PrP^{Sc} template and halting or slowing the replication cycle. If the first mechanism were engaged, then it might be reasonable to speculate that the regions of PrP bound by the inhibitory antibodies may play a direct role in forming the PrP^C-PrP^{Sc} interface. This idea is given further credence as the known binding epitopes of the inhibitory MAbs D18, 6H4 and ICSM 18 incorporate helix A (residues 143-156) of PrP^C, sequence that is positioned on the opposite face of the protein from residues Q167, Q171, T214 and Q218, which are hypothesized to participate in binding an auxiliary molecule

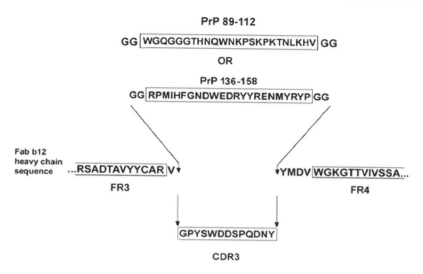

Figure 4. Schematic illustration of mouse PrP 89-112 and 136-158 peptides replacing b12 HCDR3 sequence to yield IgGs 89-112 and 136-158. The N-terminal Val residue and 4 C-terminal residues (Tyr-Met-Asp-Val) of the original b12 HCDR3 are retained; two Gly residues are added to each flank of the grafted PrP sequence.

essential to prion propagation (Kaneko *et al.*, 1998; Zulianello *et al.*, 2000; 2000). Thus D18/6H4/ICSM 18, and possibly D13, MAbs are likely to operate mechanistically by directly blocking or modifying PrP^C interaction with PrP^{Sc}, rather than by inhibiting the binding of a cofactor. In extension of this argument, PrP sequences composing the D18 and D13 antibody epitopes may directly bind PrP^{Sc} during prion replication, and thus possess an intrinsic and specific affinity for PrP^{Sc}.

4.1. PrP^{Sc}-specific Motif-grafted Antibodies

To test if sequence composing the PrP epitopes recognized by MAbs D18 (residues 133-158) and D13 (residues 96-104) did indeed interact specifically with disease-associated forms of PrP, a series of motif-grafting experiments were performed. Transplanting PrP sequence corresponding to these PrP regions into a suitable carrier molecule, such as an antibody, should impart specific recognition of PrP^{Sc}. Historically, relatively long recognition sequences have been grafted into the HCDR3 of the antibody molecule to generate desired binding properties (McLane *et al.*, 1995). Thus, PrP sequences containing

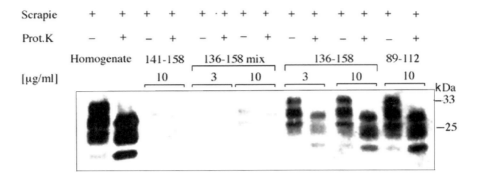

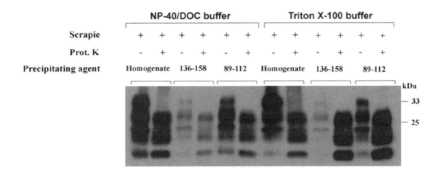

Figure 5. Immunoprecipitation of PrP^Sc and PrP27-30 from prion-infected mouse brain using PrP-grafted antibodies 89-112, 136-158, 136-158 mix and 141-158. **a.** PrP^Sc and PrP 27-30 present in crude homogenate are shown in lanes 1 and 2, respectively. No or trace amounts of PrP were precipitated by PrP-IgGs 141-158 (lanes 3 and 4) and 136-158 random (lanes 5-8), indicating that specificity for PrP^Sc and PrP 27-30 is critically dependent upon the grafted PrP 136-158 (lanes 9-12) and 89-112 (lanes 13 and 14) sequence motifs. **b.** Immunoprecipitation of PrP^Sc and PrP 27-30 in the presence of NP-40/DOC and Triton X-100. A prion-infected mouse brain was bisected laterally and one hemisphere homogenized in buffer containing 1% NP-40 and 1% DOC, and the other hemisphere homogenized in buffer containing 1% Triton X-100. Roughly equal amounts of PrP^Sc were present in each of these preparations (lanes 1, 2 and 7, 8). IgGs 89-112 and 136-158 (10 μg/ml) precipitated very similar amounts of PrP^Sc from both brain homogenates (lanes 3, 5 and 9, 11), but significantly greater amounts of PrP 27-30 were immunoprecipitated from Triton X-100 homogenate (lanes 10 and 12) than from homogenate prepared using 1% NP-40 and 1% DOC (lanes 4 and 6).

residues 96-104 and 136-158 were independently transplanted into the HCDR3 of IgG Fab b12 (Burton *et al.*, 1994), a human recombinant antibody specific for HIV-1 gp120. The b12 antibody was chosen as the recipient molecular scaffold for transplanted PrP sequence because the parental antibody possesses a relatively long HCDR3 (18 amino acids) that projects vertically from the surface of the antigen binding site (Ollmann Saphire *et al.*, 2001). To maximally distance PrP sequence from the antibody surface, each graft was placed between the first N-terminal residue and four C-terminal residues of the parental HCDR3 (Figure 4). In addition, two glycine residues were incorporated at each flank of the PrP sequence. The resulting PrP-antibodies 89-112 and 136-158 were expressed as IgGs in CHO cells.

Reactivity of the resulting PrP-grafted antibody molecules was evaluated in an immunoprecipitation assay against PrP^C, PrP^{Sc} and PrP27-30 (Figure 5). The data indicate that the antibody recipients of the PrP-sequence grafts bound specifically and robustly with PrP^{Sc} and PrP27-30 from mouse brains infected with either RML or 79a scrapie prions, but not PrP^C. To accumulate further evidence demonstrating that the specificity of the PrP^{Sc} interaction was conferred by the grafted PrP sequence, an antibody molecule in which the amino acids composing the 136-158 PrP graft were randomly reordered was created. The resulting molecule, termed 136-158 mix, showed only trace reactivity with PrP^{Sc} and PrP27-30 at a concentration of 10 µg/ml, and no reactivity at a concentration of 3 µg/ml (Figure 5). Intriguingly, the PrP 141-158 antibody also reacted only extremely weakly with PrP^{Sc} and PrP27-30, indicating that PrP sequence between amino acids 136 and 140 (inclusive) may be critical for PrP^C-PrP^{Sc} interactions (Figure 5). In previous studies, specific interaction between plasminogen and PrP^{Sc} (Fischer *et al.*, 2000) was shown to be dependent upon the presence of detergents that disrupt membrane rafts (Shaked Y. *et al.*, 2002). Importantly, the PrP-grafted antibodies bound well to PrP^{Sc} and PrP 27-30 in brain homogenates prepared using either Nonidet P-40 and sodium deoxycholate, reagents that disrupt membrane rafts, or Triton X-100, a detergent that preserves raft architecture.

The affinities of the PrP-grafted antibodies for disease-associated PrP conformations, were estimated in a series of immunoprecipitation assays in which decreasing concentrations of antibodies were used.

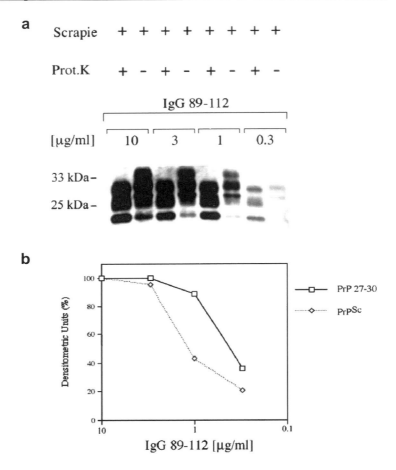

Figure 6. a. Titration of IgG 89-112 reactivity in an immunoprecipitation assay against mouse PrP^Sc (- Prot. K) and PrP27-30 (+ Prot. K). **b.** Densitometric measurement of PrP bands identified in the immunoblot in 'a'. Values are given as densitometric units, where 100% is assigned to the PrPΣχ or PrP 27–30 band produced by immunoprecipitation in the presence of 10 μg/ml of specific antibody.

The results indicated that IgG 87-112 possesses an apparent affinity constant (K_d) of approximately 2 nM for PrP 27-30 and 7 nM for PrP^Sc (Figure 6). Using a similar approach, the K_d of IgG 136-158 for abnormal PrP conformers was estimated to lie in the 20-30 nM range.

4.2. Antibodies to Misfolded PrP by Immunization

Despite considerable efforts in many laboratories, immunization and selection strategies employing infectious and highly purified preparations of PrP27-30, in addition to recombinant PrP folded in a β-sheet-rich conformation, have not yielded antibodies reacting exclusively with native (non-denatured) conformations of PrP[Sc] or PrP27-30. However, this impasse now appears to have been overcome (Paramithiotis *et al.*, 2003). Ten IgM antibodies that displayed PrP[Sc]-specific immunoprecipitation from ME7 and 139A scrapie-infected mouse brain were recovered from Balb/c mice immunized with keyhole limpet hemocyanin coupled to a CYYRRYYRYY peptide. The rationale behind this immunization strategy was that induction of β-sheet-rich structure in recombinant PrP was associated with an increased solvent accessibility of tyrosine residues, over that measured in PrP with α-helical rich conformation. Thus, antibodies directed against tyrosine motifs may bind to abnormal forms of PrP, but not the normal cellular conformer. Overall, this antibody panel was also shown to recognize hamster, sheep, bovine and human PrPs. Notably, however, in the immunoprecipitation assays MAbs were used at a concentration of 100 μg/ml, suggesting that their affinities may be relatively modest. Nonetheless, one Mab, 9A4, bound to a subpopulation of dendritic cells from scrapie-infected sheep lymphnodes.

We can be hopeful that this new generation of antibodies, specific for disease-associated PrP conformations, can shed light on mechanisms of prion replication, and bolster ongoing efforts to develop effective prion therapeutics and diagnostics.

5. Immuno-diagnostic Detection of Prions

The BSE crisis in Europe, and ongoing emergence of vCJD and CWD have demonstrated that the transmission of prion diseases is a significant international problem, with important ramifications for public health, and substantial impact in several strategically important economic sectors, particularly agriculture. For these reasons, many government agencies have introduced policies directed toward the containment and eventual eradication of prion diseases. A key component of these initiatives is the development of effective monitoring and surveillance programs that can identify prion-infected individuals and livestock. Historically, diagnosis of TSEs has relied upon relatively 'low-tech' and labour intensive immunohistochemical methodologies. These techniques are based upon the immunological detection of abnormal forms of PrP. Methodologically, tissue samples are first treated with proteinase K to remove endogenous PrP^C. Any residual misassembled PrP conformers possessing partial resistance to proteolytic degradation are then denatured via heat or chemical treatment, which results in the exposure of previously cryptic antibody epitopes. Typically, in this type of diagnostic analysis CNS tissues that contain the highest concentrations of PrP^{Sc} are used. However, in certain TSEs, prion titers accumulate to histochemically detectable titers in peripheral tissues. For example, lymphoid tissues such as the third eyelid in sheep, or tonsil in deer, have been routinely examined for diagnostic purposes.

The exponential increase in demand for prion diagnostic testing that resulted from the onset of the BSE crisis and the ensuing governmental legislation, necessitated that different approaches were required to develop more practical, rapid and low cost postmortem detection of PrP^{Sc}. To meet this demand, a large number of diagnostic testing platforms were developed in different laboratories (reviewed in Bennion and Daggett, 2002), and four of these have now been approved for use by the European Commission. Each of these testing platforms shares the same basic approach, which is to use specific antibodies to detect denatured forms of PrP^{Sc}. Of the two most widely used tests, one employs an immunoblotting approach in which proteinase K digested brain homogenates are first electrophoresed through a sodium dodecyl sulphate polyacrylamide gel, then transferred to a membrane and PrP detected via the 6H4 antibody. The second test utilizes an ELISA based sandwich assay where, after concentration and denaturation steps, PrP^{Sc} is captured and detected by two MAbs recognizing different PrP epitopes (Grassi *et al.*, 2000). The diagnostic systems described above have contributed greatly to large-scale epidemiological analysis and surveillance of BSE disease in cattle and sheep populations, most notably in Europe, North America and Japan.

The remaining challenge in prion diagnostics is the development of a commercially viable testing system to detect infectious prions in blood or possibly urine (Shaked *et al.*, 2001) of humans and animals during the preclinical stage of disease. Ante-mortem testing is required to protect the human blood supply from contamination with vCJD prions, which could be replicating subclinically in a fraction of the many millions of individuals likely to have been exposed to BSE-tainted products, and to more effectively monitor and control prion disease in domestic livestock. The recent transmission of BSE and natural scrapie via blood transfusion in sheep underscores that all available precautions must be taken to exclude prions from the blood supply (Houston *et al.*, 2000; Hunter *et al.*, 2002). However, the available evidence suggests that prion titers in blood are likely to be low and may be present only transiently.

Clearly, any diagnostic test aspiring to detect prions in blood requires extreme sensitivity, while retaining absolute accuracy. Ultimately, the detection of single units of prion infectivity is desirous, and here calibrating the sensitivity of each prototypic diagnostic test requires comparison to prion infectivity as determined by bioassay. However, a degree of caution must be exercised here, since the definition of infective titer will vary with the recipient of the infective inoculum. For example, titrating BSE infectivity in cattle indicated that earlier studies using an RIII mouse bioassay (Deslys *et al.*, 2001) had underestimated BSE prion titers by a factor of approximately 1000. Mice carrying a bovine PrP transgene (Tg(BoPrP$^{+/+}$)4092/ Prnp$^{o/o}$), however, are actually 10 times more sensitive than cattle to infection with BSE prions (Safar *et al.*, 2002). Of the many studies described in the literature (Bennion and Daggett, 2002), demonstrably the most sensitive test reported to date is the conformation dependent immunoassay (CDI) (Safar *et al.*, 1998; 2002). This test appears able to detect BSE prions in bovine brainstems with a sensitivity similar to that of end-point titrations in Tg(BoPrP) mice. In this system, antibody binding was once again measured against PrP epitopes exposed following chemical denaturtation of PrP^{Sc}. Briefly, PrP^{Sc} was first concentrated from tissue samples using sodium phosphotungstate acid, then denatured and captured with the Fab D18 antibody. Captured PrP was quantitated by measuring binding of Fab P antibody labeled with Europium chelate using a time-resolved fluorescence instrument. The technique is advantaged in that it may be applied to samples with or without removing endogenous PrP^C via proteinase K digestion. Thus any proteinase K sensitive forms of PrP^{Sc} that would otherwise be lost are retained for quantitation. Whether this approach is suitable for measuring prions in non-CNS tissues has yet to be determined, although using recombinant methodologies to evolve antibodies with improved binding kinetics and specificity for PrP antigens from different species, will likely elevate the sensitivity of the test even further.

Given the very high levels of diagnostic sensitivity required for detection of prion infectivity in blood, considerable attention was given to a report by Saborio *et al.* (2001), in which a method for the amplification of disease-associated PrP conformers was described. Here, in a cyclical process reminiscent of DNA amplification by the polymerase chain reaction, PrP^{Sc} in prion-infected brain homogenate was purported to convert PrP^C in normal brain homogenate into additional molecules of proteinase resistant PrP. Following sonication to break up newly formed prion aggregates, additional normal brain material was added, and the conversion process repeated. After 5 cycles of amplification via repeated sonication and incubation, a 97% conformational conversion of PrP^C was reported. This novel approach offers tremendous potential for enhancing the sensitivities of the immuno-diagnostic platforms, but its reproducibility and general applicability remains in question.

References

Aguzzi, A., Glatzel, M., Montrasio, F., Prinz, M., and Heppner, F.L. 2001. Interventional strategies against prion diseases. Nat. Rev. 2: 745-749.

Aguzzi, A. and Weissmann, C. 1998. Prion diseases. Haemophilia. 4: 619-627.

Anderson, R.M., Donnelly, C.A., Ferguson, N.M., Woolhouse, M.E.J., Watt, C.J., Udy, H.J., MaWhinney, S., Dunstan, S.P., Southwood, T.R.E., Wilesmith, J.W., Ryan, J.B.M., Hoinville, L.J., Hillerton, J.E., Austin, A.R., and Wells, G.A.H. 1996. Transmission dynamics and epidemiology of BSE in British cattle. Nature. 382: 779-788.

Andrews, N.J., Farrington, C.P., Ward, H.J., Cousens, S.N., Smith, P.G., Molesworth, A.M., Knight, R.S., Ironside, J.W., and Will, R.G. 2003. Deaths from variant Creutzfeldt-Jakob disease in the UK. Lancet. 361: 751-752.

Barclay, G.R., Houston, E.F., Halliday, S.I., Farquhar, C.F., and Turner, M.L. 2002. Comparative analysis of normal prion protein expression on human, rodent, and ruminant blood cells by using a panel of prion antibodies. Transfusion. 42: 517-526.

Barret, A., Tagliavini, F., Forloni, G., Bate, C., Salmona, M., Colombo, L., De Luigi, A., Limido, L., Suardi, S., Rossi, G., Auvre, F., Adjou, K.T., Sales, N., Williams, A., Lasmezas, C., and Deslys, J.P. 2003. Evaluation of quinacrine treatment for prion diseases. J. Virol. 77: 8462-8469.

Barry, R.A., Kent, S.B., McKinley, M.P., Meyer, R.K., DeArmond, S.J., Hood, L.E., and Prusiner, S.B. 1986. Scrapie and cellular prion proteins share polypeptide epitopes. J. Infect. Dis. 153: 848-854.

Barry, R.A. and Prusiner, S.B. 1986. Monoclonal antibodies to the cellular and scrapie prion proteins. J. Infect. Dis. 154: 518-521.

Bendheim, P.E., Barry, R.A., DeArmond, S.J., Stites, D.P., and Prusiner, S.B. 1984. Antibodies to a scrapie prion protein. Nature. 10: 418-421.

Bennion, B.J. and Daggett, V. 2002. Protein conformation and diagnostic tests: the prion protein. Clin. Chem. 48: 2105-2114.

Beringue, V., Mallinson, G., Kaisar, M., Tayebi, M., Sattar, Z., Jackson, G., Anstee, D., Collinge, J., and Hawke, S. (2003). Regional heterogeneity of cellular prion protein isoforms in the mouse brain. Brain. 126: 2065-2073.

Bode, V., Pocchiari, M., Gelderblom, H., and Diringer, H. 1985. Characterization of antisera against scrapie-associated fibrils (SAF) from affected hamster and cross-reactivity with SAF from scrapie-affected mice and from patients with Creutzfeldt-Jakob disease. J.Gen.Virol. 66: 2471-2478.

Boelle, P.Y., Thomas, G., Valleron, A.J., Cesbron, J.Y., and Will, R. 2003. Modelling the epidemic of variant Creutzfeldt-Jakob disease in the UK based on age characteristics: updated, detailed analysis. Stat. Methods Med. Res. 12: 221-233.

Bolton, D.C., Seligman, S.J., Bablanian, G., Windsor, D., Scala, L.J., Kim, K.S., Chen, C.M., Kascak, R.J., and Bendheim, P.E. 1991. Molecualr location of a species-specific epitope on the hamster scrapie agent protein. J. Virol. 65: 3667-3675.

Borchelt, D.R., Scott, M., Taraboulos, A., Stahl, N., and Prusiner, S.B. 1990. Scrapie and cellular prion proteins differ in their kinetics of synthesis and topology in cultured cells. J. Cell Biol. 110: 743-752.

Brandner, S., Isenmann, S., Raeber, A., Fischer, M., Sailer, A., Kobayashi, Y., Marino, S., Weissmann, C., and Aguzzi, A. 1996. Normal host prion protein necessary for scrapie-induced neurotoxicity. Nature. 379: 339-343.

Bruce, M.E. 2000. 'New varriant' Creutzfeldt-Jakob disease and bovine spongiform encephalopathy. Nat. Med. 6: 258-259.

Bruce, M.E., Will, R.G., Ironside, J.W., McConnell, I., Drummond, D., Suttie, A., McCardle, L., Chree, A., Hope, J., Birkett, C., Fraser, H., and Bostock, C.J. 1997. Transmissions to mice indicate that 'new variant' CJD is caused by the BSE agent. Nature. 389: 498-501.

Büeler, H., Fischer, M., Lang, Y., Bluethmann, H., Lipp, H.P., DeArmond, S.J., Prusiner, S.B., Aguet, M., and Weissmann, C. 1992. Normal development and behaviour of mice lacking the neuronal cell-surface PrP protein. Nature. 356: 577-582.

Burns, C.S., Aronoff-Spencer, E., Legname, G., Prusiner, S.B., Antholine, W.E., Gerfen, G.J., Peisach, J., and Millhauser, G.L. 2003. Copper coordination in the full-length, recombinant prion protein. Biochemistry. 42: 6794-6803.

Burton, D.R., Pyati, J., Koduri, R., Sharp, S.J., Thornton, G.B., Parren, P.W.H.I., Sawyer, L.S.W., Hendry, R.M., Dunlop, N., Nara, P.L., Lamacchia, M., Garratty, E., Stiehm, E.R., Bryson, Y.J., Cao, Y., Moore, J.P., Ho, D.D., and Barbas, C.F. 1994. Efficient neutralization of primary isolates of HIV-1 by a recombinant human monoclonal antibody. Science. 266: 1024-1027.

Butler, D.A., Scott, M.R.D., Bockman, J.M., Borchelt, D.R., Taraboulos, A., Hslao, K.K., Kingsbury, D.T., and Prusiner, S.B. 1988. Scrapie-infected murine neuroblastoma cells produce protease-resistant prion proteins. J. Virol. 62: 1558-1564.

Caughey, B. 2001. Interactions between prion protein isoforms: the kiss of death? Trends Biochem. Sci. 26: 235-242.

Caughey, B. and Race, R.E. 1992. Potent inhibition of scrapie-associated PrP accumulation by congo red. J. Neurochem. 59: 768-771.

Caughey, B. and Raymond, G.J. 1993. Sulfated polyanion inhibition of scrapie-associated PrP accumulation in cultured cells. J. Virol. 67: 643-650.

Caughey, B., Raymond, L.D., Raymond, G.J., Maxson, L., Silveira, J., and Baron, G.S. 2003. Inhibition of protease-resistant prion protein accumulation *in vitro* by curcumin. J. Virol. 77: 5499-5502.

Caughey, W.S., Raymond, L.D., Horiuchi, M., and Caughey, B. 1998. Inhibition of protease-resistant prion protein formation by porphyrins and phthalocyanines. Proc. Natl. Acad. Sci. USA. 95: 12117-12122.

Chabry, J., Priola, S.A., Wehrly, K., Nishio, J., Hope, J., and Chesebro, B. 1999. Species-independent inhibition of abnormal prion protein (PrP) formation by a peptide containing a conserved PrP sequence. J. Virol. 73: 6245-6250.

Chuong, C.M., McClain, D.A., Streit, P., and Edelman, G.M. 1982. Neural cell adhesion molecules in rodent brains isolated by monoclonal antibodies with cross-species reactivity. Proc. Natl. Acad. Sci. 79: 4234-4238.

Cohen, F.E., Pan, K.M., Huang, Z., Baldwin, M., Fletterick, R.J., and Prusiner, S.B. 1994. Structural clues to prion replication. Science. 264: 530-531.

Collinge, J. 2001. Prion diseases of humans and animals: their causes and molecular basis. Annu. Rev. Neurosci. 24: 519-550.

Collinge, J., Sidle, K.C.L., Meads, J., Ironside, J., and Hill, A.F. 1996. Molecular analysis of prion strain variation and the aetiology of 'new variant' CJD. Nature. 383: 685-690.

Collins, S.J., Lewis, V., Brazier, M., Hill, A.F., Fletcher, A., and Masters, C.L. 2002. Quinacrine does not prolong survival in a murine Creutzfeldt-Jakob disease model. Ann. Neurol. 52: 503-506.

Demart, S., Fournier, J.G., Creminon, C., Frobert, Y., Lamoury, F., Marce, D., Lasmezas, C., Dormont, D., Grassi, J., and Deslys, J.P. 1999. New insight into abnormal prion protein using monoclonal antibodies. Biochem. Biophys. Res. Commun. 265: 652-657.

Deslys, J.P., Comoy, E., Hawkins, S., Simon, S., Schimmel, H., Wells, G., Grassi, J., and Moynagh, J. 2001. Screening slaughtered cattle for BSE. Nature. 409: 476-478.

Donne, D.G., Viles, J.H., Groth, D., Mehlhorn, I., James, T.L., Cohen, F.E., Prusiner, S.B., Wright, P.E., and Dyson, H.J. 1997. Structure of the recombinant full-length hamster prion protein PrP(29-231): The N-Terminus is highly flexible. Proc. Natl. Acad. Sci. USA. 94: 13452-13457.

Dormont, D. 2002. Prion diseases: pathogenesis and public health concerns. FEBS Lett. 529: 17-21.

Enari, M., Flechsig, E., and Weissmann, C. 2001. Scrapie prion protein accumulation by scrapie-infected neuroblastoma cells abrogated by exposure to a prion protein antibody. Proc. Natl. Acad. Sci. USA. 98: 9295-9299. (Abstract)

Esiri, M.M., Carter, J., and Ironside, J.W. 2000. Prion protein immunoreactivity in brain samples from an unselected autopsy population: findings in 200 consecutive cases. Neuropathol. Appl. Neurobiol. 26: 273-284.

Farquhar, C., Dickinson, A., and Bruce, M. 1999. Prophylactic potential of pentosan polysulphate in transmissible spongiform encephalopathies. Lancet. 353: 117

Fischer, M.B., Roeckl, C., Parizek, P., Schwarz, H.P., and Aguzzi, A. 2000. Binding of disease-associated prion protein to plasminogen. Nature. 408: 479-483.

Gabizon, R. and Prusiner, S.B. 1990. Prion liposomes. Biochem. J. 266: 1-14.

Garnett, A.P. and Viles, J.H. 2003. Copper binding to the octarepeats of the prion protein. Affinity, specificity, folding, and cooperativity: insights from circular dichroism. J. Biol. Chem. 278: 6795-6802.

Ghani, A.C., Ferguson, N.M., Donnelly, C.A., and Anderson, R.M. 2003. Short-term projections for variant Creutzfeldt-Jakob disease onsets. Stat. Methods Med. Res. 12: 191-201.

Gilch, S., Wopfner, F., Renner-Muller, I., Kremmer, E., Bauer, C., Wolf, E., Brem, G., Groschup, M.H., and Schatzl, H.M. 2003. Polyclonal anti-PrP auto-antibodies induced with dimeric PrP interfere efficiently with PrPSc propagation in prion-infected cells. J. Biol. Chem. 278: 18524-18531.

Gonzalez, L., Martin, S., and Jeffrey, M. 2003. Distinct profiles of PrP(d) immunoreactivity in the brain of scrapie- and BSE-infected sheep: implications for differential cell targeting and PrP processing. J. Gen. Virol. 84: 1339-1350.

Grassi, J., Comoy, E., Simon, S., Creminon, C., Frobert, Y., Trapmann, S., Schimmel, H., Hawkins, S.A., Moynagh, J., Deslys, J.P., and Wells, G.A. 2001. Rapid test for the preclinical postmortem diagnosis of BSE in central nervous system tissue. Vet. Rec. 149: 577-582.

Grassi, J., Creminon, C., Frobert, Y., Fretier, P., Turbica, I., Rezaei, H., Hunsmann, G., Comoy, E., and Deslys, J.P. 2000. Specific determination of the proteinase K-resistant form of the protein using two-site immunometric assays. Application to the post-mortem diagnosis of BSE. Arch. Virol. Suppl. 197-205.

Harris, D.A. 2001. Biosynthesis and cellular processing of the prion protein. Adv. Protein Chem. 57: 203-228.

Heppner, F.L., Musahl, C., Arrighi, I., Klein, M.A., Rulicke, T., Oesch, B., Zinkernagel, R.M., Kalinke, U., and Aguzzi, A. 2001. Prevention of scrapie pathogenesis by transgenic expression of anti-prion protein antibodies. Science. 294: 178-182.

Hill, A.F., Desbruslais, M., Joiner, S., Sidle, K.C., Gowland, I., Collinge, J., Doey, L.J., and Lantos, P. 1997. The same prion strain causes vCJD and BSE. Nature. 389: 448-450.

Hilton, D.A., Ghani, A.C., Conyers, L., Edwards, P., McCardle, L., Penney, M., Ritchie, D., and Ironside, J.W. 2002. Accumulation of prion protein in tonsil and appendix: review of tissue samples. British Med J. 325: 633-634.

Houston, F., Foster, J.D., Chong, A., Hunter, N., and Bostock, C.J. 2000. Transmission of BSE by blood transfusion in sheep. Lancet. 356: 999-1000.

Hunter, N., Foster, J., Chong, A., McCutcheon, S., Parnham, D., Eaton, S., MacKenzie, C., and Houston, F. 2002. Transmission of prion diseases by blood transfusion. J. Gen. Virol. 83: 2897-2905.

Ironside, J.W. and Head, M.W. 2003. Variant Creutzfeldt-Jakob disease and its transmission by blood. J. Thromb. Haemost. 1: 1479-1486.

Ivanova, L., Barmada, S., Kummer, T., and Harris, D.A. 2001. Mutant prion proteins are partially retained in the endoplasmic reticulum. J. Biol. Chem. 276: 42409-42421.

Jeffrey, M., McGovern, G., Goodsir, C.M., Brown, K.L., and Bruce, M.E. 2000. Sites of prion protein accumulation in scrapie-infected mouse spleen revealed by immuno-electron microscopy. J. Pathol. 191: 323-332.

Kaneko, K., Zulianello, L., Scott, M., Cooper, C.M., Wallace, A.C., James, T.L., Cohen, F.E., and Prusiner, S.B. 1998. Evidence for protein X binding to a discontinuous epitope on the cellular prion protein during scrapie prion propagation. Proc. Natl. Acad. Sci. USA. 94: 10069-10074.

Kanu, N., Imokawa, Y., Drechsel, D.N., Williamson, R.A., Birkett, C.R., Bostock, C.J., and Brockes, J.P. 2002. Transfer of scrapie prion infectivity by cell contact in culture. Curr. Biol. 12: 523-530.

Kanyo, Z.F., Pan, K.M., Williamson, R.A., Burton, D.R., Prusiner, S.B., Fletterick, R.J., and Cohen, F.E. 1999. Antibody binding defines a structure for an epitope that participates in the PrP(C)-->PrP (Sc) Conformational Change. J. Mol. Biol. 293: 855-863.

Kascsak, R.J., Rubenstein, R., Merz, P.A., Tonna-DeMasi, M., Fersko, R., Carp, R.I., Wisniewski, H.M., and Diringer, H. 1987. Mouse polyclonal and monoclonal antibody to scrapie-associated fibril proteins. J. Virol. 61: 3688

Keshet, G.I., Bar-Peled, O., Yaffe, D., Nudel, U., and Gabizon, R. 2000. The cellular prion protein colocalizes with the dystroglycan complex in the brain. J. Neurochem. 75: 1889-1897.

Knight, R. and Stewart, G. 1998. The new variant form of Creutzfeldt-Jakob disease. FEMS Immunol. Med. Microbiol. 21: 97-100.

Korth, C., Stierli, B., Streit, P., Moser, M., Schaller, O., Fischer, R., Schulz-Schaeffer, W., Kretzschmar, H., Raeber, A., Braun, U., Ehrensperger, F., Hornemann, S., Glockshuber, R., Riek, R., Billeter, M., Wuthrich, K., and Oesch, B. Prion (PrP^{Sc})-specific epitope defined by a monoclonal antibody. 1997 Nature. 390: 74-77.

Korth, C., May, B.C., Cohen, F.E., and Prusiner, S.B. 2001. Acridine and phenothiazine derivatives as pharmacotherapeutics for prion disease. Proc. Natl. Acad. Sci. USA. 98: 9836-9841.

Krasemann, S., Groschup, M.H., Harmeyer, S., Hunsmann, G., and Bodemer, W. 1996. Generation of Monoclonal Antibodies against Human prion proteins in PrP$^{o/o}$ Mice. Mol. Med. 2: 725-734.

Laffling, A.J., Baird, A., Birkett, C.R., and John, H.A. 2001. A monoclonal antibody that enables specific immunohistological detection of prion protein in bovine spongiform encephalopathy cases. Neurosci. Lett. 300: 99-102.

Leclerc, E., Peretz, D., Ball, H., Sakurai, H., Legname, G., Serban, A., Prusiner, S.B., Burton, D.R., and Williamson, R.A. 2001. Immobilized prion protein undergoes spontaneous rearrangement to a conformation having features in common with the infectious form. EMBO J. 20: 1547-1554.

Leclerc, E., Peretz, D., Ball, H., Solforosi, L., Legname, G., Safar, J., Serban, A., Prusiner, S.B., Burton, D.R., and Williamson, R.A. 2003. Conformation of PrP^c on the cell-surface as probed by antibodies. J. Mol. Biol. 326: 475-483.

Li, R., Liu, T., Wong, B.S., Pan, T., Morillas, M., Swietnicki, W., O'Rourke, K., Gambetti, P., Surewicz, W.K., and Sy, M.S. 2000. Identification of an Epitope in the C terminus of normal prion protein whose expression is modulated by binding events in the N terminus. J. Mol. Biol. 301: 567-573.

Liu, T., Li, R., Wong, B.S., Liu, D., Pan, T., Petersen, R.B., Gambetti, P., and Sy, M.S. 2001a. Normal cellular prior protein is preferentially expressed on subpopulations of murine hemopoietic cells. J. Immunol. 166: 3733-3742.

Liu, T., Zwingman, T., Li, R., Pan, T., Wong, B.S., Petersen, R.B., Gambetti, P., Herrup, K., and Sy, M.S. 2001b. Differential expression of cellular prion protein in mouse brain as detected with multiple anti-PrP monoclonal antibodies. Brain Res. 896: 118-129.

May, B.C., Fafarman, A.T., Hong, S.B., Rogers, M., Deady, L.W., Prusiner, S.B., and Cohen, F.E. 2003. Potent inhibition of scrapie prion replication in cultured cells by bis-acridines. Proc. Natl. Acad. Sci. USA. 100: 3416-3421.

McLane, K.E., Burton, D.R., and Ghazal, P. 1995. Transplantation of a 17-amino acid alpha-helical DNA-binding domain into an antibody molecule confers sequence-dependent DNA recognition. Proc. Natl. Acad. Sci. USA. 92: 5214-5218.

Miller, M.W., Williams, E.S., McCarty, C.W., Spraker, T.R., Kreeger, T.J., Larsen, C.T., and Thorne, E.T. 2000. Epidemiology of chronic wasting disease in free-ranging cervids. J. Wildl. Dis. 36: 676-690.

Mironov, A.,Jr., Latawiec, D., Wille, H., Bouzamondo-Bernstein, E., Legname, G., Williamson, R.A., Burton, D.R., DeArmond, S.J., Prusiner, S.B., and Peters, P.J. 2003. Cytosolic prion protein in neurons. J. Neurosci. 23: 7183-7193.

Montrasio, F., Frigg, R., Glatzel, M., Klein, M.A., Mackay, F., Aguzzi, A., and Weissmann, C. 2000. Impaired prion replication in spleens of mice lacking functional follicular dendritic cells. Science. 288: 1257-1259.

Moya, K.L., Sales, N., Hassig, R., Creminon, C., Grassi, J., and Di Giamberardino, L. 2000. Immunolocalization of the cellular prion protein in normal brain. Microsc. Res. Tech. 50: 58-65.

Ollmann Saphire, E., Parren, P.W.H.I., Pantophlet, R., Zwick, M.B., Morris, G.M., Rudd, P.M., Dwek, R.A., Stanfield, R.L., Burton, D.R., and Wilson, I.A. 2001. Crystal structure of a neutralizing human IgG against HIV-1: A template for vaccine design. Science. 293: 1155-1159.

Pan, K.M., Baldwin, M., Nguyen, J., Gasset, M., Serban, A., Groth, D., Mehlhorn, I., Huang, Z., Fletterick, R.J., Cohen, F.E., and Prusiner, S.B. 1993. Conversion of α-helices into β-sheets features in the formation of the scrapie prion protein. Proc. Natl. Acad. Sci. USA. 90: 10926-10966.

Paramithiotis, E., Pinard, M., Lawton, T., LaBoissiere, S., Leathers, V.L., Zou, W.Q., Estey, L.A., Lamontagne, J., Lehto, M.T., Kondejewski, L.H., Francoeur, G.P., Papadopoulos, M., Haghighat, A., Spatz, S.J., Head, M., Will, R., Ironside, J., O'Rourke, K., Tonelli, Q., Ledebur, H.C., Chakrabartty, A., and Cashman, N.R. 2003. A prion protein epitope selective for the pathologically misfolded conformation. Nat. Med. 9: 893-899.

Peretz, D., Scott, M.R., Groth, D., Williamson, R.A., Burton, D.R., Cohen, F.E., and Prusiner, S.B. 2001a. Strain-specified relative conformational stability of the scrapie prion protein. Protein Sci. 10: 854-863.

Peretz, D., Williamson, R.A., Kaneko, K., Vergara, J., Leclerc, E., Schmitt-Ulms, G., Mehlhorn, I.R., Legname, G., Wormald, M.R., Rudd, P.M., Dwek, R.A., Burton, D.R., and Prusiner, S.B. 2001b. Antibodies inhibit prion propagation and clear cell cultures of prion infectivity. Nature. 412: 739-743.

Peretz, D., Williamson, R.A., Legname, G., Matsunaga, Y., Vergara, J., Burton, D.R., DeArmond, S.J., Prusiner, S.B., and Scott, M.R. 2002. A change in the conformation of prions accompanies the emergence of a new prion strain. Neuron. 34: 921-932.

Peretz, D., Williamson, R.A., Matsunaga, Y., Serban, H., Pinilla, C., Bastidas, R., Rozenshteyn, R., Papahadjopoulos, D.P., James, T.J., Houghten, R.A., Cohen, F.E., Prusiner, S.B., and Burton, D.R. 1997. A conformational transition at the N-terminus of the prion protein features in formation of the scrapie isoform. J. Mol. Biol. 273: 614-622.

Perrier, V., Wallace, A.C., Kaneko, K., Safar, J., Prusiner, S.B., and Cohen, F.E. 2000. Mimicking dominant negative inhibition of prion replication through structure-based drug design. Proc. Natl. Acad. Sci. USA. 97: 6073-6078.

Peters, P.J., Mironov, A.,Jr., Peretz, D., Van Donselaar, E., Leclerc, E., Erpel, S., DeArmond, S.J., Burton, D.R., Williamson, R.A., Vey, M., and Prusiner, S.B. 2003. Trafficking of prion proteins through a caveolae-mediated endosomal pathway. J. Cell. Biol. 162: 703-717.

Petersen, G., Song, D., Hugle-Dorr, B., Oldenburg, I., and Bautz, E.K. 1995. Mapping of linear epitopes recognized by monoclonal antibodies with gene-fragment phage display libraries. Mol. Gen. Genet. 249: 425-431.

Pocchiari, M., Schmittinger, S., and Masullo, C. 1987. Amphotericin B delays the incubation period of scrapie in intracerebrally inoculated hamsters. J. Gen. Virol. 68: 219-223.

Priola, S.A., Raines, A., and Caughey, W.S. 2000. Porphyrin and phthalocyanine antiscrapie compounds. Science. 287: 1503-1506.

Prusiner, S.B. 1998a. The prion diseases. Brain Pathol. 8: 499-513.

Prusiner, S.B. 1998b. Prions. Proc. Natl. Acad. Sci. USA. 95: 13363-13383.

Prusiner, S.B., Groth, D., Serban, A., Koehler, R., Foster, D., Torchia, M., Burton, D.R., Yang, S.L., and DeArmond, S.J. 1993. Ablation of the prion protein (PrP) gene in mice prevents scrapie and facilitates production of anti-PrP antibodies. Proc. Natl. Acad. Sci. USA. 90: 10608-10612.

Race, R. and Chesebro, B. 1998. Scrapie infectivity found in resistant species. Nature. 392: 770

Race, R., Meade-White, K., Raines, A., Raymond, G.J., Caughey, B., and Chesebro, B. 2002. Subclinical scrapie infection in a resistant species: persistence, replication, and adaptation of infectivity during four passages. J. Infect. Dis. 186 (Suppl 2): S166-S170

Race, R., Raines, A., Raymond, G.J., Caughey, B., and Chesebro, B. 2001. Long-term subclinical carrier state precedes scrapie replication and adaptation in a resistant species: Analogies to bovine spongiform encephalopathy and variant creutzfeldt-jakob disease in humans. J. Virol. 75: 10106-10112.

Raymond, G.J., Bossers, A., Raymond, L.D., O'Rourke, K.I., McHolland, L.E., Bryant, P.K.3., Miller, M.W., Williams, E.S., Smits, M., and Caughey, B. 2000. Evidence of a molecular barrier limiting susceptibility of humans, cattle and sheep to chronic wasting disease. EMBO J. 19: 4425-4430.

Riek, R., Hornemann, S., Wider, G., Billeter, M., Glockshuber, R., and Wuthrich, K. 1996. NMR structure of the mouse prion protein domain PrP (121-231). Nature. 382: 180-182.

Rubenstein, R., Merz, P.A., Kascsak, R.J., Carp, R.I., Scalici, C.L., Fama, C.L., and Wisniewski, H.M. 1987. Detection of scrapie-associated fibrils (SAF) and SAF proteins form scrapie-affected sheep. J. Infect. Dis. 156: 36-42.

Saborio, G.P., Permanne, B., and Soto, C. 2001. Sensitive detection of pathological prion protein by cyclic amplification of protein misfolding. Nature. 411: 810-813.

Safar, J., Wille, H., Itri, V., Groth, D., Serban, H., Torchia, M., Cohen, F.E., and Prusiner, S.B. 1998. Eight prion strains have PrP^{Sc} molecules with different conformations. Nature Medicine. 4: 1157-1165.

Safar, J.G., Scott, M., Monaghan, J., Deering, C., Didorenko, S., Vergara, J., Ball, H., Legname, G., Leclerc, E., Solforosi, L., Serban, H., Groth, D., Burton, D.R., Prusiner, S.B., and Williamson, R.A. 2002. Measuring prions causing bovine spongiform encephalopathy or chronic wasting disease by immunoassays and transgenic mice. Nat. Biotechnol. 20: 1147-1150.

Sethi, S., Lipford, G., Wagner, H., and Kretzschmar, H. 2002. Postexposure prophylaxis against prion disease with a stimulator of innate immunity. Lancet. 360: 229-230.

Shaked, G.M., Shaked, Y., Kariv-Inbal, Z., Halimi, M., Avraham, I., and Gabizon, R. 2001. A protease-resistant prion protein isoform is present in urine of animals and humans affected with prion diseases. J. Biol. Chem. 276: 31479-31482.

Shaked. Y., Engelstein, R., and Gabizon, R. (2002). The binding of prion proteins to serum components is affected by detergent extraction conditions. J. Neurochem. (Abstract) 82: 1-5.

Sigurdson, C.J., Barillas-Mury, C., Miller, M.W., Oesch, B., van Keulen, L.J., Langeveld, J.P., and Hoover, E.A. 2002. PrP^{CWD} lymphoid cell targets in early and advanced chronic wasting disease of mule deer. J. Gen. Virol. 83: 2617-2628.

Soto, C., Kascsak, R.J., Saborío, G.P., Aucouturier, P., Wisniewski, T., Prelli, F., Kascsak, R., Mendez, E., Harris, D.A., Ironside, J., Tagliavini, F., Carp, R.I., and Frangione, B. 2000. Reversion of prion protein conformational changes by synthetic β-sheet breaker peptides. Lancet. 355: 192-197.

Souan, L., Margalit, R., Brenner, O., Cohen, I.R., and Mor, F. 2001. Self prion protein peptides are immunogenic in Lewis rats. J. Autoimmun. 17: 303-310.

Spraker, T.R., Zink, R.R., Cummings, B.A., Sigurdson, C.J., Miller, M.W., and O'Rourke, K.I. 2002. Distribution of protease-resistant prion protein and spongiform encephalopathy in free-ranging mule deer (*Odocoileus hemionus*) with chronic wasting disease. Vet. Pathol. 39: 546-556.

Supattapone, S., Nguyen, H.O., Cohen, F.E., Prusiner, S.B., and Scott, M.R. 1999. Elimination of prions by branched polyamines and implications for therapeutics. Proc. Natl. Acad. Sci. USA. 96: 14529-14534.

Supattapone, S., Wille, H., Uyechi, L., Safar, J., Tremblay, P., Szoka, F.C., Cohen, F.E., Prusiner, S.B., and Scott, M.R. 2001. Branched polyamines cure prion-infected neuroblastoma cells. J. Virol. 75: 3453-3461.

Tagliavini, F., McArthur, R.A., Canciani, B., Giaccone, G., Porro, M., Bugiani, M., Lievens, P.M.J., Bugiani, O., Peri, E., Dall'Ara, P., Rocchi, M., Poli, G., Forloni, G., Bandiera, T., Varasi, M., Suarato, A., Cassutti, P., Cervini, M.A., Lansen, J., Salmona, M., and Post, C. 1997. Effectiveness of anthracycline against experimental prion disease in Syrian hamsters. Science. 276: 1119-1122.

Takahashi, K., Shinagawa, M., Doi, S., Sasaki, S., Goto, H., and Sato, G. 1986. Purification of crapie agent from infected animal brains and raising of antibodies to the purified fraction. Mol. Immunol. 30: 123-131.

Tal, Y., Souan, L., Cohen, I.R., Meiner, Z., Taraboulos, A., and Mor, F. 2003. Complete Freund's adjuvant immunization prolongs survival in experimental prion disease in mice. J. Neurosci. Res. 71: 286-90.

Wadsworth, J.D.F., Joiner, S., Hill, A.F., Campbell, T.A., Desbruslais, M., Luthert, P.J., and Collinge, J. 2001. Tissue distribution of protease resistant prion protein in variant creutzfeldt-jakob disease using highly sensitive immunoblotting assay. Lancet. 358: 171-180.

Weissmann, C., Enari, M., Klohn, P.C., Rossi, D., and Flechsig, E. 2002. Transmission of prions. J. Infect. Dis. 186 (Suppl 2): S157-S165

White, A.R., Enever, P., Tayebi, M., Mushens, R., Linehan, J., Brandner, S., Anstee, D., Collinge, J., and Hawke, S. 2003. Monoclonal antibodies inhibit prion replication and delay the development of prion disease. Nature. 422: 80-83.

Wiley, C.A., Burrola, P.G., Buchmeier, M.J., Wooddell, M.K., Barry, R.A., Prusiner, S.B., and Lampert, P.W. 1987. Immuno-gold localization of prion filaments in scrapie-infected hamster brains. Lab. Invest. 57: 646-656.

Williamson, R.A., Peretz, D., Pinilla, C., Ball, H., Bastidas, R.B., Rozenshteyn, R., Houghten, R.A., Prusiner, S.B., and Burton, D.R. 1998. Mapping the prion protein using recombinant antibodies. J. Virol. 72: 9413-9418.

Williamson, R.A., Peretz, D., Smorodinsky, N., Bastidas, R., Serban, A., Mehlhorn, I., DeArmond, S., Prusiner, S.B., and Burton, D.R. 1996. Circumventing tolerance in order to generate autologous monoclonal antibodies to the prion protein. Proc. Natl. Acad. Sci. USA. 93: 7279-7282.

Wisniewski, T., Brown, D.R., and Sigurdsson, E.M. 2002. Therapeutics in alzheimer's and prion diseases. Biochem. Soc. Trans. 30: 574-578.

Wüthrich, K. and Riek, R. 2001. Three-dimensional structures of prion proteins. Adv. Protein Chem. 57: 55-82.

Zhang, H., Stockel, J., Melhorn, I., Groth, D., Baldwin, M.A., Prusiner, S.B., James, T.L., and Cohen, F.E. 1997. Physical studies of conformational plasticity in a recombinant prion protein. Biochemistry. 36: 3543-3553.

Zulianello, L., Kaneko, K., Scott, M., Erpel, S., Han, D., Cohen, F.E., and Prusiner, S.B. 2000. Dominant-negative inhibition of prion formation diminished by deletion mutagenesis of the prion protein. J. Virol. 74: 4351-4360.

From: Prions and Prion Diseases: Current Perspectives
Edited by: Glenn C. Telling

Chapter 9

Prnd and the Doppel Protein

David Westaway

Abstract

Doppel (Dpl) is a PrP-related protein encoded by the *Prnd* gene downstream of *Prnp*. Like PrP^C, Dpl is a GPI-anchored protein with two N-glycosylation sites and a globular C-terminus containing three α-helices and two short β-strands, and both proteins bind copper ions *in vitro*. Dpl is distinguished from PrP^C by a shorter N-terminal region (lacking both octarepeats and an optional transmembrane region), a kink in helix B, and a different location and stoichiometry of Cu binding site. *In vivo*, Dpl is apparently unable to be converted to a PrP^{Sc}-like form. Also, whereas mouse PrP^C is expressed abundantly in the CNS and is present in several peripheral tissues, Dpl expression in adult is restricted almost exclusively to the testis. When expressed in the CNS, however, Dpl produces neurodegeneration with similarities to phenotypes produced by some mutant forms of PrP^C, and furthermore, the neurotoxic effects of Dpl are blocked by expression of wild type PrP^C. These findings, as well as an unexpected parallel between Familial Creutzfeldt-Jakob disease mutations and conserved residues in Dpl suggest cross-talk between Dpl and PrP^C in certain physiological and pathogenic situations. Defining the basis for the overlapping and complementary functions of Dpl and PrP isoforms comprises an important challenge for future research.

1. Introduction

1.1. Discovery of the Prion Protein Gene, *Prnp*

Prior to 1985, the genetic origin of the infectious isoform of the prion protein was unknown: indeed, some had speculated that PrP would prove to be encoded by the genomic nucleic acid of hypothetical "slow virus". Using technologies championed in the laboratory of Leroy Hood, elucidation of the N-terminal sequence of highly infectious preparations of prions containing PrP27-30 allowed the synthesis of degenerate oligonucleotide probes and the subsequent identification of a cDNA clone that encoded all but the first 10 amino acids of the mature prion protein (Oesch *et al.*, 1985). This cDNA clone was then used as a hybridization probe to interrogate denatured preparations of purified infectious prions and genomic DNA and total RNA preparations isolated from the brains of healthy and prion-infected hamsters (Oesch *et al.*, 1985). These analyses established that, whereas DNA or RNA molecules encoding prion proteins could not be detected in denatured prion preparations, prion gene sequences were present in the genomic DNA of hamsters. This chromosomal gene, cloned and mapped the following year, is transcribed in both healthy and infected hamsters to generate a 2.1 kb PrP mRNA. Subsequent work identified the translation product of this mRNA in healthy animals as the protease-sensitive alpha-helical glycoprotein PrP^C, setting the stage in turn for conformational hypotheses of prion replication.

1.2. The Prion Gene Complex (*Prn*)

Although discerning the exact function of PrP^C has proven challenging, it is abundantly clear that prion protein genes (designated "*Prnp*") are ubiquitous in mammals. Analysis of genomic clones and hybridization analyses to genomic DNA failed to provide any evidence for additional prion protein genes. Thus, on one hand, from a practical point of view, PrP has long been discerned as a single-copy gene. On the other hand, linkage analysis of segregating crosses between mouse inbred strains (NZW and I/LnJ) with differing scrapie incubation times was at one time used to define a prion incubation time gene, *Prn-i*, and, using flanking region polymorphisms, *Prn-i* was deduced to lie closely adjacent to the prion protein structural gene (Carlson *et al.*, 1986). These data therefore suggested the potential existence of a prion gene complex (*Prn*), encoding proteins (other than PrP^C) also fulfilling a role in prion replication. To distinguish between these contrasting views *Prnp* molecular clones were obtained from NZW and I/LnJ mice. Nucleotide sequencing defined the existence of two allelic coding region polymorphisms (L108F, T189V, found in *cis*) within a constellation of flanking region polymorphisms defining a *Prnp*^{b} haplotype (Carlson *et al.*, 1988) distinguishing inbred mouse strains, and suggesting that *Prn-i* was actually *Prnp* itself (Westaway *et al.*, 1987). This inference was eventually established by producing altered scrapie disease incubation times in "knock-in" mice expressing the 108F, 189V *Prnp* "b" allele (*Prnp*^{b}) (Moore *et al.*, 1998). By defining congruence between *Prnp* and *Prn-i*, these experiments appeared to dispel the notion of a prion gene complex. In fact, this situation proved fleeting, with large-scale sequencing and bioinformatic projects (described below) revealing a gene called *Prnd*, lying downstream of *Prnp*. Although having no clear relationship to the determination of incubation periods for experimental scrapie, this novel gene encodes a protein with intriguing properties and comprises the subject of this chapter (Moore *et al.*, 1999). Very recently database mining has revealed a possible third member of the *Prn* gene complex, *Prnt*, lying 3 kb 3' to *Prnd*. (Makrinou *et al.*, 2002). Since information concerning the nascent *Prnt* gene and a putative 94 amino acid translation product is still scant, this gene will not be considered further here.

The *Prnd* gene was discovered during the course of sequencing a genomic DNA cosmid clone isolated from the aforementioned I/LnJ inbred strain of mice (Lee *et al.*, 1998; Moore *et al.*, 1999; Westaway *et al.*, 1991). This DNA sequencing project was undertaken with the express purpose of finding genes adjacent to *Prnp*, since transgenic mice created with the I/LnJ-4 cosmid did not behave not in accord with expectations for the dominant scrapie incubation time allele present in the I/LnJ strain (Westaway *et al.*, 1991). Although subsequent studies provided an explanation for the paradoxical scrapie incubation time behavior of transgenic mice harboring the I/LnJ-4 cosmid (sensitivity of prion replication to PrP^C expression levels outweighing effects of mis-matched allelic type (Carlson *et al.*, 1994)) this cosmid clone was of interest as it extended further 3' than the extant human, hamster and sheep *Prnp* cosmid clones. When sequenced, the 3' flanking of this clone revealed an open reading frame (ORF) predicted to encode a protein related to PrP itself. Subsequent work has established that this ORF encompasses N- and C-terminal signal peptides and encodes a cell-surface protein called Doppel ("Doppel" is corruption of the acronym derived from <u>do</u>wnstream <u>p</u>rion <u>p</u>rotein-<u>l</u>ike gene.

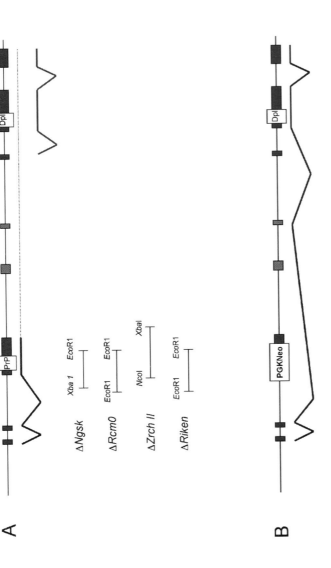

Figure 1. Structure and transcription of the mouse *Prn* complex: effect of *Prnp* deletions. Panel A shows the wt mouse *Prn* gene complex with the protein coding exons ("PrP", "Dpl") as thick boxes and non-coding exons as thinner boxes. Intergenic exons are cross-hatched. Spliced mRNAs are shown below the genes, with a dotted line extending beyond the *Prnp* mRNA indicating "leaky" transcriptional termination that may be germane to the genesis of intergenic splicing. Endpoints of the deletions (defined by restriction endonuclease sites) used to create four $Prnp^{0/0}$ alleles associated with cerebellar ataxia are shown below the panel. Panel B depicts the structure of a mutant *Prnp* gene complex containing a *Prnp* knock-out allele generated by interposition of a selectable marker "PGKNeo". Note that the *Prnp* coding exon (exon 3) splice acceptor is absent from the targeted allele. A chimeric mRNA resulting from intergenic splicing is depicted below the targeted allele. Modified from Mastrangelo and Westaway (2001).

2. Physiology and Biochemistry of the Doppel Protein

2.1. Origin and Properties of the *Prnd* Gene

The Dpl coding sequence is conserved in different strains of mice, humans (Moore *et al.*, 1999) sheep and cows (Tranulis *et al.*, 2001) and, apparently from database entries of partial coding sequences, in a number of other mammals including elephants and whales. The mouse doppel gene *Prnd* lies ~16 kb downstream of the *Prnp* coding region (Figure 1A), whereas the human gene lies 27 kb (or perhaps closer (Makrinou *et al.*, 2002)) downstream of *PRNP* (Moore *et al.*, 1999). The bovine Dpl gene lies 16 kb downstream of *Prnp* (Comincini *et al.*, 2001), with the analogous distance for the ovine gene reported as 20 or 52 kb (Comincini *et al.*, 2001; Essalmani *et al.*, 2002). Thus the prion gene complex, *Prn*, comprises *Prnp* and *Prnd*, and presumably arose by duplication of a "proto" prion protein gene (and possibly even triplication followed by a gene inversion, depending the outcome of future studies on the putative *Prnt* gene (Makrinou *et al.*, 2002). Insofar as no nucleic acid homology remains between the *Prnp* and *Prnd* genes, and the respective proteins only exhibit 24% sequence identity, we may infer that duplication or triplication from a proto-prion gene was an ancient event. Indeed, presence of a *Prn* gene complex in mice, humans, rats, cows and sheep suggests this gene complex may have existed prior to the speciation of mammals.

Domain maps of mammalian (mouse) PrP and Dpl are presented in Figure 2A. PrP genes encoding proteins with a basic residues at the N-terminus, conserved central hydrophobic regions encoding the amino acids Met/ValAlaGlyAlaAlaAlaAlaGlyAla, and S-S linked α-helical C-terminal domains have been isolated from diverse species including turtles, birds and the Africa clawed toad, *Xenopus laevis* (Gabriel *et al.*, 1992; Simonic *et al.*, 2000; Strumbo *et al.*, 2001). None of these genes are predicted to encode proteins with two disulphide bonds and are thus more PrP-like. A cDNA isolated recently from the pufferfish *Fugu rubipes* encodes a protein with the first two of these features, but not the third (Suzuki *et al.*, 2002). Domain maps of PrP-like genes from amphibians and fish are presented in Figure 2B.

A

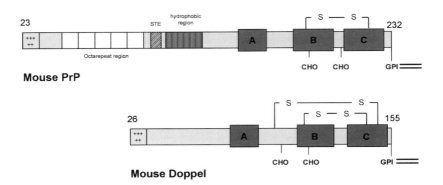

B

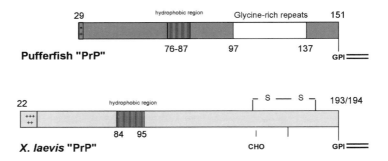

Figure 2. Structure of cellular prion proteins *A*. Schematic of modern-day Dpl and PrP exemplified by analysis of mouse molecular clones and proteins. This panel shows the positions of the predicted features of mouse Dpl and its relationship to experimentally demonstrated features of mammalian PrP, including posttranslational modifications and secondary structure (modified from Moore *et al.*, 1999 (Moore *et al.*, 1999). Three α-helices found in PrP and Dpl are shown as gray boxes A, B, and C (Mo *et al.*, 2001). The residue numbers for the predicted disulfides and Asn-linked glycosylation sites (CHO) are also shown. The box with vertical stripes shows the transmembrane region containing the highly conserved sequence motif AlaGlyAlaAlaAlaAlaGlyAla found in all PrPs, and the stop transfer effector region (STE) is represented by a box filled with diagonal stripes. The numbering of both sequences is that of the mouse. The + symbols indicate a cluster of basic residues as per panel A. Copper binding regions are bracketed. *B*. Structure of PrP-like proteins inferred from the sequence analysis of Pufferfish and *X. laevis* genomic DNA and cDNA clones. Alpha helical regions of these proteins have not yet been determined by experimental analysis. A hydrophobic region is denoted by a striped box and the "+" symbols indicate a cluster of basic residues.

2.2. *Prnd* mRNA and Protein Expression

In mice *Prnd* is interrupted by introns, and is transcribed into two major mRNA species (Moore *et al.*, 1999) (Figure 1A). Although expressed in embryos, *Prnd* is expressed in a more restricted manner in adult mice than the broad pattern observed for *Prnp* (Makrinou *et al.*, 2002; Moore *et al.*, 1999). By northern and western blot analysis, Dpl mRNA and protein are found abundantly in adult testis, at lower levels in heart, but are effectively absent from the adult CNS using standard methods of northern and western blot analysis (Moore *et al.*, 1999; Silverman *et al.*, 2000). Whether Dpl expression can be induced under conditions of cellular stress remains unknown.

Recent studies have focused upon the types of cells in the testis that express Dpl protein. We have used the antibody E6977 (Moore *et al.*, 2001) directed against recombinant doppel protein to perform immunohistochemistry of seminiferous tubules. Exploiting *Prnd* knockout mice and pre-adsorption with rDpl to delineate and/or block non-specific binding we have localized doppel expression to developing spermatids (P. Mastrangelo, P. Horne, S. Varmuza, R. Moore *et al.*, in preparation). Using the mutagen busulfan to depopulate seminiferous tubules of germ cells we infer that expression levels in Sertoli cells (support cells needed for spermatid maturation) are much lower than in developing spermatids. However, other groups have reported Dpl expression in Sertoli cells of humans, using testis from a Sertoli-deficient syndrome as a point of reference (Peoc'h *et al.*, 2002). Immunostaining on the tails of mature human sperm was also described, thereby indicating differences in the cellular distribution of Dpl between humans and mice.

2.3. Biochemical Properties of the Dpl Protein

Sequence inspection reveals Dpl as an N-terminally truncated version of PrP^C, lacking hexarepeat motifs, octarepeat motifs, and the stop-transfer-effector (STE) and transmembrane regions lying in the center of the molecule. In contrast to this marked difference at the front end of the molecule, the C-terminal domain is remarkably PrP-like, considering that the proteins have only ~24% sequence identity. Spectroscopic analyses of recombinant Dpl reveals an α-helical signature, virtually superimposable with that of rPrP (Silverman *et al.*, 2000). The disulphide bond of PrP linking residues 178 and 212

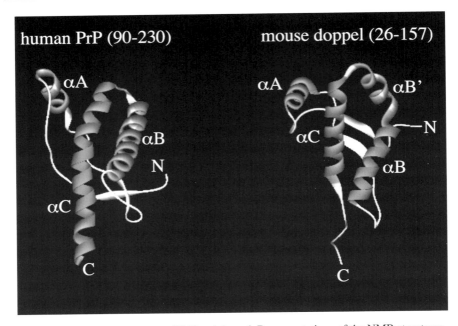

Figure 3. NMR structures of PrP and doppel. Representations of the NMR structures derived from recombinant human PrP 90-230 (left) (Zahn *et al.*, 2000) and mouse Dpl 26-157 (right) (Mo *et al.*, 2001). Alpha-helices (αA, αB or αB' and αC) are shown in red whereas short beta strand motifs are highlighted in turquoise. The assigned human PrP structure corresponds to residues L125 to R228 and the assigned mouse Dpl structure corresponds to residues to R51 to A157. This figure was compiled from the protein database (pdb) files 1QM0 and 1I17 using Web Lab™ Viewer light version 3.2. The protein structures are orientated with the C-termini (site a GPI anchor addition and of membrane attachment) at the bottom. Despite a striking overall similarity, the Dpl structure differs from PrP in the division of helix αB into αB and αB', displacement of the beta strands by one or two residues, and an altered plane of the beta strands versus the axes of helices αB and αC. Modified from Westaway and Carlson, 2002 (Westaway and Carlson, 2002).

has a close cognate in Dpl (residues 109 and 143), though this is augmented by a second disulphide bond lying between residues 95 and 148. Somewhat similar results have been obtained for human recombinant doppel (Lu *et al.*, 2000). These low-resolution structural studies have now been superceded by an NMR structure (Mo *et al.*, 2001). This displays a PrP-like fold of 3 α-helices and two short β-strand motifs, even though many of the individual amino acid residues within analogous portions of the structure are not identical between PrP and Dpl. Also, there are differences in the tertiary structure of PrP and Dpl that may be functionally significant. The angle of helix A versus helices B and C is different, helix B has a marked interruption ("kink") and the loop between helices B and C is shorter and is "naked", lacking an N-glycosylation site (Figure 3).

Studies from many laboratories have shown that the octarepeat region in PrP^C has the property of binding copper in a selective and multivalent manner (Hornshaw, 1995; Stöckel *et al.*, 1998; Viles *et al.*, 1999). These octarepeat motifs lie within unstructured region of the protein and presumably mediate the ability of full-length PrP^C to be purified from cell lysates by immobilized metal affinity chromatography (IMAC), using copper or nickel charged resins (Pan *et al.*, 1992). Since Dpl lacks octarepeat motifs one might have predicted that this protein would not bind to copper ions *in vitro*, and would not be enriched by IMAC chromatography with Cu-charged affinity resin. In fact only one of the predictions has proven to be correct. Exploiting a full-length recombinant version of mouse Dpl and an "internal" synthetic peptide (Dpl 101-145), Qin *et al.* used four techniques to probe metal binding capabilities (Qin *et al.*, 2003). This series of experiments demonstrated that Dpl has one selective Cu binding site with a K_d of less than or equal to 0.5µM, and hence comparable to highest affinity Cu-binding sites within PrP^C. This unique site is located within the α-helical domain of doppel, with the Dpl 101-145 encompassing the αB-loop-αC sub-region exhibiting similar binding properties to Dpl 27-154. Although IMAC chromatography was not used by Qin *et al.*, it is possible that Dpl cannot be purified efficiently by IMAC chromatography (Shaked *et al.*, 2002) because (unlike PrP^C), the binding site is monovalent and is not located in a flexible "arm", but is instead upon one face of a planar, triangular sub-domain defined by the αB/αB'loop-αC helical region.

Like PrP^C, doppel is anchored to the external face of the plasma membrane (PM) by a GPI anchor (Silverman *et al.*, 2000). There is an emerging consensus that PrP^C locates to raft-like sub-domains of the PM (Vey *et al.*, 1996), characterized by their distinct lipid content, and hence detergent solubility properties, but no consensus as to whether these rafts are always associated with marker proteins such as caveolins. The biochemical function performed by PrP^C at this particular cellular locale(s) is, of course, controversial. Insofar as this issue has been reviewed extensively elsewhere we will not revisit it here, but it is of interest the note that Dpl may localize to a yet different sub-domain of the PM, as defined by failure to observe co-immunoprecipitation of the two proteins under detergent conditions where rafts are inferred to remain intact (Shaked *et al.*, 2002). A caveat is that these studies were performed upon testis homogenates and assume that there is an abundant cellular population that expresses both PrP^C and Dpl, such that they could both incorporate into the same

A Human Dpl

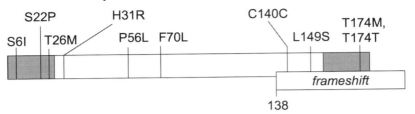

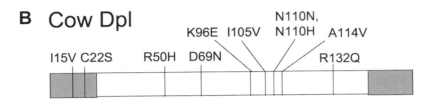

B Cow Dpl

C Sheep Dpl

Figure 4. Polymorphisms in mammalian *PRND* genes. **A.** The human *PRND* open reading frame (ORF) is indicated by a rectangle with N- and C-terminal signal peptides denoted by stippled boxes. Point mutation polymorphisms in the human *PRND* gene are shown above the ORF while a frameshift mutation is shown as an open rectangle below and to the right of the ORF. **B.** Multiple single nucleotide polymorphisms detected in the bovine *Prnd* gene (representation as per panel A. **C.** In contrast to B, missense polymorphisms are absent from the sheep *Prnd* gene alleles examined to date. Data is from references (Comincini *et al.*, 2001; Essalmani *et al.*, 2002; Mead *et al.*, 2000; Peoc'h *et al.*, 2000).

rafts. As discussed below, the situation with respect to co-expression of these two proteins in the male reproductive tract remains to be clarified.

2.4. Genetics and the Physiological Function of Dpl

What are the consequences of inactivation of the Dpl gene? In humans, besides a number of silent or missense polymorphisms (Figure 4A), heterozygosity for a frameshift allele was detected in one patient in a survey of Creutzfeldt Jakob disease (CJD) patients and normal subjects. Whether or not this allele predisposes to familial CJD (F-CJD) in heterozygous form remains to be established as linkage analyses were not presented. Inspection of the translated sequence of the frameshifted protein predicts a protein that it lacks the C-terminal cysteine residues of both disulphide bonds (Figure 4B). We speculate that this protein would be highly unstable, and that the frameshift mutation would therefore comprise a null allele. Further investigation of the frequency of this putative *PRND* null allele in the human population and the consequence of homozygosity will therefore be of interest.

With respect to *Prnd* null alleles in rodents, several laboratories have generated lines of Dpl k/o (*Prnd*$^{0/0}$) mice. In all cases homozygous null animals derived from the mating of heterozygotes are viable, develop normally and have apparently normal neuroanatomy: however, the male *Prnd*$^{0/0}$ mice have markedly reduced fertility. This phenotype stands in contrast to phenotypes seem in *Prnp*$^{0/0}$ mice, which are subtle, disputed, or require exposure to an exogenous stimulus. One possibility is that proto cellular prion proteins evolved first to facilitate spermatogenesis, a role quite different from our typical perception of these proteins from their involvement in neurodegenerative diseases. Alternatively, subsequent to a duplication of a neurally-expressed progenitor gene, the expression tropism of Dpl diverged such that it eventually came to perform an essential function with regards to male reproductive ability.

Although a behavioral deficit could underlie the inability of male *Prnd*$^{0/0}$ mice to yield offspring, given data that the cells expressing Dpl are spermatids, it is more plausible that infertility has a physiological basis and is a direct result of a deficit in male gametes. Studies are underway to pinpoint this lesion and thereby gain insight into Dpl's biochemical attributes. In one instance reproductive disability has been equated with abnormalities in sperm cell ultrastructure, in the ability of sperm to penetrate the zona pellucida of the oocyte, and in limited viability of eggs once fertilized by *Prnd*$^{0/0}$ sperm. However, the precise biochemical defect in the Zürich *Prnd*$^{0/0}$ mice remains to be pinpointed and, furthermore, it is unclear if other lines of *Prnd*$^{0/0}$ mice derived from different gene targeting strategies share the exact same deficit. Possible effects of *Prnd* targeting constructs upon expression of the *Prnt* transcription unit (which, in one human genomic clone is closely juxtaposed with *Prnd*) may have to be considered in this equation. It is also of interest to note that PrP^C expression has been reported at a later stage in gametogenesis, on mature sperm (Shaked *et al.*, 1999). Insofar as *Prnp*$^{0/0}$ males are not sterile, PrP^C's role (being present at a later stage in gametogenesis) may be ancillary and it is unlikely that the two prion proteins collaborate physically to contribute to the production of healthy gametes. Nonetheless, it will be useful to determine whether expression of PrP^C at an earlier stage in gametogenesis, i.e., under the control of the *Prnd* promoter, can rescue the sterility of *Prnd*$^{0/0}$ mice.

3. The Pathobiology of Doppel

3.1. Genetics and the Pathobiology of Dpl

A mooted Dpl null allele has been noted above but what of activating mutations? In the case of the prion protein gene, point mutations cause familial prion diseases and overexpression causes a neuromyopathic syndrome (Westaway *et al.*, 1994). Currently, we do not know whether point mutations in the human Dpl gene cause "doppelopathies" but given the expression tropism of Dpl it is reasonable to assume phenotypes affecting the male reproductive tract.

Like PrP^C, overexpression of wt Dpl is pathogenic. Insights into this effect arose from close scrutiny of a puzzling feature of PrP^C-deficient *Prnp*$^{0/0}$ mice (Moore *et al.*, 1999; 2001; Rossi *et al.*, 2001; Sakaguchi *et al.*, 1996). Seven lines of *Prnp*$^{0/0}$ mice have been constructed, yielding two phenotypes, phenotypically normal in middle age or succumbing to a cerebellar ataxia marked by loss of Purkinje cell neurons (Table 1). The solution to this puzzle arose from the discovery of a class of *Prnd* mRNAs initiated from the *Prnp* promoter (Moore *et al.*, 1999; Westaway *et al.*, 1994). Though these chimeric *Prnp/Prnd* mRNAs generated by intergenic splicing are present at low levels in the CNS of wt mice they are strongly up-regulated in lines of *Prnp*$^{0/0}$ mice exhibiting ataxia (see Figure 1B) (Moore *et al.*, 1999; 2001; Silverman *et al.*, 2000). A simple explanation for the divergent behaviour *Prnp*$^{0/0}$ mice is that gene targeting construct deletions which remove the *Prnp* exon 3 splice acceptor site favor ectopic expression of *Prnd* from the *Prnp* promoter (Figure 1, Table 1). To test this hypothesis, the *Prnd* coding region was placed under the control of the (hamster) *Prnp* promoter using the cos.Tet expression vector (Scott *et al.*, 1992). Subsequent to

Table 1. PrP-ablated mice, ataxia, and Dpl overexpression

Prnp$^{0/0}$ Tg line	Ataxia in middle-age?	Deletion of *Prnp* exon 3 splice acceptor site?	Expression of Dpl in the CNS?	Reference
(Wt mouse)	No	No	No	n.a.
Zrch I	No	No	No	(Büeler *et al.*, 1992)
NPU	No	No	No	(Manson *et al.*, 1994)
Ngsk	Yes	Yes	Yes	(Li *et al.*, 2000; Moore *et al.*, 1999; Sakaguchi *et al.*, 1996)
Rcm0	Yes	Yes	Yes	(Moore *et al.*, 1999)
ZrchII	Yes	Yes	Yes	(Rossi *et al.*, 2001; Weissmann and Aguzzi, 1999)
Riken	yes	yes	Yes (S. Itohara, pers. comm.)	(Yokoyama *et al.*, 2001)

n.a., not applicable

pronuclear microinjections, a variety of Dpl "founder" Tg mice were identified. Though some founders did not breed or died in adolescence, several stable Tg lines were established. Of note, 4 independent Tg lines exhibiting high levels of Dpl CNS expression exhibited an ataxic phenotype, often in early adult life. This was characterized by loss of cerebellar granule and Purkinje cells. Apoptosis was notable in the granule cell population, accompanied by loss of Purkinje cells, and these changes were accompanied by a florid activation of astrocytes (Moore *et al.*, 2001). These data provide compelling evidence that the ataxia seen in Dpl-overexpressing *Prnp*$^{0/0}$ mice is a direct consequence of a toxic effect of this PrP-like protein. This conclusion is also supported by an earlier study from Rossi *et al.* (2001). Here Dpl overexpression was noted in the ZrchII (Zurich II) line of *Prnp*$^{0/0}$ mice with ataxia, and this effect was exacerbated by introduction of a *Prnd*-expressing cosmid clone. However, while the neurotoxicity of Dpl is now established beyond reasonable doubt, the neuropathology associated with Dpl-overexpression is distinct from that attributable to PrP^{C} overexpression or replication of rodent adapted prions introduced by intracerebral inoculation.

3.2. Dpl, Conformational Change, and Prion Replication

If Dpl is a protein with PrP^C-like properties, can it be converted to an alternative, pathogenesis-associated conformer? Can Dpl participate in or modulate prion disease marked by accumulation of PrP^{Sc}, for example in infections such as scrapie or CJD, or in familial prion disease such as F-CJD? Although this possibility has not been exhaustively tested, several observations already argue against this notion. First, Dpl mRNAs are not readily detected in CNS neurons (a preferred site for prion replication) without recourse to sensitive PCR techniques to amplify cDNAs (RT-PCR) (Moore *et al.*, 1999). Furthermore, the NMR structure of mouse Dpl (Mo *et al.*, 2001) suggests structural parameters that may preclude Dpl from undergoing a pathogenic conformational transformation. These are (i) the lack of an apparent equivalent to the conformationally "plastic" region of PrP^C, (ii) two disulphide bonds constraining the alpha-helical domain, whereas PrP^C has only one, (iii) a positively charged residue at the analogue of position 171, which causes a dominant negative effect upon prion replication in sheep PrP^C and in transgenic mice (Kaneko *et al.*, 1997; O'Rourke *et al.*, 1997; Perrier *et al.*, 2002; Westaway *et al.*, 1994) and (iv) the existence of partially unfolded intermediates in urea-induced denaturation studies in the case of PrP, but not in the case of Dpl (Nicholson *et al.*, 2002).

In the genetic realm, although *Prnd* polymorphisms are known (Figure 4), with one possible exception, none have been suggested as being associated with altered susceptibility to CJD (Mead *et al.*, 2000; Peoc'h *et al.*, 2000). More recently, transplantation experiments have been used to demonstrate that Dpl-deficient neural grafts can support prion replication (Behrens *et al.*, 2001). Lastly, modest (non-toxic) overexpression of Dpl in the CNS does not impact incubation times for the Rocky Mountain Laboratory (RML) isolate of experimental scrapie (Moore *et al.*, 2001). The attendant conclusions are that Doppel neither undergoes pathogenic conformational changes, nor modulates pathogenic conformational in PrP^C. However, this view is shaped in part by data deriving from intracerebral inoculation of rodent-adapted prion isolates and it is important to bear in mind that the supply and availability of alpha-helical PrP^C is an important factor in the genesis of PrP^{Sc}, and also that prion diseases exist with familial or sporadic (as well as infectious) etiologies. In this light, interactions between PrP^C and Doppel, and unexpected parallels between expression of Dpl in the CNS and some genetic prion disease bear further comment.

4. Intersections in the Activities of PrP^C and Doppel

4.1. Interactions Between Dpl and PrP^C in Neurons

As noted above, sequential expression of Dpl and PrP^C in the process of gametogenesis may be inferred from preliminary histochemical analyses and may prove interesting at the biochemical and cell biological level. Another intersection in the biology of these two proteins arises from consideration of ataxic *Prnp*$^{0/0}$ mice. Prior to the discovery of the *Prnd* gene, ataxia in *Prnp*$^{0/0}$ lines such as the *"Ngsk"* (Nagasaki) line was originally attributed to a deficiency of PrP^C, that, for reasons unknown, did not manifest itself in the ZrchI line of *Prnp*$^{0/0}$ mice. Accordingly, an attempt was made to "rescue" the ataxic phenotype by reintroduction of a wt PrP transgene. The outcome of this experiment was abrogation of the ataxic phenotype, leading to the erroneous conclusion that ataxia was caused by a lack of PrP^C. Two points in this "paradigm" bear elaboration.

First, the rescue phenomenon is robust. Besides the foregoing experiment, a protective effect of PrP transgenes was noted in hippocampal cell-lines derived from Riken *Prnp*$^{0/0}$ mice bearing a deletion spanning the *Prnp* exon 3 splice acceptor (Kuwahara *et al.*, 1999). More recently, the phenomenon has been validated by crossing a wt hamster PrP transgene to ataxia-prone Tg-Dpl mice (Moore *et al.*, 2001). In some settings Dpl's lethal effect upon neurons is antagonized by wt PrP even when the PrP gene is present at only one copy per diploid genome (Rossi *et al.*, 2001). Second, rescue is not accomplished by a gene regulatory effect as levels of *Prnd* mRNA and Dpl protein are unaffected by PrP transgenes (Moore *et al.*, 1999; 2001). Rather, these data are compatible with a posttranslational mechanism, presumably representing competing biological activities of the two mature proteins. Similar biochemical properties of PrP and Doppel that could underlie "competition" would include a positively charged N-terminal region, synthesis in the secretory pathway and cell-surface presentation, a similarly folded C-terminal domain and a shared predilection for copper ions. Some potential cellular mechanisms are shown in Figure 4.

One possibility for the "sibling rivalry" between Dpl and PrP^C is competitive binding for a common ligand or receptor, with the similarly structured α-helical domains of Dpl and PrP^C comprising a putative binding region (Figure 5, panels A and C). While the identity

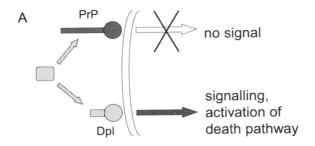

A

PrP

no signal

Dpl

signalling,
activation of
death pathway

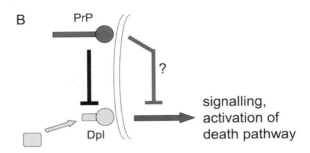

B

PrP

?

Dpl

signalling,
activation of
death pathway

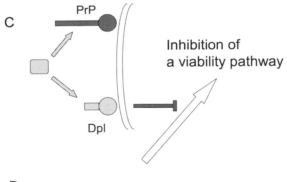

C

PrP

Dpl

Inhibition of
a viability pathway

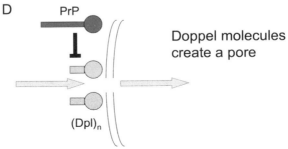

D

PrP

Doppel molecules
create a pore

$(Dpl)_n$

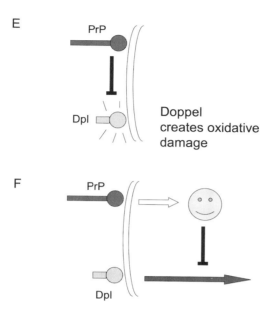

PrP has a non-specific protective effect

Figure 5. Models for neurotoxicity of Doppel. Speculative models are represented in panels A-F. PrP is represented by a turquoise ball (C-terminal alpha-helical domain) and stick (N-terminal unstructured domain. Dpl is represented in a similar fashion but in a lighter color. A) PrP has no direct effect on a death pathway activated by Doppel but competes for a protein or inorganic co-activator of Dpl (rectangle with rounded corners), for example, a ligand or Cu^{2+} ions. B) PrP inhibits a death pathway by physical interaction with Dpl at the cell-surface (black symbol indicating inhibition) or interferes with an intracellular relay in the signaling pathway (grey symbol indicating inhibition. C) As per A but instead of Dpl initiating a "new", inappropriate signaling cascade it undercuts an endogenous pathway needed for cell viability. D) Homomeric assemblies of Dpl ("(Dpl_n") create pores allowing a toxic imbalance of metabolites (flux indicated by yellow arrows. PrP interferes with the formation of the homomeric protein complex resulting in assemblies without pores. E) Dpl creates free radicals that cause oxidative damage (dashes) to proteins or membrane components, and PrP (perhaps by a scavenging effect involving copper chelation) antagonizes this effect and in consequence is neuroprotective. F) In a mechanism lacking a specific interaction with Dpl and PrP^C, PrP is broadly neuroprotective ("smiley face") and thereby counteracts a toxic pathway initiated by Dpl.

of the target protein is unknown, it is not unreasonable to suggest that it might play a role in a signal transduction cascade. Provocatively, both Dpl and PrP^C in the form of Fab fusion proteins bind to cerebellar granule cells (Legname *et al.*, 2002), a target for the toxic activity of Dpl and "Shmerling" PrP (see below. Another possibility is that by binding to Dpl, PrP^C might perturb a particular assembly state of Dpl (e.g. monomer, homodimer) required to initiate the same signaling cascade (Figure 5B). In both of these scenarios Dpl would play a positive role in activating a "new" signal transduction pathway inappropriate for CNS neurons, and leading to apoptosis. Conversely, ectopic Dpl could have a negative effect, for example by disrupting a pathway necessary for cell viability (Figure 5C). Dpl could also be neurotoxic without requiring interactions with other proteins. For example, Dpl could multimerize to form a membrane pore (Figure 5D) and allow inappropriate ingress or egress of cellular contents or toxic substances: PrP^C might intercalate into these assemblies and thereby disrupt their neurotoxicity (Behrens and Aguzzi, 2002) although, since Dpl lacks a potential transmembrane domain like the central region of PrP^C, it is difficult to conceive how Dpl might form a pore in the first place. Yet, another possibility, prompted in part by the notion that PrP^C is protective and perhaps a metalloprotein or even a metalloenzyme, is that PrP^C and Dpl have antithetical properties with respect to oxidative metabolism (Figure 5E) (Wong *et al.*, 2001). A problem with this scenario is that our analyses of Dpl Tg mice have failed to reveal elevated levels of protein oxidation relative to non-Tg controls (Qin *et al.*, 2003). Also, it is now clear that Dpl may itself exist in Cu-metalled forms, suggesting that Dpl and PrP^C apoproteins co-expressed in the CNS could compete for Cu co-factors (i.e., a variant of Figure 5A).

Though the foregoing scenarios imply specificity in the interaction between PrP and Dpl, a rather different type of "interaction" might also have to be considered. Extrapolating from the position that PrP possesses anti-apoptotic properties, the "specificity" implied by titration of PrP and Doppel levels might be misleading (Brown *et al.*, 2002). For example, Dpl might be pro-apoptotic by activation of an inappropriate pathway, with PrP blocking this effect via a broad-ranging protective (and presumably anti-apoptotic) effect, but yet without any physical interactions between the two varieties of GPI-anchored protein (Figure 5F). In any event, discerning between these possibilities at the bench will probably be useful as it will also clarify PrP^C's physiological activities. As noted before (Westaway and Carlson, 2002), since PrP^C is a modulator of CNS doppel disease, "modifier" locus screens for Tg mice expressing Doppel might yield proteins that interact with PrP or GPI-linked cell signaling assemblies.

4.2. "Shmerling Syndrome" and CNS Dpl Expression

The effects of CNS expression of Doppel are similar to "Shmerling syndrome", an ataxia associated with an artificial interstitial deletion of mouse PrP (Shmerling *et al.*, 1998). Shmerling *et al.* noted that while ZrchI $Prnp^{0/0}$ mice exhibit normal development, ZrchI mice expressing transgenes encoding PrPΔ32-121 or PrPΔ32-134 developed cerebellar ataxia. With regards to neuropathology, this was marked by loss of granule cells accompanied by prominent astrogliosis, as well as spongiosis in the cerebellar white matter tracts. Purkinje cells were unaffected. Pathology in these mice could be completely abrogated by reintroduction of full-length wt PrP^C. Indeed, the many similarities in the genetics (PrPΔ32-121 or PrPΔ32-134 are N-terminally reduced forms of PrP and thus Dpl-like, both syndromes are rescued by wt PrP expression) and pathology of "Shmerling syndrome" and the ectopic Dpl expression syndrome begs the question as to whether these disorders reflect a shared pathogenic pathway (Moore *et al.*, 1999; Weissmann and Aguzzi, 1999). While one difference is the lack of involvement of Purkinje cells in mice expressing PrPΔ32-121 or PrPΔ32-134, this discrepancy may have a technical origin as the vector used to express PrPΔ32-121 or PrPΔ32-134 may lack a Purkinje cell enhancer (Fischer *et al.*, 1996). Also, in Tg(Dpl) mice made using the hamster PrP cosmid vector cos.Tet, the balance of pathology is slightly skewed from that seen in ataxic $Prnp^{0/0}$ mice, such that granule cell death is also prominent in addition to Purkinje cell death (Moore *et al.*, 2001). This blurs an apparent distinction between the two syndromes. One limitation in the striking convergence between Shmerling and Dpl syndromes is that these particular deletions (PrPΔ32-121 or PrPΔ32-134) deletions have no equivalents in human prion diseases. Also, other internal deletions in PrP give rise to a yet again distinct pathology (i.e., distinct from the pathologies of wt Dpl CNS expression, wt PrP overexpression, and prion infections), namely that of a lysosomal storage disease (Muramoto *et al.*, 1997).

4.3. *PRNP* Mutations and Dpl

Similarities between "Shmerling" and "doppel" syndromes lead us to ask whether any mutant human PrPs found in familial prion diseases might resemble Dpl more closely than wt PrP. No trend was noted in the new amino acid residues produced by the missense mutations found in GSS and FFI and the identity of the equivalent amino acid residues in the Dpl protein. Of note, however, 4 familial Creutzfeldt-Jakob disease (F-CJD) missense mutations result in amino acids identical to or of similar charge to residues in Dpl. For example, the lysine residue produced by the mutation E200K is mirrored in a conserved lysine residue in the Dpl protein at position 129. These four F-CJD mutations (V180I, E196K, E200K and E211Q) occur in a cluster in the C-terminal region of PrP (Mastrangelo *et al.*, 2002). This concordance is unlikely to have a trivial explanation to do with gene structure and evolution as (i) there is no nucleotide homology between the PrP and Dpl genes (Moore *et al.*, 1999) and (ii) single base substitutions in these codons are capable of producing 5-7 alternative amino acids that do not match the Dpl sequence. While the exact significance of this finding remains open, it begs the question as to whether a boundary between PrP^C and Dpl attributes is eroded by these F-CJD mutations, perhaps contributing to the pathogenesis of this particular sub-variety of prion disease. One possibility is that the primary effect of F-CJD mutations is not thermodynamic destabilization of PrP^C, but is to make the mutant PrPs weak mimetics of Dpl. This could cause the mutant PrPs to be exposed to trafficking or protein:protein interactions different from those of wt PrP and that (while benign in the context of Dpl) would jeopardize the mutant PrP^C to misfolding.

5. Concluding Remarks

Discovery of the Dpl protein has provided an unexpected sibling for PrP and a new point of entry into the physiology of prion proteins. Whether or not the *Prn* gene complex also includes a third *Prnp*-related gene denoted *Prnt* remains to be rigorously established. The toxicity of Doppel for CNS neurons is striking (given that it is modulated by PrP^C) and may provide insights into control of neuronal apoptosis. Since Doppel deficient mice have a phenotype in a developmental system (gametogenesis) that lends itself to experimental analysis, it seems possible that the function of the doppel protein will be deciphered. It will be of interest to see how the interactions of PrP^C and Dpl play out in these two tissues and whether physiological conformational changes feature in the biology of the two cellular prion proteins.

6. Acknowledgments

This work was supported by the Canadian Institutes of Health Research grants MOP363377 and MME54190, the Bayer Blood Partnership Research Fund, and the Alzheimer Society of Ontario.

References

Behrens, A., and Aguzzi, A. 2002. Small is not beautiful: antagonizing functions for the prion protein PrP^C and its homologue Dpl. Trends Neurosci. 25: 150-154.

Behrens, A., Brandner, S., Genoud, N., and Aguzzi, A. 2001. Normal neurogenesis and scrapie pathogenesis in neural grafts lacking the prion protein homologue Doppel. EMBO Rep. 2: 347-352.

Brown, D. R., Nicholas, R. S., and Canevari, L. 2002. Lack of prion protein expression results in a neuronal phenotype sensitive to stress. J. Neurosci. Res. 67: 211-224.

Büeler, H., Fischer, M., Lang, Y., Bluethmann, H., Lipp, H.-P., DeArmond, S. J., Prusiner, S. B., Aguet, M., and Weissmann, C. 1992. Normal development and behaviour of mice lacking the neuronal cell-surface PrP protein. Nature. 356: 577-582.

Carlson, G. A., Ebeling, C., Yang, S.-L., Telling, G., Torchia, M., Groth, D., Westaway, D., DeArmond, S. J., and Prusiner, S. B. 1994. Prion isolate specified allotypic interactions between the cellular and scrapie prion proteins in congenic and transgenic mice. Proc. Natl. Acad. Sci. USA. 91: 5690-5694.

Carlson, G. A., Goodman, P. A., Lovett, M., Taylor, B. A., Marshall, S. T., Peterson-Torchia, M., Westaway, D., and Prusiner, S. B. 1988. Genetics and polymorphism of the mouse prion gene complex: the control of scrapie incubation time. Mol. Cell. Biol. 8: 5528-5540.

Carlson, G. A., Kingsbury, D. T., Goodman, P. A., Coleman, S., Marshall, S. T., DeArmond, S. J., Westaway, D., and Prusiner, S. B. 1986. Linkage of prion protein and scrapie incubation time genes. Cell. 46: 503-511.

Comincini, S., Foti, M. G., Tranulis, M. A., Hills, D., Di Guardo, G., Vaccari, G., Williams, J. L., Harbitz, I., and Ferretti, L. 2001. Genomic organization, comparative analysis, and genetic polymorphisms of the bovine and ovine prion Doppel genes (PRND. Mamm. Genome. 12: 729-733.

Essalmani, R., Taourit, S., Besnard, N., and Vilotte, J. L. 2002. Sequence determination and expression of the ovine doppel-encoding gene in transgenic mice. Gene. 285: 287-290.

Fischer, M., Rulicke, T., Raeber, A., Sailer, A., Moser, M., Oesch, B., Brandner, S., Aguzzi, A., and Weissmann, C. 1996. Prion protein (PrP) with amino-proximal deletions restoring susceptibility of PrP knockout mice to scrapie. EMBO J. 15: 1255-1264.

Gabriel, J.-M., Oesch, B., Kretzschmar, H., Scott, M., and Prusiner, S. B. 1992. Molecular cloning of a candidate chicken prion protein. Proc. Natl. Acad. Sci. USA. 89: 9097-9101.

Hornshaw, M. P., McDermott, J.R., and Candy, J.M. 1995. Copper binding to the N-terminal tandem repeat region of mammalian and avian prion protein. Biochem. Biophys. Res. Comm. 207: 621-629.

Kaneko, K., Zulianello, L., Scott, M., Cooper, C. M., Wallace, A. C., James, T. L., Cohen, F. E., and S.B., P. 1997. Evidence for protein X binding to a discontinuous epitope on the cellular prion protein during scrapie prion propagation. Proc. Natl. Acad. Sci. USA. 94: 10069-10074.

Kuwahara, C., Takeuchi, A. M., Nishimura, T., Haraguchi, K., Kubosaki, A., Matsumoto, Y., Saeki, K., Yokoyama, T., Itohara, S., and Onodera, T. 1999. Prions prevent neuronal cell-line death. Nature. 400: 225-226.

Lee, I., Westaway, D., Smit, A. F. A., Wang, K., Seto, J., Chen, L., Acharya, C., Ankener, M., Baskin, D., Cooper, C., Yao, H., Prusiner, S. B., and Hood, L. 1998. Complete genomic sequence and analysis of the prion protein gene region from three mammalian species. Genome Research. 8: 1022-1037.

Legname, G., Nelken, P., Guan, Z., Kanyo, Z. F., DeArmond, S. J., and Prusiner, S. B. 2002. Prion and doppel proteins bind to granule cells of the cerebellum. Proc. Natl. Acad. Sci. USA. 99: 16285-16290.

Li, A., Sakaguchi, S., Atarashi, R., Roy, B. C., Nakaoke, R., Arima, K., Okimura, N., Kopacek, J., and Shigematsu, K. 2000. Identification of a novel gene encoding a PrP-like protein expressed as chimeric transcripts fused to PrP exon 1/2 in ataxic mouse line with a disrupted PrP gene [In Process Citation]. Cell. Mol Neurobiol 20: 553-67.

Lu, K., Wang, W., Xie, Z., Wong, B. S., Li, R., Petersen, R. B., Sy, M. S., and Chen, S. G. 2000. Expression and structural characterization of the recombinant human doppel protein. Biochem. 39: 13575-13583.

Makrinou, E., Collinge, J., and Antoniou, M. 2002. Genomic characterization of the human prion protein (PrP) gene locus. Mamm. Genome. 13: 696-703.

Manson, J. C., Clarke, A. R., hooper, M. L., aitchison, L., McConnel, I., and Hope, J. 1994. 129/Ola mice carryinga null mutation in PrP that abolishes mRNA production are developmentally normal. Mol. Neurobiol. 8: 121-127.

Mastrangelo, P., Serpell, L., Dafforn, T., Lesk, A., Fraser, P., and Westaway, D. 2002. A cluster of familial Creutzfeldt-Jakob Disease mutations recapitulate conserved residues in Doppel: a case of molecular mimicry? FEBS Letters. 532: 21-24.

Mastrangelo, P., and Westaway, D. 2001. The prion gene complex encoding PrP^C and doppel: insights from mutational analysis. Gene. 275: 1-18.

Mead, S., Beck, J., Dickinson, A., Fisher, E. M., and Collinge, J. 2000. Examination of the human prion protein-like gene doppel for genetic susceptibility to sporadic and variant Creutzfeldt-Jakob disease. Neurosci. Lett. 290: 117-120.

Mo, H., Moore, R. C., Cohen, F. E., Westaway, D., Prusiner, S. B., Wright, P. E., and Dyson, H. J. 2001. Two different neurodegenerative diseases caused by proteins with similar structures. Proc. Natl. Acad. Sci. USA. 98: 2352-2357.

Moore, R., Lee, I., Silverman, G. S., Harrison, P., Strome, R., Heinrich, C., Karunaratne, A., Pasternak, S. H., Chishti, M. A., Liang, Y., Mastrangelo, P., Wang, K., Smit, A. F. A., Katamine, S., Carlson, G. A., Cohen, F. E., Prusiner, S. B., Melton, D. W., Tremblay, P., Hood, L. E., and Westaway, D. 1999. Ataxia in prion protein (PrP) deficient mice is associated with upregulation of the novel PrP-like protein *doppel*. J. Mol. Biol. 293: 797-817.

Moore, R., Mastrangelo, P., Bouzamondo, E., Heinrich, C., Legname, G., Prusiner, S. B., Hood, L., Westaway, D., DeArmond, S., and Tremblay, P. 2001. Doppel-induced cerebellar degeneration in transgenic mice. Proc. Natl. Acad. Sci. USA. 98: 15288-15293.

Moore, R. C., Hope, J., McBride, P. A., McConnell, I., Selfridge, J., Melton, D. W., and Manson, J. C. 1998. Mice with gene targetted prion protein alterations show that *Prnp*, *Sinc* and *Prni* are congruent. Nature Genetics. 18: 118-125.

Muramoto, T., DeArmond, S. J., Scott, M., Telling, G. C., F.E., C., and Prusiner, S. 1997. Heritable disorder resembling neuronal storage disease in mice expressing prion protein with deletion of an alpha-helix. Nature Medicine. 3: 750-755.

Nicholson, E. M., Mo, H., Prusiner, S. B., Cohen, F. E., and Marqusee, S. 2002. Differences between the Prion Protein and its Homolog Doppel: A Partially Structured State with Implications for Scrapie Formation. J. Mol. Biol. 316: 807-815.

O'Rourke, K. I., Holyoak, G. R., Clark, W. W., Mickelson, J. R., Wang, S., Melco, R. P., Besser, T. E., and Foote, W. C. 1997. PrP genotypes and experimental scrapie in orally inoculated Suffolk sheep in the United States. J. Gen. Virol. 78: 975-978.

Oesch, B., Westaway, D., Wälchli, M., McKinley, M. P., Kent, S. B. H., Aebersold, R., Barry, R. A., Tempst, P., Teplow, D. B., Hood, L. E., Prusiner, S. B., and Weissmann, C. 1985. A cellular gene encodes scrapie PrP 27-30 protein. Cell. 40: 735-746.

Pan, K.-M., Stahl, N., and Prusiner, S. B. 1992. Purification and properties of the cellular prion protein from Syrian hamster brain. Protein Sci. 1: 1343-1352.

Peoc'h, K., Guerin, C., Brandel, J. P., Launay, J. M., and Laplanche, J. L. 2000. First report of polymorphisms in the prion-like protein gene (*PRND*): implications for human prion diseases. Neurosci. Lett. 286: 144-148.

Peoc'h, K., Serres, C., Frobert, Y., Martin, C., Lehmann, S., Chasseigneaux, S., Sazdovitch, V., Grassi, J., Jouannet, P., Launay, J. M., and Laplanche, J. L. 2002. The human "prion-like" protein Doppel is expressed in both Sertoli cells and spermatozoa. J. Biol. Chem. 277: 43071-43078.

Perrier, V., Kaneko, K., Safar, J., Vergara, J., Tremblay, P., DeArmond, S. J., Cohen, F. E., Prusiner, S. B., and Wallace, A. C. 2002. Dominant-negative inhibition of prion replication in transgenic mice. Proc. Natl. Acad. Sci. USA. 97: 6073-6078.

Qin, K., Coomaraswamy, J., Mastrangelo, P., Yang, Y., Lugowski, S., Petromilli, C., Prusiner, S. B., Fraser, P. E., Goldberg, J. M., Chakrabartty, A., and Westaway, D. 2003. The PrP-like protein Dpl binds copper. J. Biol. Chem. 278: 8888-8896.

Rossi, D., Cozzio, A., Flechsig, E., Klein, M. A., Rulicke, T., Aguzzi, A., and Weissmann, C. 2001. Onset of ataxia and Purkinje Cell. loss in PrP null mice inversely correlated with Dpl level in brain. EMBO J. 20: 694-702.

Sakaguchi, S., Katamine, S., Nishida, N., Moriuchi, R., Shigematsu, K., Sugimoto, T., Nakatani, A., Kataoka, Y., Houtani, T., Shirabe, S., Okada, H., Hasegawa, S., Miyamoto, T., and Noda, T. 1996. Loss of cerebellar purkinje cells in aged mice homozygous for a disrupted PrP gene. Nature. 380: 528-531.

Scott, M. R., Köhler, R., Foster, D., and Prusiner, S. B. 1992. Chimeric prion protein expression in cultured cells and transgenic mice. Protein Sci. 1: 986-997.

Shaked, Y., Hijazi, N., and Gabizon, R. 2002. Doppel and PrP(C) do not share the same membrane microenvironment. FEBS Lett 530: 85-88.

Shaked, Y., Rosenmann, H., Talmor, G., and Gabizon, R. 1999. A C-terminal-truncated PrP isoform is present in mature sperm. J. Biol. Chem. 274: 32153-32158.

Shmerling, D., Hegyi, I., Fischer, M., Blattler, T., Brandner, S., Gotz, J., Rulicke, T., Flechsig, E., Cozzio, A., C., v. M., Hangartner, C., Aguzzi, A., and Weissmann, C. 1998. Expression of amino-terminally truncated PrP in the mouse leading to ataxia and specific cerebellar lesions. Cell. 93: 203-214.

Silverman, G. L., Qin, K., Moore, R. C., Yang, Y., Mastrangelo, P., Tremblay, P., Prusiner, S. B., Cohen, F. E., and Westaway, D. 2000. Doppel is an N-glycosylated, glycosylphosphatidylinositol-anchored protein: expression in testis and ectopic production in the brains of $Prnp^{0/0}$ mice predisposed to Purkinje cell loss. J. Biol. Chem. 275: 26834-26841.

Simonic, T., Duga, S., Strumbo, B., Asselta, R., Ceciliani, F., and Ronchi, S. 2000. cDNA cloning of turtle prion protein. FEBS Lett. 469: 33-38.

Stöckel, J., Safar, J., Wallace, A. C., Cohen, F. E., and Prusiner, S. B. 1998. Prion protein selectively binds copper(II) ions. Biochemistry 37, 7185-93.

Strumbo, B., Ronchi, S., Bolis, L. C., and Simonic, T. 2001. Molecular cloning of the cDNA coding for Xenopus laevis prion protein. FEBS Lett. 508: 170-174.

Suzuki, T., Kurokawa, T., Hashimoto, H., and Sugiyama, M. 2002. cDNA sequence and tissue expression of Fugu rubripes prion protein- like: a candidate for the teleost orthologue of tetrapod PrPs. Biochem. Biophys. Res. Commun. 294: 912-917.

Tranulis, M. A., Espenes, A., Comincini, S., Skretting, G., and Harbitz, I. 2001. The PrP-like protein Doppel gene in sheep and cattle: cDNA sequence and expression. Mamm. Genome. 12: 376-379.

Vey, M., Pilkuhn, S., Wille, H., Nixon, R., DeArmond, S. J., Smart, E. J., Anderson, R. G., Taraboulos, A., and Prusiner, S. B. 1996. Subcellular colocalization of the cellular and scrapie prion proteins in caveolae-like membranous domains. Proc. Natl. Acad. Sci. USA. 93: 14945-14949.

Viles, J. H., Cohen, F. E., Prusiner, S. B., Goodin, D. B., Wright, P. E., and Dyson, H. J. 1999. Copper binding to the prion protein: structural implications of four identical cooperative binding sites. Proc. Natl. Acad. Sci. USA. 96: 2042-2047.

Weissmann, C., and Aguzzi, A. 1999. Perspectives: neurobiology. PrP's double causes trouble [published erratum appears in Science 1999 Dec 10;286(5447):2086]. Science. 286: 914-915.

Westaway, D., and Carlson, G. A. 2002. Mammalian prion proteins: enigma, variation and vaccination. Trends Biochem. Sci. 27: 301-307.

Westaway, D., Cooper, C., Turner, S., Da Costa, M., Carlson, G. A., and Prusiner, S. B. 1994. Structure and polymorphism of the mouse prion protein gene. Proc. Natl. Acad. Sci. USA. 91: 6418-6422.

Westaway, D., DeArmond, S. J., Cayetano-Canlas, J., Groth, D., Foster, D., Yang, S.-L., Torchia, M., Carlson, G. A., and Prusiner, S. B. 1994. Degeneration of skeletal muscle, peripheral nerves, and the central nervous system in transgenic mice overexpressing wild-type prion proteins. Cell. 76: 117-129.

Westaway, D., Goodman, P. A., Mirenda, C. A., McKinley, M. P., Carlson, G. A., and Prusiner, S. B. 1987. Distinct prion proteins in short and long scrapie incubation period mice. Cell. 51: 651-662.

Westaway, D., Mirenda, C. A., Foster, D., Zebarjadian, Y., Scott, M., Torchia, M., Yang, S.-L., Serban, H., DeArmond, S. J., Ebeling, C., Prusiner, S. B., and Carlson, G. A. 1991. Paradoxical shortening of scrapie incubation times by expression of prion protein transgenes derived from long incubation period mice. Neuron. 7: 59-68.

Westaway, D., Zuliani, V., Cooper, C. M., Da Costa, M., Neuman, S., Jenny, A. L., Detwiler, L., and Prusiner, S. B. 1994. Homozygosity for prion protein alleles encoding glutamine-171 renders sheep susceptible to natural scrapie. Genes Dev. 8: 959-969.

Wong, B. S., Liu, T., Paisley, D., Li, R., Pan, T., Chen, S. G., Perry, G., Petersen, R. B., Smith, M. A., Melton, D. W., Gambetti, P., Brown, D. R., and Sy, M. S. 2001. Induction of ho-1 and nos in doppel-expressing mice devoid of prp: implications for doppel function. Mol. Cell Neurosci. 17: 768-775.

Yokoyama, T., Kimura, K. M., Ushiki, Y., Yamada, S., Morooka, A., Nakashiba, T., Sassa, T., and Itohara, S. 2001. In vivo conversion of cellular prion protein to pathogenic isoforms, as monitored by conformation-specific antibodies. J. Biol. Chem. 276: 11265-11271.

Zahn, R., Liu, A., Luhrs, T., Riek, R., von Schroetter, C., Lopez Garcia, F., Billeter, M., Calzolai, L., Wider, G., and Wuthrich, K. 2000. NMR solution structure of the human prion protein. Proc. Natl. Acad. Sci. USA. 97: 145-150.

From: Prions and Prion Diseases: Current Perspectives
Edited by: Glenn C. Telling

Chapter 10

Cellular Control of Prion Formation and Propagation in Yeast

Yury O. Chernoff

Abstract

Yeast prions provide a model for understanding both the molecular mechanisms of mammalian amyloidoses and the general principles of protein-based inheritance. In yeast, initial prion formation is induced by protein overproduction and facilitated in a cascade-like fashion by pre-existing prion isoforms of unrelated proteins, containing prion-forming domains of similar amino acid composition. Pre-existing yeast prions also increase aggregation and toxicity of the heterologous polyglutamine proteins, thus manifesting themselves as susceptibility factors for other aggregation disorders. Formation and propagation of yeast prions is modulated by various cellular regulatory networks, including the stress-defense systems (heat shock proteins [Hsps] and the ubiquitin pathway), cytoskeleton, and functional partners of the specific prion-forming proteins. The very ability of yeast prions to spread in cell divisions or by cytoplasmic exchanges, that turns them into "malignant" subcellular "tumors" and distinguishes prions from non-perpetuating protein aggregates, requires a certain level of Hsp activity. Thus, prions have "learned" how to use the cellular stress-defense systems to their own advantage. Stresses and chemical agents can cure yeast cells of prions by altering the Hsp levels or activities. The mechanism of prion curing by guanidine hydrochloride is specifically discussed. Curing yeast cells of prions *via* chaperone manipulations suggests an approach for development of new potential anti-prion treatments.

1. Introduction

The behavior of infectious agents of transmissible encephalopathies in mammals could not be explained completely within the framework of the "central dogma" of molecular biology ("DNA-RNA-protein"). Most experts agree that these agents, termed "prions" by S.B. Prusiner (Prusiner, 1982), are composed of an abnormally conformed protein that replicates without nucleic acid. A few remaining "dissenters" continue to criticize certain aspects of this model (e.g. see Chesebro, 1997) but have so far failed to present an equally convincing alternative explanation. The prion model promoted the revolutionary idea that proteins can transmit information directly from one protein molecule onto another protein molecule.

The prion concept was further strengthened by the discovery of prion proteins in yeast (Wickner, 1994) and other fungi (Coustou *et al.*, 1997). Yeast prions are non-Mendelian elements that control certain phenotypically detectable traits and are transmitted *via* cytoplasm rather than by extracellular infection as in mammals (see Figure 1A). While this generally follows the mode of transmission of yeast viruses, which are also not capable of extracellular infection, such a mechanism establishes yeast prions as protein-based genetic elements

Figure 1. Yeast prions. A - Non-Mendelian behavior of yeast prions. Designations: - nuclear chromosomes (different homologs are indicated by different thickness); ▮ - prion isoform; ⬤ – non-prion isoform. In contrast to nuclear genes, yeast prions are usually transmitted to all or most of the meiotic progeny in a non-Mendelian fashion, as prion protein can convert non-prion protein into a prion. B - Structural and functional organization of the yeast prion proteins known to date. Sup35 (eRF3) protein: PFD – prion-forming domain, A – aminoacyl-tRNA binding site, G – GTP binding sites. Two regions of PFD that have been proven by mutational analysis to specifically affect prion propagation are shown, namely QN – QN-rich stretch, and R – region of oligopeptide repeats. Position of the $\Delta 22/69$ deletion that specifically affects prion aggregate "shearing" (Borchsenius *et al.*, 2001) is indicated. See comments in the text. Other yeast prion proteins: regions containing QN-rich stretches and proven to be responsible for the prion propagation (prion-forming domains, or PFDs) are shown as filled boxes. Numbers correspond to the amino acid positions.
* The designation [NU$^+$] was initially proposed for the prion formed by chimeric protein that contained New1 PFD (Santoso *et al.*, 2000; Osherovich and Weissman, 2001). It has not yet been reported whether complete New1 protein is capable of forming a prion on its own.

A

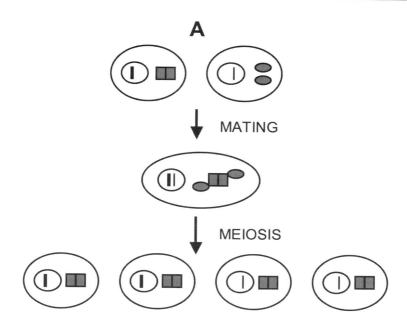

NUCLEAR GENES = 2:2
PRION : NON-PRION = 4:0

B

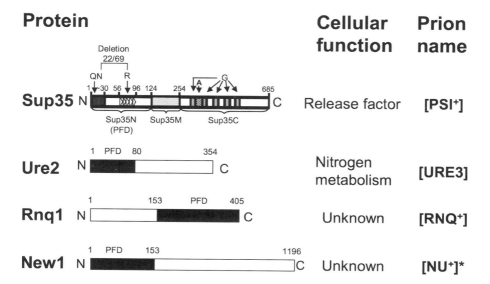

Protein	Cellular function	Prion name
Sup35	Release factor	[PSI⁺]
Ure2	Nitrogen metabolism	[URE3]
Rnq1	Unknown	[RNQ⁺]
New1	Unknown	[NU⁺]*

responsible for certain phenotypic traits. As in the case of mammalian prion protein (PrP), yeast prion-forming proteins are coded by nuclear genes, so that prions are not completely independent of the nuclear genome. However, differences between alternative phenotypic traits are not determined by differences in DNA sequence. Cells containing a prion may have the same sequence of nucleotides in DNA as in the non-prion cells; nevertheless they exhibit alternative phenotypic traits that are inherited in cell generations.

Although yeast prion proteins are not homologous to mammalian PrP, they exhibit a surprisingly similar mode of behavior. Moreover, some yeast prion proteins possess homologs in mammals, although it is not yet known whether these homologs exhibit prion properties as well. Domains of yeast proteins that are responsible for prion formation and propagation, called prion-forming domains, or PFDs, are characterized by high glutamine (Q) and/or asparagine (N) contents. This makes yeast prion proteins similar to the proteins involved in so-called polyQ disorders, such as Huntington's disease in humans. These patterns, combined with the power of yeast genetic analysis, make yeast an excellent experimental model for studying the general mechanisms of prion diseases and other aggregation-related disorders.

According to the "nucleated polymerization" model (Lansbury and Caughey, 1995), the prion is an insoluble protein polymer, which can "seed" polymerization of the normally soluble cellular protein, thus converting it into a prion. While it remains to be proven whether aggregation plays a causative role in prion conversion, it is clear that protein molecules in the prion form tend to generate β-sheet rich fiber-like aggregates, called amyloids. This resembles other amyloidoses and neural inclusion diseases, such as Alzheimer's disease, Parkinson's disease and Huntington's disease. While a variety of proteins are capable of forming amyloids *in vitro*, only a small subset of proteins generates self-perpetuating aggregates *in vivo*. Clearly, cells are capable of counteracting the amyloid-forming potential of most proteins. The following questions arise: 1) How do cells counteract the amyloid-forming processes? 2) How are cell defense systems overcome by prions?

This chapter reviews the recent progress in our understanding of the cellular control of prion formation and propagation that comes almost exclusively from the experiments on yeast models. Other aspects of yeast prions and *in vitro* behavior of yeast prion proteins have been reviewed elsewhere (*e.g.,* Serio and Lindquist, 2000; Chernoff, 2001; Derkatch and Liebman, 2001; Wickner *et al.*, 2001; Chernoff *et al.*, 2002; Zhouravleva *et al.*, 2002) and will be mentioned here only briefly when it is necessary for understanding the materials of the given chapter.

2. Yeast Prions

2.1. [PSI⁺]/Sup35 System

Yeast non-Mendelian factor [PSI⁺], first described by B. Cox (Cox, 1965), is a prion isoform of the essential protein Sup35, a yeast counterpart of the translational termination factor eRF3 (see Zhouravleva *et al.*, 2002 for review). [PSI⁺] cells are partially defective in termination of translation. This could be detected by nonsense read-through, that is visualized by growth or color on the specific media in the specially designed yeast strains containing nonsense-reporter constructs (see Chernoff *et al.*, 2002). For example, yeast strains containing the *ade1-14* allele, in which the open reading frame (ORF) of *ADE1* gene is interrupted by a UGA mutation, do not grow on medium without adenine (-Ade) and accumulate a red pigment on complete medium. Presence of the [PSI⁺] prion in the *ade1-14* cells leads to UGA read-through (nonsense-suppression) that partially restores growth on –Ade and decreases accumulation of the red pigment. Biochemical analysis confirms that Sup35 protein is soluble in [psi⁻] (non-prion) cells but forms insoluble aggregates in [PSI⁺] cells (Patino *et al.*, 1996; Paushkin *et al.*, 1996). Different "variants" ("strains") of [PSI⁺] are inherently different from each other by both level of nonsense-suppression and efficiency of mitotic transmission in cell generations (Derkatch *et al.*, 1996). These "variants" of [PSI⁺] are reminiscent of so-called "strains" of mammalian prions (Dickinson and Meikle, 1971), and probably correspond to different self-perpetuating states of prion aggregates.

Yeast Sup35 protein consists of three regions with distinct functions (Figure 1B; see Zhouravleva *et al.*, 2002 for more detail). The C-proximal region (Sup35C) shows a significant homology with the translation elongation factor eEF1-A and is required and sufficient for translation termination and cell viability. The N-proximal region (Sup35N) is a prion-forming domain (PFD), not essential for viability and termination but required for prion formation (Derkatch *et al.*, 1996) and propagation (Doel *et al.*, 1994; Ter-Avanesyan *et al.*, 1994). The charged M (middle) region (Sup35M) is required for neither viability and termination nor [*PSI*⁺] formation and propagation. Sup35 PFD is "transferable": when fused to another protein, such as green fluorescent protein (GFP) (Patino *et al.*, 1996), or rat glucocorticoid receptor (Li and Lindquist, 2000), Sup35N converts this protein into an aggregated state in the [PSI⁺] background. *In vitro* data confirm that Sup35N or NM regions (but not M, C or MC regions) form amyloid fibrils *in vitro* (Glover *et al.*, 1997; King *et al.*, 1997). The Sup35N region exhibits an unusual amino acid (aa) composition. It is rich in Q and N residues (45%), and also in glycine (G) and tyrosine (Y) residues (33%) (Kushnirov *et al.*, 1988), compared to 10% and 8% average in the yeast proteome, respectively (Young *et al.*, 2000). Two subregions of Sup35N specially influence prion propagation: QN-rich stretch (aa 6-33), and R, region of oligopeptide repeats (aa 41-97). The presence of G-rich oligopeptide repeats (consensus sequence PQGGYQQYN), which are similar to those found at approximately the same position of mammalian PrP (consensus sequence PHGGGWGQ), is one of the most intriguing features of the Sup35 PFD (Cox, 1994; Kushnirov *et al.*, 1995). In contrast to PrP, the oligopeptide repeats of Sup35 are required for prion propagation (Parham *et al.*, 2001).

2.2. Heterologous [PSI⁺] Prions

Comparative study of the *SUP35* genes from 10 distantly related yeast species (see below) revealed a higher degree of variability among the Sup35N (27-43% of identity) and Sup35M (24-56% of identity) sequences, in contrast to a highly conserved (67-88% of identity) Sup35C (see Zhouravleva *et al.*, 2002 for review). However, the QN and R regions remain most conserved within Sup35N (Kushnirov *et al.*, 1990; Santoso *et al.*, 2000; Nakayashiki *et al.*, 2001). Oligopeptide repeats were present in all Sup35 proteins of budding yeasts, although they were somewhat divergent in sequence.

Chimeric Sup35 proteins in which the Sup35N (or NM) region of *S. cerevisiae* was substituted for the corresponding region from *Pichia methanolica*, *Candida albicans*, *Candida maltosa*, *Debaryomyces hansenii*, *Kluyveromyces lactis*, *Yarrowia lipolytica* or *Zygosaccharomyces rouxii*, were all able to form the prion state in *S. cerevisiae* (Chernoff *et al.*, 2000; Kushnirov *et al.*, 2000a; Santoso *et al.*, 2000; Nakayashiki *et al.*, 2001). The full-size Sup35 protein from *P. methanolica* was also capable of forming a prion in the *S. cerevisiae* host (Zadorskii *et al.*, 2000). Like mammalian prions, yeast prions exhibit a "species barrier", meaning that interspecies conversion from endogenous *S. cerevisiae* Sup35 to chimeric Sup35 containing PFD from different species is usually inefficient (Chernoff *et al.*, 2000; Kushnirov *et al.*, 2000a; Santoso *et al.*, 2000). However, the strict "species barrier" was not observed in some combinations or conditions (Chernoff *et al.*, 2000; Nakayashiki *et al.*, 2001). It is likely that prion behaviour of the Sup35 PFD homologs in *S. cerevisiae* reflects their ability to form a prion in the homologous systems. Indeed, *K. lactis* Sup35 homolog is capable of forming aggregates, associated with translational suppression, in *K. lactis* cells (Nakayashiki *et al.*, 2001). Therefore, it is clear that the prion-forming ability of Sup35 is conserved across budding yeast.

Higher eukaryotes (mice and humans) contain the Sup35 homologs with highly conserved Sup35C region but greatly divergent Sup35N and M regions, compared to yeast (see Zhouravleva *et al.*, 2002). The N domains of the higher eukaryotic eRF3 proteins show no obvious sequence similarity to the corresponding domains of lower eukaryotes and contain no QN-rich stretches and oligopeptide repeats, although they still exhibit an unusual aa composition, being enriched in proline (P), serine (S) and G residues. This suggests the possibility of high structural flexibility. Prion formation by non-yeast Sup35 homologs has not been proven yet.

2.3. [URE3]/Ure2 System

The yeast non-Mendelian element [URE3], discovered by F. Lacroute (Lacroute, 1971; Aigle and Lacroute, 1975), is a prion form of the protein Ure2, a posttranslational regulator in the nitrogen metabolism pathway. Due to partial inactivation of Ure2, [URE3] enables yeast cells to uptake ureidosuccinic acid from the medium, and therefore confers to the yeast *ura2* mutants the ability to utilize ureidosuccinic acid as a source of uracil. Other detection assays for [URE3] are also available in the specially designed yeast strains (*e.g.*, Schlumpberger *et al.*, 2001). Like for [PSI+], distinct self-perpetuating "variants" ("strains") have been isolated for [URE3] (Schlumpberger *et al.*, 2001). As in case of Sup35, Ure2 protein can be subdivided into the N-proximal QN-rich prion-forming domain and C-proximal enzymatic domain (Masison and Wickner, 1995; see Figure 1B). However, Ure2 also contains a secondary prion-forming domain that could be activated by some internal deletions (Maddelein *et al.*, 1999). The prion isoform of Ure2 protein is protease-resistant and forms aggregates (Masison and Wickner, 1995; Edskes *et al.*, 1999a), and Ure2 PFD forms amyloids *in vitro* (Taylor *et al.*, 1999; Thual *et al.*, 1999; Schlumpberger *et al.*, 2000). In contrast to Sup35, Ure2 is not conserved in mammals, although Ure2 homologs were identified in various yeast and fungal species.

2.4. Other Prions and Prion Candidates in Yeast

Prion-forming domains of both Sup35 and Ure2 are characterized by high QN-rich contents. Moreover, this feature is conserved in the distant yeast homologs of Sup35, despite the high level of overall divergence of aa sequences. It has previously been proposed that QN-rich sequences may play a role of "polar zippers" involved in protein-protein interactions (Perutz, 1996). A number of proteins with QN-rich stretches were identified in the yeast genome, as well as in the other eukaryotic genomes (Michelitsch and Weissman, 2000). Some of these proteins also contain oligopeptide repeats, another characteristic feature of Sup35 PFD(see Chernoff *et al.*, 2000; Michelitsch and Weissman, 2000; Chernoff, 2001). At least in two cases, it has been proven that a QN-rich domain of the other protein retains the ability to convert Sup35 into a prion when it is substituted for the Sup35 PFD. This applies to proteins New1 (Santoso *et al.*, 2000) and Rnq1 (Sondheimer and Lindquist, 2000) (see Figure 1B). In case of Rnq1, it has also been confirmed that this protein is capable of spontaneously forming a prion called [RNQ⁺] (Sondheimer *et al.*, 2001). New1 and Rnq1 proteins are not essential, and their normal cellular functions are not known. However, [RNQ⁺] prion has recently been identified (Derkatch *et al.*, 2001) with the previously described (Derkatch *et al.*, 1997) non-Mendelian element [PIN⁺], that is detectable phenotypically *via* its effect on *de novo* appearance of [PSI⁺] (see below).

It should be noted that candidate prions are by no means confined to QN-rich proteins. Neither mammalian PrP nor the prion protein Het-s, found in the fungus *Podospora* (Coustou *et al.*, 1997), contain QN-rich stretches. Several non-Mendelian elements of peculiar behavior recently described in yeast and other fungi, such as [*kil-d*] (Talloczy *et al.*, 2000) or [ISP⁺] (Volkov *et al.*, 2002), could eventually turn out to be prions as well. Taken together, these data suggest that prion-like phenomena are widespread in nature and may play an important role not only in disease, but also in inheritance of phenotypic traits.

3. Initiation of Aggregation and Prionization *In Vivo*

The prion concept explains spread of infection by the protein-converting ability of the prion isoform. Within the framework of the "nucleated polymerization" model, this is achieved *via* "seeded" polymerization of the soluble protein by pre-existing prion aggregates. However, the mechanism of initial formation of prion aggregates from non-prion protein in mammalian systems remains a mystery. It has been proposed that initial prion isoforms could appear due to somatic mutations in PrP protein, that make it more likely to convert into the prion isoform (see Prusiner, 1994). Mutations with high prion-forming potential are indeed associated with heritable prion diseases, but whether such mutations contribute to the initial appearance of prions in the case of sporadic prion diseases, remains an open question. No mammalian experimental model has been described thus far that would be capable of answering this question.

Due to large numbers, high proliferation speed, powerful genetic techniques, and simple phenotypic detection assays available for at least some prions, yeast provides an ideal experimental model for studying the mechanism of initial prion formation. Indeed, experiments in yeast have shown that initial prion aggregation could be modulated by protein levels rather than by somatic mutations.

3.1. Prion Induction by Protein Overproduction

One of the foundation stones for the prion model of [PSI+] and [URE3] (Wickner, 1994) was the observation that each of these non-Mendelian elements can be induced *de novo* by transient overproduction of the specific protein. Indeed, multicopy plasmids bearing the *SUP35* gene converted a fraction of the initially [psi-] cells into a [PSI+] state, which was maintained after the loss of the plasmid (Chernoff *et al.*, 1993). The same result was observed when single-copy *SUP35* gene was overexpressed from an inducible promoter; moreover, *de novo* [PSI+] induction required overproduction of Sup35 protein, rather than just an increase in mRNA level (Derkatch *et al.*, 1996). Likewise, transient overproduction of Ure2 protein induced [URE3] (Wickner, 1994). Prion induction by overproduction required the presence of PFD region in the overproduced protein (Masison and Wickner, 1995; Derkatch *et al.*, 1996). Moreover, transient overproduction of the fragments encompassing the prion-forming domain but lacking the enzymatic C-proximal region induced prion formation even more efficiently than did overproduction of the complete protein. In some cases, overproduction increased frequencies of *de novo* prion formation 10^4-10^5 fold, reaching levels of 10-50%. Prions formed by Sup35 proteins with heterologous PFDs (Chernoff *et al.*, 2000; Kushnirov *et al.*, 2000a; Santoso *et al.*, 2000; Nakayashiki *et al.*, 2001) or PFDs originating from other proteins, such as New1 (Santoso *et al.*, 2000) and Rnq1 (Sondheimer and Lindquist, 2000), were also induced by overproduction of the corresponding PFD-containing region in each case. *Podospora* prion [Het-s] was induced by overproduction of the Het-s protein (Coustou *et al.*, 1997). Prion induction by protein overproduction confirms that initial events, leading to prion formation in yeast, occur at the "post-DNA" level and should be considered as protein rather than DNA "mutations". One can not rule out a possibility that accumulation of the aberrant misfolded translation products, which initiate a process of prionization, is caused or facilitated by transcription errors or transcript alterations, similar to so-called "transcriptional mutagenesis" (Viswanathan *et al.*, 1999). However, it is also likely that spontaneous misfolding and aggregation simply become more probable with an increase of the cellular concentration of the corresponding protein. The overproduction may also shift the balance between the newly synthesized protein and chaperone helpers, assisting in proper protein folding. This might facilitate protein misfolding, resulting in prion formation (see below).

It is possible that naturally occurring variations in protein levels also contribute to prion formation in the natural conditions. However, this should be noted that while PrP overproduction may lead to the appearance of certain symptoms of prion disease in mice (Bueler *et al.*, 1992), no formation of infectious prion protein in result of PrP overproduction in non-prion containing mammalian cells has been detected thus far. It is therefore of interest to further consider cellular factors that influence efficiency of overproduction-induced prion formation in yeast.

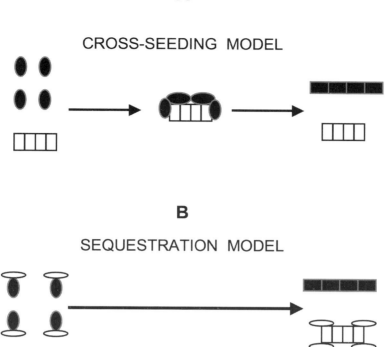

Figure 2. Two models for "prion cascades" in yeast. Designations: ▉ and ● - prion and non-prion isoforms of one and the same protein, respectively; ▢ - pre-existing prion isoform of another protein; ○ - protein factor(s) that normally prevents conversion of the soluble protein into a prion. **A)** "Cross-seeding" model. Pre-existing prion aggregates "seed" aggregation of the unrelated overproduced protein. Aggregation results in prion formation. **B)** "Sequestration" model. Prion aggregates of one protein sequester the factor that normally prevents conversion of the other protein into a prion. Based on data and hypotheses presented by: Derkatch *et al.*, 2001; Osherovich and Weissman, 2001. See comments in the text.

3.2. Prion Cascades

Further investigations have demonstrated that *de novo* [PSI⁺] induction by overproduced Sup35 is efficient only in the *S. cerevisiae* strains bearing a non-Mendelian element that was named [PIN⁺], from [PSI⁺] inducibility (Derkatch *et al.*, 1997). Although spontaneous [PSI⁺] formation in the absence of Sup35 overproduction usually occurs at very low level (about 10^{-6} – for example, see Chernoff *et al.*, 1999) even in [PIN⁺] strains, it appears to be even less frequent in the isogenic [pin⁻] strains. Neither translational suppression nor maintenance of the pre-existing [PSI⁺] state is affected by [PIN⁺]. [PIN⁺] was distinct from [PSI⁺], as [PIN⁺] transmission did not require PFD of Sup35 (Derkatch *et al.*, 1997).

Subsequently, the initially characterized [PIN⁺] element was identified with [RNQ⁺], the prion isoform of Rnq1 protein (Derkatch *et al.*, 2001). It was also shown that other QN-rich prions, such as [URE3] or prion isoform of the chimeric protein New1-Sup35, can substitute for [RNQ⁺] in this role (Derkatch *et al.*, 2001; Osherovich and Weissman, 2001). Moreover, several other QN-rich yeast proteins have been identified that promote [PSI⁺] formation in the [pin⁻] strains when they are co-overproduced with Sup35 (Derkatch *et al.*, 2001). It is possible that these proteins form prions, or prion-like aggregates, on their own, that in turn induces "prionization" of Sup35. Thus, pre-existing QN-rich prion aggregates promote formation of the new prions, generated by different (but also QN-rich) proteins. This demonstrates the existence of prion "cascades" in the yeast cell.

Two models have been proposed to explain the [PIN⁺] ("prion-inducing") effect of pre-existing prion aggregates (Derkatch *et al.*, 2001; Osherovich and Weissman, 2001). The "cross-seeding" model (Figure 2A) suggests that pre-existing prion aggregates of one protein serve as "seeds" for *de novo* formation of prion aggregates by the other protein. Although mature prion aggregates do not co-localize, it is possible that such co-localization could be detected at the early stages of prion formation, but is difficult to catch. The alternative "sequestration" model (Figure 2B) suggests that pre-existing prion aggregates sequester some protein co-factors that are involved in either proper folding or clearance of the other protein and are normally counteracting its aggregate-forming potential. Some proteins playing such a role are indeed known (see below), but so far, it was not possible to relate them to the [PIN⁺] phenomenon. Further experiments are needed to distinguish between the models.

Some Sup35 fragments, bearing the Sup35N (prion-forming) domain with certain non-Sup35 extensions but lacking the Sup35C region, can induce [PSI⁺] even in the [pin⁻] strains (Derkatch *et al.*, 1997, 2000). It remains unclear how the "[pin⁻] barrier" is overcome in these cases. It is worth noting that the extensions with [PIN⁺]-independent effect that are characterized to date contain a high percentage of the hydrophobic aa residues, and the resulting proteins are characterized by low yield, indicative of their low proteolytic stability (Derkatch *et al.*, 2000). It is possible that short fragments of Sup35 PFD, generated by proteolysis of these constructs, possess increased aggregation potential and are capable of generating the initial prion aggregates even in the absence of pre-existing "seeds".

It should be noted that in addition to the "co-induction" phenomenon, antagonistic interactions between some yeast prions have also been reported. For example, despite the fact that [PSI$^+$] and [URE3] facilitate *de novo* appearance of each other, they also inhibit each other's propagation in the cells where both prions are already present (Bradley *et al.*, 2002; Schwimmer and Masison, 2002). Overall, the complexity of the prion "networks" existing in the yeast cell cytoplasm is still poorly understood and requires further investigation.

3.3. Role of Prions in Aggregation of the Other Proteins

Recent data show that in addition to affecting each other, yeast prions also influence aggregation and toxicity of some foreign proteins, expressed in the yeast cell. The chimeric protein containing the exon 1 of human huntingtin (Htt), fused in frame to the jellyfish green fluorescent protein (GFP), has been used in these experiments. Huntingtin is associated with Huntington's disease, a neurodegeneration disorder resulting from expansion of the poly-Q stretch within the region encoded by exon 1. Poly-Q expanded huntingtin forms huge intracellular aggregates in mammalian cells that are suspected to be related to huntingtin's toxicity (see Waelters *et al.*, 2001; Ross, 2002). Poly-Q-expanded Htt-GFP construct also aggregated and caused toxicity in the yeast cells containing the [RNQ$^+$] prion (Meriin *et al.*, 2002). In the [rnq$^-$] or *rnq1Δ* cells, only rare aggregates with different characteristics were observed, and no toxicity was detected. This agrees with the previous observation that aggregation of the other mammalian poly-Q protein, MJD, is facilitated in the [PIN$^+$] or [NU$^+$] yeast cells (Osherovich and Weissman, 2001). [PSI$^+$] also appears to be able to partly substitute for [RNQ$^+$] in regard to Htt-GFP aggregation and toxicity (G. Newnam and Y. Chernoff, unpublished). It looks like at least in the case of [RNQ$^+$], Rnq1 protein directly interacts with Htt-GFP, possibly promoting its aggregation (Meriin *et al.*, 2003). Taken together, these data confirm that cytoplasmic prions, not necessarily toxic on their own, may cause susceptibility to other aggregation-related disorders. As a variety of

QN-rich proteins exist in the mammalian genomes, this notion could be relevant to the mammalian aggregation-related disorders as well.

4. Role of Hsp104 in Prion Propagation

Although it was proposed that propagation of mammalian prions may require participation of auxiliary protein(s) such as the hypothetical protein X (reviewed in: Prusiner, 1997; Harrison *et al.*, 1997), the first direct evidence of a chaperone role in prion propagation has come from the yeast model (Chernoff *et al.*, 1995). Propagation of yeast prions requires components of the cellular stress-response system. In yeast, the proteins affecting prion maintenance *in trans* were first identified by studying the effects of transient protein overproduction, the same tool that helped to identify Sup35 and Ure2 as prion carriers. Conventional genetic procedures allowing for inactivation of the corresponding genes were used on further stages.

4.1. Effects of Hsp104 on Yeast Prions

The first cellular protein shown to affect [PSI⁺] propagation *in trans* was the yeast chaperone protein Hsp104. Increased Hsp104 levels inhibit nonsense-suppression by [PSI⁺] (Chernoff and Ono, 1993; Chernoff *et al.*, 1995). Moreover, yeast cells can be cured of [PSI⁺] as a result of transient Hsp104 overproduction (Chernoff *et al.*, 1995). Surprisingly, deletion of the *HSP104* gene also resulted in loss of [PSI⁺]. The "[PSI⁺] no more" effect of the *hsp104Δ* deletion was recessive, while mutations inactivating the ATP-binding domains of Hsp104 protein exhibited a dominant negative effect on [PSI⁺] (Chernoff *et al.*, 1995). This could be due to an oligomeric organization of the Hsp104 protein (see below). Apparently, "mutant" monomers inactivate the whole oligomeric unit. Indeed, Hsp104 mutations that exhibit a dominant negative effect on [PSI⁺] propagation also inhibit Hsp104-dependent thermotolerance (see below) in a dominant fashion (Schirmer *et al.*, 2001; Wegrzyn *et al.*, 2001). These data confirm that the Hsp104 chaperone plays a unique role in [PSI⁺] propagation. If the balance between Hsp104 and the [PSI⁺]-forming protein Sup35 is shifted, reproduction of the prion state is impaired.

Hsp104 is also required for propagation of the yeast prions other than [PSI⁺], specifically [RNQ⁺], or [PIN⁺] (Derkatch *et al.*, 1997),

[NU$^+$] (Santoso *et al.*, 2000) and [URE3] (Moriyama *et al.*, 2000). However, these prions are not cured by overproduction of Hsp104. It has also been reported that heterologous [PSI$^+$]^{PS} prion, formed by the chimeric Sup35 protein with the Sup35N domain from *Pichia methanolica*, is less sensitive to overproduced Hsp104 than endogenous *S. cerevisiae* [PSI$^+$] (Kushnirov *et al.*, 2000a,b), although the curing effect of excess Hsp104 on [PSI$^+$]^{PS} was still observed at higher Hsp104 concentrations (E. Lewitin and Y. Chernoff, unpublished data). Interestingly, the prion formed by modified *S. cerevisiae* Sup35, in which M region was substituted by charged linker region from the human topoisomerase gene, was not cured by Hsp104 alterations (Liu *et al.*, 2002). Likewise, the heterologous prion formed in *S. cerevisiae* by the Sup35 protein with oligopeptide repeats containing no QN residues did not appear to require Hsp104, although maintenance of this prion was so far confirmed only at relatively high levels of the heterologous Sup35 protein (Crist *et al.*, 2002).

4.2. Hsp104 and Protein Disaggregation

The Hsp104 chaperone belongs to the evolutionarily conserved ClpB/Hsp100 family, members of which participate in various cellular processes (see Schirmer *et al.*, 1996). In yeast, Hsp104 is responsible for so-called induced thermotolerance, that is, adaptation to severe heat shock induced by pre-incubation at mild heat shock conditions (Sanchez and Lindquist, 1990). Hsp104 is also involved in the response to some other environmental stresses, as well as in the control of spore viability and long term viability of starving vegetative cells (Sanchez *et al.*, 1992). In molecular terms, Hsp104 is an ATPase (Parsell *et al.*, 1991) shown to promote solubilization of aggregated heat damaged proteins *in vivo* (Parsell *et al.*, 1994) and *in vitro* (Glover and Lindquist, 1998). This "disaggregation" function of Hsp104 is achieved in cooperation with other chaperones, Hsp70-Ssa (Sanchez *et al.*, 1992; Glover and Lindquist, 1998) and Hsp40-Ydj1 (Glover and Lindquist, 1998).

Certain features of the Hsp104 effect on protein aggregates could be understood from a comparison with its *E. coli* homolog, ClpB. The bi-chaperone network, consisting of the ClpB protein and DnaK-DnaJ-GrpE complex (a prokaryotic counterpart of the eukaryotic Hsp70-Hsp40), promotes *in vitro* re-solubilization and refolding of misfolded proteins, previously aggregated in the absence of chaperones (Goloubinoff *et al.*, 2000). It is thought that ClpB directly binds protein aggregates and, as a result of ATP hydrolysis, undergoes structural changes which increase the hydrophobic exposure of the aggregates and allow the DnaK-DnaJ-GrpE complex to bind and promote further disaggregation and refolding. Therefore, Hsp104 is required for initiation of the disaggregation process. It is possible that Hsp104 protein and some members of the Hsp70 and Hsp40 families act on the yeast protein aggregates by a similar mechanism. As the process of disaggregation can not be initiated in the absence of Hsp104, we will further refer to Hsp104 as to "disaggregase", even though its activity alone is apparently not sufficient for the completion of this process.

4.3. Of Hsp104's Requirement for Prion Propagation

Initiation of protein "disaggregation" by Hsp104 could explain the fact that an excess of this protein interferes with propagation of some prion aggregates. Indeed, the Sup35 protein was shifted from insoluble (prion) to the soluble (non-prion) fraction in the [PSI+] yeast cells overproducing Hsp104 (Patino *et al.*, 1996; Paushkin *et al.*, 1996). However, more difficult to explain is the observation that loss of Hsp104 activity eliminates [PSI+]. It has initially been hypothesized that Hsp104 activity could be involved in generation of the intermediate, a partially misfolded conformer of Sup35, that serves as a source for prion conversion (Chernoff *et al.*, 1995). For example, it is possible that alterations of the Hsp104 shift the balance between completely folded Sup35 protein and partly unstructured polypeptide which serves as a substrate for the process of aggregation. The role of unstructured polypeptide in the process of Sup35 aggregation is in an agreement with *in vitro* data (Serio *et al.*, 2000). Another model suggests that Hsp104 could be involved in the process of prion conversion *per se* (Patino *et al.*, 1996). One more model (Paushkin *et al.*, 1996) proposes that Hsp104 is needed for initiation of the "seeding" reaction: in the absence of Hsp104, prion aggregates can grow but can not be broken down into the oligomeric "seeds" which initiate new rounds of prion reproduction. As a result, huge aggregates are eventually diluted and lost in cell divisions. In either model, lack of Hsp104 activity blocks prion reproduction rather than removes prion conformers instantly. Therefore, Hsp104 inactivation should result in the generation-dependent loss of a prion. Indeed, this is the case for

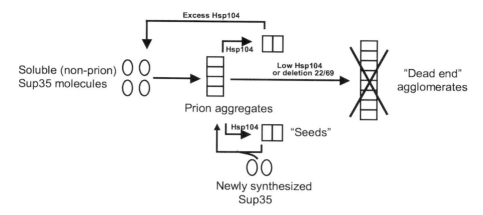

Figure 3. Role of Hsp104 in the [PSI⁺] propagation cycle. Designations: ☐ – prion isoform; ○ – non-prion isoform. Moderate levels of Hsp104 are responsible for initiation of aggregate "shearing", leading to formation of the new "seeds". High levels of Hsp104 disaggregate prion polymers into monomers that refold to the non-prion conformation. Lack or low level of Hsp104, or deletion of the region between amino acid residues 22 and 69, impairing the "shearing" capability of Sup35 aggregates, result in uncontrollable aggregate growth and accumulation of large agglomerates incapable of proliferation.

[PSI⁺] and [PIN⁺] (Wegrzyn *et al.*, 2001). However, [PSI⁺] loss after Hsp104 inactivation appears to be more rapid than would be expected if Hsp104 simply blocks [PSI⁺] proliferation without altering its structure. Together with the observation that the presence of Hsp104 influences two-hybrid interactions between Sup35N and some other proteins (Bailleul *et al.*, 1999), this suggests a more complex role of Hsp104 in [PSI⁺] maintenance.

Recent data (Wegrzyn *et al.*, 2001) indicate that most likely, the major role of Hsp104 in [PSI⁺] propagation is to break the huge aggregates down into the small "seeds". In the experiments that used [PSI⁺] cells with GFP-tagged Sup35, loss of Hsp104 activity was accompanied by rapid accumulation of huge cytologically detectable Sup35-GFP aggregates. Moreover, the proportion of cells containing these large aggregates correlated with the proportion of [psi⁻] colonies, obtained after plating these cultures onto solid medium. This shows that large aggregates are not capable of efficiently propagating the prion state. Formation of the large aggregates in response to Hsp104 inactivation was also detected by secondary immunofluorescence in the initially [PSI⁺] cells containing non-tagged Sup35, confirming that these aggregates do not represent an artifact related to the presence of GFP tag.

Apparently, Hsp104-dependent disaggregating activity keeps the aggregate size at a level that is optimal for prion propagation (see Figure 3). If there is no Hsp104 present, rapid uncontrolled growth of an aggregate leads to the loss of its prion-propagating activity. If prion aggregates are composed of amyloid-like fibers (which seems likely), the inverse correlation between average length of a fiber and efficiency of amyloid propagation is expected, as only ends of the fibers are active in immobilizing protein molecules from the solution. In addition to this, large cytologically visible clumps could represent the complexes of the bundled fibers, in which the internally located components are screened from interaction with the soluble protein floating in cytoplasm. It is also possible that large aggregates are more difficult to transmit from mother cells to the buds during cell divisions, although direct evidence for such an asymmetry is lacking thus far. Some or all of these factors working together apparently turn large aggregates into "dead ends" of prion propagation. Indeed, some conditions other than Hsp104 inactivation increase aggregate size and also exhibit an inhibitory effect on prion propagation. For example, Sup35 overproduction in prion-containing cells results in appearance of the large cytologically visible aggregates (see Chernoff *et al.*, 2002 for review) and also causes detectable loss of a prion (K. Allen and Y. Chernoff, unpublished data).

It should be noted that formation of the cytologically detectable clumps by GFP-tagged prion proteins has been used to detect yeast prions in the cells (Patino *et al.*, 1996). This tool remains useful but should be employed with caution. While visible aggregates certainly result from pre-existing prions, the data discussed above suggest that such aggregates could hardly be identified with proliferating prion units themselves. Indeed, some isolates of [URE3] do not form cytologically visible Ure2 clumps (Fernandez-Bellot *et al.*, 2002).

The "aggregate-shearing" role of Hsp104 is supported by the experiments with the deletion derivative of Sup35, Sup35-Δ22/69 (previously known as Sup35-ΔBstEII), that lacks a region between aa residues 22 and 69. This derivative, initially described as "[PSI$^+$] no more" (Ter-Avanesyan *et al.*, 1994), in fact retains the ability to both induce *de novo* prion formation upon overproduction and turn into a prion state (Borchsenius *et al.*, 2001). However, the resulting prion (called [PSI$^+$]$^{\Delta22/69}$) is very unstable in non-selective conditions. Moreover, prion aggregates formed by the Sup35-Δ22/69 protein are characterized by extremely large size and unusual morphology. These data suggest that the Δ22/69 deletion inhibits aggregate shearing. This decreases production of the new prion seeds and converts a heritable ("infectious") prion into one that is rapidly growing but almost inactive in prion conversion, and therefore produces mitotically unstable protein aggregate. In contrast to its effect on "conventional" [PSI+], moderate excess of the Hsp104 disaggregase increases the frequency and longevity of the [PSI+]$^{Δ22/69}$ derivatives (Borchsenius *et al.*, 2001), apparently by promoting their shearing and generating larger number of the active seeds.

Interestingly, the heterologous prion [PSI+]^{PS} described above, that is relatively insensitive to the curing effect of excess Hsp104, is also characterized by the larger proportion of protein in the fast precipitating fraction corresponding to the large aggregates (Kushnirov *et al.*, 2000a,b). It is possible that the relatively low mitotic stability of [PSI+]^{PS}, relative to the endogenous [PSI+] prion (Chernoff *et al.*, 2000; Kushnirov *et al.*, 2000a), is partly explained by its low sensitivity to the "shearing" defect of Hsp104.

Taken together, these data suggest that a major distinction between the prion aggregates and "non-infectious" (non-propagating) aggregates is the mode of their response to the chaperone system of the cell. *In vivo*, the "shearing" activity of Hsp104 essentially makes yeast prion a prion.

4.4. Hsp104's Role in Prion Curing by Environmental Stresses and Chemical Agents

Certain chemical agents and environmental treatments are known to cure yeast cells of prions. Most detailed analysis of prion-curing treatments has been performed in case of [PSI+] (see Cox *et al.*, 1988 for review). Since Hsp104 levels are increased by at least some of these treatments, it was logical to suggest that curing yeast cells of [PSI+] in such cases occurs by modifying levels or activity of Hsp104 (Chernoff *et al.*, 1995). Indeed, Hsp104 is induced by heat shock, ethanol and osmotic stress (Sanchez *et al.*, 1992), treatments known to cure yeast cells of [PSI+] to a certain degree (see Cox *et al.*, 1988 for review). The role of Hsp104 induction in [PSI+] curing by heat shock has been confirmed by recent investigations (Y. Chernoff, G. Newnam, L. Ozolins and J. Birchmore, in preparation). Moreover, Hsp104 is also expressed in response to UV or radiation induced DNA damage (T. Magee and K. McEntee, personal communication, quoted in: Chernoff *et al.*, 1995). Therefore, it has been proposed that a [PSI⁺]-curing effect of UV is mediated by induction of Hsp104 (Chernoff *et al.*, 1995). This removes a major obstacle to the "protein only" model of the [PSI⁺] phenomenon, that is, the observation that a "[PSI⁺]-curing" effect of UV can be partially reversed by photoreactivation (Cox *et al.*, 1988). Indeed, photoreactivation is a light-dependent DNA repair process that decreases remaining DNA damage and may in turn result in decreased Hsp104 induction. Thus, the effect of some prion-curing agents could be mediated by chaperones in the same way as effects of DNA mutagens are mediated by DNA repair systems.

The most extensively characterized [PSI⁺] curing agent, guanidine hydrochloride or GuHCl (Tuite *et al.*, 1981), also increases the levels of Hsp104 expression (Chernoff *et al.*, 1995; Lindquist *et al.*, 1995). However, it is unlikely that the GuHCl effect on [PSI⁺] is explained by Hsp104 induction. First, GuHCl also cures yeast cells of [URE3] (Wickner, 1994) and [PIN⁺], or [RNQ⁺] (Derkatch *et al.*, 1997), prions that are not sensitive to excess Hsp104. Second, while the Sup35-solubilizing (Paushkin *et al.*, 1996) and [PSI⁺]-curing (Wegrzyn *et al.*, 2001) effects of overproduced Hsp104 are relatively fast, prion curing by GuHCl is strictly generation-dependent (Eaglestone *et al.*, 2000). The stress treatments increasing Hsp104 levels, such as ethanol stress or heat shock, do not increase the [PSI⁺]-curing effect of GuHCl (Ferreira *et al.*, 2001). Moreover, Hsp104 ATPase activity is inhibited *in vitro* by millimolar concentrations of GuHCl (Glover and Lindquist, 1998), the same concentrations that are used to cure yeast cells of [PSI⁺]. This means that even though Hsp104 levels are elevated in the presence of GuHCl, the total Hsp104 activity is unlikely to be increased. In fact, recent data show that the Hsp104-dependent component of thermotolerance is strongly inhibited by millimolar concentrations of GuHCl, suggesting a significant decrease of Hsp104 activity in the presence of Hsp104 (Ferreira *et al.*, 2001; Jung *et al.*, 2001). It is therefore likely that GuHCl-induced loss of prions occurs due to inactivation (rather than induction) of Hsp104. Indeed, mutant alleles of the *HSP104* gene have been identified that confer resistance to the prion-curing effect of GuHCl (Jung *et al.*, 2002). This confirms an essential role of Hsp104 inactivation in the process of GuHCl-induced prion curing.

However, certain kinetic parameters of [PSI⁺] curing by GuHCl treatment and by Hsp104 inactivation are different from each other.

Loss of [PSI⁺] in the presence of GuHCl is observed only after a long lag period of about 4-6 generations, depending on the [PSI⁺] isolate (Eaglestone *et al.*, 2000). This has been interpreted as evidence that GuHCl does not affect pre-existing prion "seeds" but rather blocks "seed" proliferation, resulting in subsequent loss of a prion by dilution in cell divisions. Length of the lag period in the presence of GuHCl has therefore been used as a tool to calculate the average number of the actively proliferating [PSI⁺] "seeds" per cell (Eaglestone *et al.*, 2000). Indeed, such a mode of action for GuHCl would be in agreement with the role of Hsp104 as an "aggregate-shearing" agent that is required for initiation of the process generating new "seeds" from pre-existing aggregates (Wegrzyn *et al.*, 2001). However, it turned out that direct inactivation of Hsp104 by genetic manipulations cures yeast cells of [PSI⁺] faster than growth in the presence of GuHCl.

When the [PSI⁺] yeast diploid, heterozygous by *hsp104Δ* deletion, has been sporulated and dissected, the [PSI⁺] loss in *hsp104Δ* spores has been detected after the short 2-division lag, in contrast to 5-division lag, detected in the same prion isolate during GuHCl treatment (Wegrzyn *et al.*, 2001). Moreover, the actual lag in case of *hsp104Δ* must be even shorter than the observed one, as Hsp104 is a very stable protein whose levels are decreased in the cells that have lost the HSP104 gene only by dilution in subsequent cell divisions. Thus, 2 cell divisions are needed to dilute Hsp104 down to 25% of the normal levels. As independent experiments confirm that 35-40% of wild-type Hsp104 activity are sufficient to maintain [PSI⁺] (Wegrzyn *et al.*, 2001), it appears that [PSI⁺] loss begins almost immediately after the cellular concentrations of Hsp104 drop below this level. Indeed, the alternative approach, employing the dominant negative mutant allele of the *HSP104* gene (*HSP104-KT*) under the inducible promoter, also detected rapid loss of [PSI⁺] loss at least in some strains and experimental conditions (Wegrzyn *et al.*, 2001). This should also be noted that abovementioned prion, formed by the modified Sup35 protein with topoisomerase linker instead of M region and incurable by *hsp104* deletion, was still cured by GuHCl (Liu *et al.*, 2002).

To deal with at least some of these discrepancies, Ness *et al.* (2002) proposed that rapid loss of [PSI⁺] in the sporulation/dissection experiments could be due to dilution of the number of prion seeds during meiosis, when four haploid cells are generated from one initial diploid cell. If spores have a much smaller number of "seeds" to begin with, this could explain the shorter lag. As for the induction of

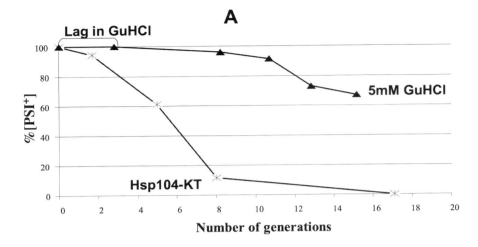

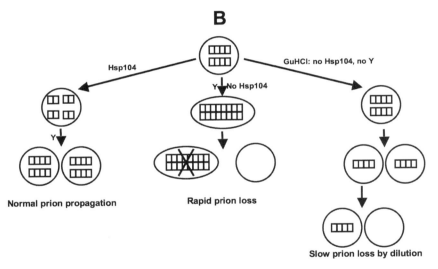

Figure 4. Comparison of the [PSI⁺]-curing effects of GuHCl treatment and Hsp104 inactivation. A – [PSI⁺] curing by Hsp104 inactivation occurs faster than [PSI⁺] curing by GuHCl. ▲ - growth in the presence of 5 mM GuHCl; ✳- galactose-induced expression of the dominant negative Hsp104-KT mutant allele. Data are from Wegrzyn *et al.*, 2001. Percentages are recalculated with mosaic colonies counted as [PSI⁺], as suggested by Ness *et al.*, 2002. [PSI⁺] curing by Hsp104 inactivation does not show a long lag, in contrast to [PSI⁺] curing by GuHCl in the same conditions. B - Model for the prion-curing effect of GuHCl. Hsp104 is responsible for aggregate "shearing" and formation of the new "seeds". The unknown factor Y assists in aggregate growth. Selective inactivation of Hsp104 results in uncontrollable aggregate growth in result of Y action that leads to the rapid loss of prion-converting activity. GuHCl inactivates both Hsp104 and Y, that results in "freezing" the aggregates and their subsequent generation-dependent loss by dilution in cell divisions. See also comments in the text.

Hsp104-KT, the same authors suggested that seemingly rapid loss of [PSI⁺] could be an artifact caused by the mutant Hsp104 that remains in the cells after induction is turned off, and continues to act during formation of a colony in the non-selective conditions.

However, neither explanation appears to remove the discrepancies completely. By Ness *et al.*'s (2002) own data, even cells in which the number of "seeds" is diluted down to 1 in the presence of GuHCl, amplify to the normal number (in these experiments, about 60) within less than one generation in the absence of GuHCl. As levels of Hsp104 in the *hsp104Δ* spores originating from the Hsp104⁺ strain initially remain normal (see above), there is no reason to expect that they could not restore the "seed" copy number in the same way. Moreover, if "seed" numbers in the germinating spores remain so low that "dilution" of seeds takes an effect almost immediately, this should lead to at least occasional spontaneous loss of [PSI⁺] in the first mitotic deletions of the normal Hsp104⁺ spores as well. Such a phenomenon has never been observed.

To distinguish between the immediate and "postponed" action of Hsp104-KT in the induction experiments, Ness *et al.* (2002) recommended to count only complete [psi⁻] colonies as ones originated from the instantly cured cells, and consider "mosaic" (sectored [PSI⁺]/[psi⁻]) colonies as ones generated by the cells remaining [PSI⁺] by the end of the induction period. Such an approach produced more or less comparable kinetics of [PSI⁺] loss on GuHCl and after Hsp104-KT induction. However, the same calculation technique applied to the data of Wegrzyn *et al.* (2001) confirms that the lag period for Hsp104-KT in these experiments was still much shorter than that for GuHCl (Figure 4A; K. Bapat, R. Wegrzyn and Y. Chernoff, unpublished data). It should also be noted that rapid increase in size of the detectable Sup35 aggregates, observed after Hsp104 inactivation due to induction of Hsp104-KT (Wegrzyn *et al.*, 2001), has never been reported for GuHCl.

One could suggest that longer lag in GuHCl experiments is due to incomplete or slow inactivation of Hsp104 by GuHCl. This, in turn, would mean that the actual number of prion "seeds" per cell is much less than the number calculated from GuHCl experiments. If there are in fact only about 3-4 seeds per cell rather than 60, this would satisfactorily explain almost immediate effect of Hsp104 inactivation.

However, such a low number of seeds seem highly unlikely, as it does not agree with high mitotic stability of the prion isolates used in these experiments.

It seems more likely that GuHCl inhibits both Hsp104 (factor, initiating aggregate shearing) and another unknown factor (or factors) promoting aggregate growth (Figure 4B). Therefore, prion aggregates are "frozen" and eventually lost by dilution in cell divisions. In the case of selective inactivation of Hsp104, the rapid uncontrollable growth of aggregates occurs that results in much more rapid loss of prion-propagating activity due to large aggregate size, as discussed above. In the case of Hsp104-insensitive prions, GuHCl could either inhibit another disaggregase substituting for Hsp104, or reduce prion proliferation by inhibiting the growth-promoting cofactors While not questioning the important (and at least for some prions, essential) role of Hsp104 inactivation in prion curing by GuHCl, this model suggests that GuHCl treatment could not be used as a simple tool for selective Hsp104 inactivation, as effects of this chemical on cells in general and prions in particular are more complex.

4.5. Are there Hsp104's Homologs/Analogs in Animal Systems?

It appears that Hsp104 also influences some aggregated proteins other than prions. For example, efficient aggregation of polyQ-expanded huntingtin in yeast requires functional Hsp104 (Krobitsch and Lindquist, 2000; Meriin *et al.*, 2002), although it is likely that Hsp104 inactivation prevents polyQ aggregation indirectly by eliminating the pre-existing endogenous yeast prions (in these experiments, [RNQ⁺]), that promote polyQ aggregation (Meriin *et al.*, 2002). It was shown that artificially introduced yeast Hsp104 can counteract polyQ aggregation in the heterologous systems, *e. g.* in the nematode *Caenorhabditis elegans* (Satyal *et al.*, 2000). *In vitro*, Hsp104 promoted conversion of mammalian PrP protein into the aggregated proteinase-resistance isoform (DebBurman *et al.*, 1997). However, it is not likely that an endogenous Hsp104 ortholog plays a major role in propagation of prions or polyQ aggregates in the mammalian cells. First of all, despite the fact that Hsp104 is conserved among fungi and plants, and its homolog (called ClpB) is also found in prokaryotes (see Schirmer *et al.*, 1996), no Hsp104 orthologs were identified in mammalian systems to date. Neither were Hsp104 orthologs found in the sequenced genomes of other animals, *Caenorhabditis elegans* or *Drosophila melanogaster*. This suggests that functions performed by Hsp104 in yeast have to be overtaken by the other chaperones in the mammalian systems. The obvious candidates are proteins of the Hsp70 and Hsp40 families that interact with Hsp104 in solubilization of the heat-damaged aggregated proteins, and are conserved across the kingdoms. Indeed, members of these families are also involved in the propagation of yeast prions.

5. Role of Other Hsps in Prion Formation and Propagation

It is clear that [PSI⁺] behavior in various growth conditions can not be solely explained by variations in Hsp104 levels or activity. For example, [PSI⁺] is not eliminated at any detectable level by growth at increased but still permissive temperature ("mild heat shock", in contrast to severe heat shock) or by incubation in the stationary phase (see Cox *et al.*, 1988 for review), although these conditions induce Hsp104 (Sanchez *et al.*, 1992). Apparently, other proteins or physiological changes associated with these conditions could interfere with Hsp104 effects on [PSI⁺]. Indeed, further experiments confirmed that the chaperones of Hsp70 and Hsp40 families, previously shown to interact with Hsp104 in disaggregating and refolding of heat damaged proteins, are also involved in the cellular control of yeast prions.

5.1. Effects of Hsp70-Ssa on Yeast Prions

The cytoplasmic proteins of Hsp70 family are represented in yeast by two major subfamilies, Ssa and Ssb. Ssa subfamily includes four proteins Ssa1, 2, 3, and 4 (Werner-Washburne *et al.*, 1987). While Ssa1 and Ssa2 are highly homologous to each other, Ssa3 and Ssa4 are somewhat diverged from them. The total level of Ssa proteins in the cell is increased in response to high temperature and other stresses, stationary phase, and sporulation, although different members of the subfamily exhibit different modes of regulation. Ssa2 is constitutively expressed at high levels, while Ssa1 is expressed at moderate levels but induced by stress; Ssa3 and Ssa4 are strictly stress-inducible proteins that are not expressed in the vegetative cells in most genotypic backgrounds. Deletion of *SSA1* and *SSA2* genes usually leads to compensatory induction of the remaining members of the subfamily. Simultaneous deletion of all four *SSA* genes is lethal.

As discussed above, the Ssa1 protein (and possibly other members of the Ssa subfamily), together with Hsp104 and Ydj1, participate in disaggregating and refolding of heat damaged protein agglomerates (Glover and Lindquist, 1998). One could expect that excess Ssa1 would facilitate [PSI⁺] loss in the presence of excess Hsp104. Surprisingly, we have shown that overproduction of Ssa1 prevents efficient [PSI⁺] "curing" by overexpressed Hsp104 and counteracts solubilization of the Sup35 aggregates in [PSI⁺] strains in the presence of Hsp104 (Newnam *et al.*, 1999). Moreover, increased Ssa1 levels lead to increased efficiency of nonsense-suppression in the [PSI⁺] strains, apparently due to increased proliferation of the [PSI⁺] prion. It appears that effects of Hsp70-Ssa on aggregates of heat-damaged proteins and on prion aggregates diverged from each other.

A mutation in the *SSA1* gene, *ssa1-21*, was shown to impair [PSI⁺] stability, most severely in the strains lacking the major constitutively expressed member of the family, *SSA2* (Jung *et al.*, 2000). In the double *ssa1 ssa2* deletion, *ssa1-21* behaved as "[PSI⁺] no more." Interestingly, double *ssa1 ssa2* deletion did not cure [PSI⁺] by itself. It is possible that Hsp70-Ssa is essential for [PSI⁺] maintenance, this function of Ssa1 is impaired by *ssa1-21* mutation, and compensatory induction of Ssa3 and Ssa4 proteins is sufficient to compensate for this function. Indeed, in the presence of Ssa1-21 compensatory induction does not occur, as this mutant is not defective in the other functions of Hsp70-Ssa. However, it is also possible that *ssa1-21* is a gain-of-function or hyperfunction allele, and this is why it affects [PSI⁺] in a dominant or semi-dominant function. Indeed, several more *ssa1* mutations antagonizing [PSI⁺] have been described recently, as well as second site mutations that compensate for the anti-prion effect of *ssa1-21* (Jones and Masison, 2003). Effects of mutations on [PSI⁺] did not correlate with their growth phenotypes, indicating that they did not result from the simple loss of the Ssa1 function. Moreover, location of second site suppressors suggested that they affect substrate-trapping interaction, indicating that the initial effect of *ssa1-21* could result from more efficient, rather than less efficient, interaction with the substrate. By itself, the "[PSI⁺] no more" phenotype of some *ssa1* mutants does not yet prove that Hsp70-Ssa is essential for [PSI⁺] propagation.

Interestingly, effects of Hsp70-Ssa on prions are strain-specific and could be protein-specific. In some strains, excess Ssa1 is antagonistic to [PSI⁺] (Chernoff *et al.*, 1995; Lindquist *et al.*, 1995; Kushnirov *et al.*, 2000b). Excess Ssa1 (but not excess Ssa2) also cures [URE3] (Schwimmer and Masison, 2002). Such a difference between two proteins is surprising, taking into account the high level of homology between them. It appears that Ssa2 acts on [PSI⁺] in the same way as Ssa1 (R. Wegrzyn and Y. Chernoff, unpublished data). Ssa3 and Ssa4 have not yet been checked directly for their effects on prions.

While excess Ssa1 "protects" [PSI⁺] from curing by excess Hsp104, it increases efficiency of [PSI⁺] curing by the dominant negative Hsp104 mutant that acts by inactivating cellular Hsp104 (Wegrzyn *et al.*, 2001). Excess Ssa1 also antagonizes the shearing-defective prion [PSI⁺]$^{\Delta 22/69}$ (Borchsenius *et al.*, 2001) and the heterologous [PSI⁺]^{PS} prions (Kushnirov *et al.*, 2000b) with suspected shearing defect (see above). Ssa1 overproduction increases efficiency of [PSI⁺] induction by excess Sup35 in the [psi⁻ PIN⁺] background, while simultaneous overproduction of Ssa1 and Sup35 in [PSI⁺] background leads to increased loss of [PSI⁺] (K. Allen, R. Wegrzyn and Y. Chernoff, unpublished data). Taken together, these data suggest that excess Ssa1 works as a "[PSI⁺] helper" when aggregate size is small, but becomes a [PSI⁺] antagonist if the aggregate size is increased. Such an observation would be consistent with the aggregation-stimulating role of Ssa1 (and possibly Ssa2). This would facilitate propagation of the small "seeding-proficient" aggregates, but further increase the size and therefore decrease prion-converting ability of the larger aggregates (*e.g.*, aggregates formed in Hsp104-depleted cells, [PSI⁺]$^{\Delta 22/69}$ aggregates, or aggregates formed as a result of Sup35 overproduction in the [PSI⁺]-containing cells), converting them to the "dead end" agglomerates.

Why should a chaperone protein stimulate aggregate growth? One possibility is that excess Ssa1 increases aggregate size indirectly, by inhibiting the activity of Hsp104 or preventing it from binding the aggregates. It is possible that overproduced Ssa1 interacts with Hsp104, but due to shortage of other, as yet unknown cofactors (*e.g.*, members of the Hsp40 family) this complex remains inefficient in disaggregating prions. In the absence of excess Ssa1, other proteins (possibly, other members of the Hsp70 family) may assist Hsp104 in its disaggregating function. Indeed, a certain level of inhibition of Hsp104 activity in thermotolerance has been observed in the cells overproducing Ssa1 (Newnam *et al.*, 1999). However, in contrast to its effect on [PSI⁺], this effect of Ssa1 was overcome by a further increase in Hsp104 levels.

Another explanation is that due to the high order structure of prion aggregates, Ssa1 protein recognizes them as legitimate subcellular structures rather than agglomerates of misfolded proteins. Therefore, Ssa1 helps to convert Sup35 protein back to the prion state, thus reversing the Hsp104 effect. This could be achieved by stabilizing misfolded proteins, generated due to Hsp104 action, in the intermediate state that facilitates their conversion back into a prion. In this way, Hsp70-Ssa may actually promote prion propagation *in vivo*. Indeed, mammalian Hsp70 homologs have been shown to be involved in the protection of high order structures (*e.g.*, cytoskeletal networks) during heat shock, suggesting that Hsp70 may assist in reassembly of the multiprotein complexes (Liang and McRae, 1997).

Although Hsp104 levels are increased during growth at high temperature and in the stationary phase, prion loss has not been observed in these conditions. As levels of Hsp70-Ssa (and specifically, levels of Ssa1) are also increased in response to the same treatments, it is likely that Hsp70-Ssa plays a major role in protection of [PSI⁺] from the curing effect of Hsp104.

5.2. Hsp70-Ssb as a Prion Antagonist

Another yeast cytosolic Hsp70 subfamily, Ssb, includes two essentially identical proteins, Ssb1 and Ssb2. In contrast to the Ssa subfamily, the Ssb subfamily is constitutively expressed, is not induced in response to high temperature, and is not essential for viability (Nelson *et al.*, 1992). Biochemical data suggest that Ssb proteins are associated with the translating ribosomes and nascent polypeptides (Nelson *et al.*, 1992; Pfund *et al.*, 1998). Alterations of Ssb levels also affect ubiquitin-dependent proteolysis (Ohba, 1994, 1997). This suggests that one of Ssb functions is the "proofreading" of the newly folded polypeptides, that is, identification of the misfolded ones and either refolding them or targeting them for degradation.

We have observed that increased levels of Ssb enhance the "[PSI⁺] curing" effect of excess Hsp104, while deletion of both *SSB* genes decreases the efficiency of [PSI⁺] "curing" by excess Hsp104 and significantly increases the rate of spontaneous [PSI⁺] formation in the [PIN⁺] background (Chernoff *et al.*, 1999). Some weak

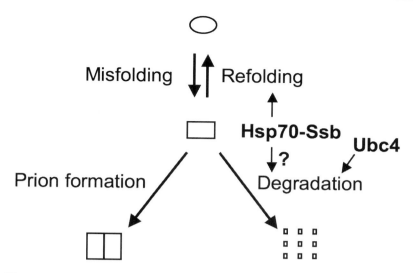

Figure 5. Model to explain the "protein mutator" phenotypes of deletions *ssb1 ssb2* and *ubc4*. Hsp70-Ssb catalyzes refolding and possibly degradation, and Ubc4 catalyzes ubiquitination (leading to degradation) of the misfolded protein, that serves as a source of prion formation. In case when refolding or degradation is impaired, prion formation is increased. See comments in the text.

[PSI$^+$] isolates generated in the absence of Ssb are antagonized by reintroduction of the *SSB1* gene (Chernoff *et al.*, 1999). Moreover, in some strains, continuous overproduction of Ssb alone is sufficient to cause detectable loss of [PSI$^+$] (Kushnirov *et al.*, 2000b; Chacinska *et al.*, 2001). Therefore, Ssb protein consistently behaves as a [PSI$^+$] antagonist. Possibly Ssb recognizes the misfolded polypeptides, which could potentially be converted into prions or the partially unfolded intermediates generated as a result of Hsp104-mediated disaggregation. Ssb either catalyzes refolding these misfolded proteins into the non-prion conformation or targets them for degradation, thus eliminating the potential source for prion generation or recovery (Figure 5). Therefore, the double *ssb1 ssb2* deletion is the first example of a "protein mutator", that increases heritable variability at the protein level by acting *in trans*.

5.3. Effects of Hsp40 Proteins on Yeast Prions

The third component of the yeast disaggregating/refolding chaperone complex, Hsp40, corresponds to the prokaryotic DnaJ. Several members of the yeast Hsp40 family have been identified, and at least two of them are shown to influence prions. Excess of Ydj1, thought to be a co-factor of Hsp70-Ssa, causes loss of [URE3] (Moriyama *et al.*, 2000) and some variants of [RNQ⁺] (Bradley *et al.*, 2002), and antagonizes certain isolates of [PSI⁺], that are sensitive to excess Ssa1 as well (Kushnirov *et al.*, 2000b). One possibility could be that Ydj1 effects are due to activation of Hsp70-Ssa. However, protection of [PSI⁺] from excess Hsp104 by excess Ssa1 does not require Ydj1 (K. Allen and Y. Chernoff, unpublished). Another member of the Hsp40 family, Sis1, is essential for viability and can not be eliminated. However, internal deletions in *SIS1* gene have been generated and shown to eliminate [RNQ⁺] (Sondheimer *et al.*, 2001). Excess of Sis1 or another protein of the Hsp40 family, YNL077W, also antagonizes [PSI⁺]^{PS}, a heterologous prion isolate formed by the chimeric *Pichia-Saccharomyces* Sup35 protein (Kryndushkin *et al.*, 2002). It is an intriguing possibility that Hsp40 proteins could be involved in determining the specificity of interactions between Hsp104/Hsp70 proteins and various prions. However, further experiments have to be undertaken in order to prove or disprove this hypothesis.

5.4. Other Chaperones or Hsp-Inducing Proteins That Affect Prions

The heterologous [PSI⁺]^{PS} prion, characterized by low sensitivity to the curing effect of excess Hsp104, low mitotic stability and probably by large aggregate size (see above), has been used to screen for the other proteins, whose overproduction exhibits an antagonistic effect on [PSI⁺] (Kryndushkin *et al.*, 2002). Among those, the chaperone Sti1, transcriptional factors Sfl1 and Ssn8, and the acidic ribosomal protein Rpp0 were identified. Overproduction of the latter three proteins activates at least some stress-related promoters, suggesting that these proteins might act via Hsp induction.

5.5. Comparison of the Effects of Hsps on Yeast Prions and Other Protein Aggregates

Hsp70 and Hsp40 proteins also influence protein aggregates other than yeast prions. For instance, Hsp70 suppresses polyQ aggregation in the *Drosophila* model (Chai *et al.*, 1999; Warrick *et al.*, 1999). Excess Hsp70-Ssa appears to antagonize assembly of the polyQ proteins into amyloid-like fibers both *in vitro* and the yeast cell (Muchowski *et al.*,

2000). Interestingly, the double *ssa1 ssa2* deletion also decreased aggregation and toxicity of the polyQ expanded huntingtin-based construct (Htt-GFP) in the yeast model (Meriin *et al.*, 2002). Hsp40 and Hsp70 proteins were also shown to antagonize polyQ aggregation in yeast and bind polyQ aggregates (Muchowski *et al.*, 2000). On the other hand, mutations in the Hsp40 homolog suppressed polyglutamine toxicity in *Drosophila* (Kazemi-Esfarjani and Benzer, 2000), while mutation in the *YDJ1* gene decreased polyQ aggregation and toxicity in yeast (Meriin *et al.*, 2002). It appears that certain levels of some Hsp70 and Hsp40 proteins facilitate polyQ aggregation, while excess Ssa becomes antagonistic to polyQ aggregates. To a certain extent, this recapitulates the effects of Ssa proteins on [PSI$^+$] system, especially if one takes into consideration that polyQ aggregates resemble the large aggregates of shearing-defective prions such as [PSI$^+$]$^{\Delta 22/69}$ (see above).

6. Other Cellular Proteins Influencing Prions

6.1. Cytoskeletal Networks

In addition to the chaperones, some other cellular proteins influence prion formation and propagation. One example is the cytoskeletal assembly protein Sla1. The known function of Sla1 is to assist in "nucleation" of microfilaments of actin cortical cytoskeleton (Holtzman *et al.*, 1993), although Sla1 is neither essential for this process nor is proven to interact directly with actin. Sla1 was shown to affect distribution of the other actin assembly protein, Sla2 (Ayscough *et al.*, 1999), playing an important role in assembly of the cortical actin patches and endocytic vesicles (see Ayscough, 2000 for review). The two-hybrid screen has uncovered interactions between the C-terminal domain of Sla1 and the N-terminal (prion-forming) domain of Sup35 (Bailleul *et al.*, 1999). Quite remarkably, the two-hybrid interaction between Sup35N and Sla1C was inhibited by Hsp104 inactivation. Moreover, the mutant derivatives of the Sup35N bearing certain *PNM* ("[PSI$^+$] no more") mutations were unable to interact efficiently with Sla1C. In the [psi$^-$] strain lacking the Sla1 protein, *de novo* [PSI$^+$] formation induced by Sup35 (or Sup35N) overproduction was significantly decreased, compared to the isogenic Sla1$^+$ strain (Bailleul *et al.*, 1999). While the [PSI$^+$] strains lacking Sla1 protein were still able to maintain a prion, the "[PSI$^+$] curing" effect of agents such as dimethylsulfoxide or overproduced Hsp104 was significantly increased in such strains. These results indicate that Sla1 protein plays an important, although not essential, role in prion formation and "recovery" from prion-curing treatments.

The molecular mechanisms of the effect of Sla1 on [PSI⁺] remain unclear. There is evidence of direct interactions between the actin cytoskeleton and components of the translational machinery, including elongation factor EF-1α (Yang *et al.*, 1990; Edmonds, 1993; Suda *et al.*, 1999), which is partially homologous to the C-terminal (release factor) domain of Sup35 (Kushnirov *et al.*, 1988; Wilson and Culbertson, 1988). Depletion of functional Sup35 results in defects of the cytoskeletal assembly (Valouev *et al.*, 2002). Prolonged incubation of yeast cells in the presence of latrunculin A, an actin cytoskeleton disrupting agent, causes loss of [PSI⁺] (Bailleul-Winslett *et al.*, 2000). However, fluorescence microscopy assays failed to detect a direct association between the Sup35^{PSI+} aggregates and the cortical actin cytoskeleton (Bailleul-Winslett *et al.*, 2000).

Interestingly, Sla1 also interacts with polyQ expanded huntingtin in the two-hybrid assay (Bailleul *et al.*, 1999), while Sla2 is a yeast homolog of mammalian Hip1, a huntingtin-interacting protein (Kalchman *et al.*, 1997). Prion "nucleated polymerization" possesses striking similarities to the assembly of the highly ordered protein complexes, such as cytoskeletal networks. Multiple alterations of the cytoskeletal structures were detected in patients with prion diseases and other aggregation-related disorders (Gray, 1986; Guiroy *et al.*, 1989; Liberski, 1994). A number of proteins involved in formation of cytoskeletal or surface structures possess the PFD-like QN-rich domains, *e. g.* Sla2 (Michelitsch and Weissman, 2000), and/or oligopeptide repeats not dissimilar from those found in PrP and Sup35 (see above), *e. g.* Sla1 (Holtzman *et al.*, 1993). It is possible that these structural elements mediate interactions with (or within) the structural networks of the cell. Intriguingly, computer analysis suggests that in addition to its role in translation, Sup35 could be involved in protein sorting (Marcotte *et al.*, 1999), making its association with components of the endocytic pathway likely. Quite remarkably, *sla1Δ* increases prion-dependent toxicity of both overproduced Sup35 and polyQ extended huntingtin-based construct (Htt-GFP) in yeast (Meriin *et al.*, 2003). One possibility is that protein complexes, involved in the formation of cortical actin patches and endocytic vesicles, facilitate nucleation of toxic misfolded proteins into less detrimental protein aggregates, possibly with the intent to target them for degradation by the endosomal/vacuolar system. This process, initially aimed at protecting cells from toxic effects of protein aggregation, inadvertently stimulates initial prion formation and prion recovery from the prion-curing treatments in the case when aggregating proteins possess a prion-forming potential. Experiments are currently underway that are aimed at testing this hypothesis.

6.2. Proteolytic Machinery

Misfolded proteins can be either refolded with the help of chaperones or eliminated by the proteolytic machinery. There are two major pathways of protein degradation in the yeast cell, specifically the proteasome pathway (see Hochstrasser *et al.*, 1999 for review) and the vacuolar pathway, analogous to the mammalian lysosome (see Van Den Hazel *et al.*, 1996 for review). Soluble proteins are usually degraded by the proteasome, while membrane proteins are targeted to the vacuole *via* endocytosis. Both pathways may include ubiquitination of the misfolded protein *via* attachment of the 76 aa protein domain, called ubiquitin (see Wilkinson, 1997; Hochstrasser *et al.*, 1999; Shaw *et al.*, 2001 for reviews). In case of proteasome-dependent degradation, polyubiquitination serves as a major signal targeting misfolded proteins to the proteasome.

By eliminating the potential source for prion conversion, proteolytic pathways may potentially counteract prion formation. Mutations in the components of ubiquitination pathway were indeed shown to influence the Parkinson's disease in humans that is related to amyloid formation (Leroy *et al.*, 1998; Shimura *et al.*, 2000). Some aggregating proteins (including huntingtin) form large inclusion bodies in mammalian cells that are called aggresomes and sequester components of the proteasome system (Kopito, 2000; Waelters *et al.*, 2001). This apparently occurs due to inability of the proteasome to destroy huge aggregates. However, no direct connections between ubiquitination and prion formation has been observed in mammalian systems to date.

As discussed above, some yeast chaperones affecting [PSI$^+$], for example Ssb, could be involved in targeting the misfolded proteins for proteolytic degradation as well as in refolding. The actin assembly protein Sla1, affecting [PSI$^+$], also plays a role in endocytosis

(Howard et al., 2002, Warren et al., 2002), an important step in the proteolytic degradation of membrane proteins. However, our recent data demonstrate an even more direct connection between prions and proteolytic pathways in yeast. Lack of the ubiquitin-conjugating enzyme, Ubc4, results in increased spontaneous[PSI⁺] formation (K. Allen, T. Chernova, E. Tennant-Clegg, K.Wilkinson, and Y. Chernoff, in preparation), that resembles a "protein mutator" phenotype of the *ssb1 ssb2* deletion (Figure 5). This suggests that the ubiquitination pathway normally counteracts prion generation.

6.3. Functional Partners of the Prion-Forming Proteins

Other proteins that physically or functionally interact with the potential prion-forming proteins in the course of their normal cellular activities sometimes influence prion formation as well. For example, efficiency of [PSI⁺] induction by the overproduced Sup35 protein is also modified by variations in levels of another component of the yeast release factor complex, Sup45 (eRF1). Increased levels of the Sup45 protein antagonize [PSI⁺] induction by the overproduced Sup35 protein (Derkatch *et al.*, 1998). Apparently, interactions with Sup45 molecules, which physically bind Sup35 (Paushkin *et al.*, 1997; Ebihara and Nakamura, 1999) and represent the normal partners of Sup35 in translation termination reaction (Stansfield *et al.*, 1995; Zhouravleva *et al.*, 1995), prevent self-aggregation and prion conversion of overproduced Sup35 protein. In contrast, Sup45 levels do not affect propagation and suppression efficiency of pre-existing [PSI⁺] (Derkatch *et al.*, 1998). Moreover, it has been reported that at least in some [PSI⁺] strains, the Sup45 molecules appear to be included in the Sup35^{PSI+} aggregates and sequestered from the translational machinery (Paushkin *et al.*, 1997).

Initial formation of [URE3] prion is also influenced by the Mks1 protein which functionally interacts with Ure2 (Edskes and Wickner, 2000). Mks1 protein is a negative posttranscriptional regulator of Ure2 activity (Edskes *et al.*, 1999b). It is possible that Mks1 physically interacts with Ure2, although direct proof for that is missing thus far. In the strains lacking Mks1 the *de novo* formation of [URE3] is blocked, while propagation of pre-existing [URE3] is not affected. The Mks1 protein itself is regulated by the Ras-cAMP pathway. Quite remarkably the Ras2^{Val19} mutation, which increases cAMP production, also blocks [URE3] formation, apparently due to inactivation of Mks1

(Edskes and Wickner, 2000). These results point to the complexity of the cellular control of [URE3] formation.

Not surprisingly, effects of the interacting partners are prion-specific. These effects are detected thus far only at the step of initial prion formation. Taken together, these data suggest that protein molecules actively involved in the cellular processes are somewhat protected from initial prion aggregation, although this protection is overcome if prion aggregates are already present in the cell.

7. Concluding Remarks: Prion Propagation as a By-Product of the Cellular Stress-Defense Machinery

The experimental evidence accumulated from the experiments on yeast models demonstrates that *in vivo* prion formation and propagation is controlled by other cellular proteins. The major role is played by the cellular machinery that is involved in stress-defense and response to accumulation of misfolded proteins. In genetic terms, this system could be interpreted as a "protein repair" system, whose errors result in "protein mutations" (prion formation). To further expand this analogy, one could say that protein aggregates actually behave as subcellular "molecular tumors". While most aggregated proteins form "non-malignant tumors", prions form the "malignant tumors" that can spread to the other cells, either in cell divisions or via cytoplasmic exchange (in yeast), or by extracellular infection (in mammals). Recent data, considered in this chapter, show that the ability of yeast prions to spread and propagate is modulated by chaperone helpers, controlling, for instance, the production of the infectious "seeds". The "prion helper" effects of Hsp104 (at intermediate levels) and Hsp70-Ssa1 indicate that yeast prions have "learned" how to use cellular stress defense systems to their own advantage, in the same way as viruses and transposons employ the cellular DNA replication and repair machinery for their own reproduction. Chaperones could become potential targets in counteracting the prion diseases and other aggregation-related disorders.

8. Acknowledgments

I am grateful to G. Newnam for the help in preparing this manuscript for publication, and to G. Telling for editing the manuscript. I thank K. Allen, K. Bapat, J. Birchmore, T. Chernova, E. Lewitin, A. Meriin, L. Ozolins, M. Sherman, R. Wegrzyn and K. Wilkinson for providing the unpublished data used in the manuscript. This work was supported in parts by grants from the National Institutes of Health (R01GM58763) and Huntington's Disease Society of America.

References

Aigle, M., and Lacroute, F. 1975. Genetical aspects of [URE3], a non-mitochondrial cytoplasmically inherited mutation in yeast. Mol. Gen. Genet. 136: 327-335.

Ayscough, K.R. 2000. Endocytosis and the development of cell polarity in yeast require a dynamic F-actin cytoskeleton. Curr. Biol. 10:1587-90.

Ayscough, K.R., Eby, J.J., Lila, T., Dewar, H., Kozminski, K.G., and Drubin, D.G. 1999. Sla1p is a functionally modular component of the yeast cortical actin cytoskeleton required for correct localization of both Rho1p-GTPase and Sla2p, a protein with talin homology. Mol. Biol. Cell. 10: 1061-1075.

Bailleul, P.A., Newnam, G.P., Steenbergen, J.N., and Chernoff, Y.O. 1999. Genetic study of interactions between the cytoskeletal assembly protein Sla1 and prion-forming domain of the release factor Sup35 (eRF3) in *Saccharomyces cerevisiae*. Genetics. 153: 81-94.

Bailleul-Winslett, P.A., Newnam, G.P., Wegrzyn, R.D., and Chernoff, Y.O. 2000. An antiprion effect of the anticytoskeletal drug latrunculin A in yeast. Gene Expression. 9: 145-156.

Borchsenius, A.S., Wegrzyn, R.D., Newnam, G.P., Inge-Vechtomov, S.G., and Chernoff, Y.O. 2001. Yeast prion protein derivative defective in aggregate shearing and production of new "seeds". EMBO J. 20: 6683-6691

Bradley, M.E., Edskes, H.K., Hong, J.Y., Wickner, R.B., and Liebman, S.W. 2002. Interactions among prions and prion "strains" in yeast. Proc. Natl. Acad. Sci. USA. 4: 16392-16399.

Bueler, H., Fischer, M., Lang, Y., Bluethmann, H., Lipp. H.P., DeArmond, S.J., Prusiner, S.B., Aguet, M., and Weissmann, C. 1992. Normal development and behaviour of mice lacking the neuronal cell-surface PrP protein. Nature. 356: 577-582

Chacinska, A., Szczesniak, B., Kochneva-Pervukhova, N.V., Kushnirov, V.V., Ter-Avanesyan, M.D., and Boguta, M. 2001. Ssb1 chaperone is a [PSI+] prion-curing factor. Curr. Genet. 39: 62-67.

Chai, Y., Koppenhafer, S.L., Bonini, N.M., and Paulson, H.L. 1999. Analysis of the role of heat shock protein (Hsp) molecular chaperones in polyglutamine disease. J. Neurosci. 19: 10338-10347.

Chernoff, Y.O. 2001. Mutation processes at the protein level: is lamarck back? Mutat. Res. 488: 39-64.

Chernoff, Y.O., and Ono, B. 1993. Dosage-dependent modifiers of psi-dependent omnipotent suppression in yeast. In: Protein Synthesis and Targeting in Yeast. A.J.P. Brown, M.F. Tuite, and J.E.G. McCarthy, Eds. NATO ASI, Ser. H: Cell Biology. Springer Verlag. p. 101-110.

Chernoff, Y.O., Derkach, I.L., and Inge-Vechtomov, S.G. 1993. Multicopy *SUP35* gene induces de-novo appearance of psi-like factors in the yeast *Saccharomyces cerevisiae*. Curr. Genet. 24: 268-270.

Chernoff, Y.O., Lindquist, S.L., Ono, B., Inge-Vechtomov, S.G., and Liebman, S.W. 1995. Role of the chaperone protein Hsp104 in propagation of the yeast prion-like factor [psi+]. Science. 268: 880-884.

Chernoff, Y.O., Newnam, G.P., Kumar, J., Allen, K., and Zink, A.D. 1999. Evidence for a protein mutator in yeast: role of the Hsp70-related chaperone Ssb in formation, stability, and toxicity of the [PSI] prion. Mol. Cell. Biol. 19: 8103-8112.

Chernoff, Y.O., Galkin, A.P., Lewitin, E., Chernova, T.A., Newnam, G.P., and Belenkiy, S.M. 2000. Evolutionary conservation of prion-forming abilities of the yeast Sup35 protein. Mol. Microbiol. 35: 865-876.

Chernoff, Y.O., Uptain, S.M., and Lindquist, S.L. 2002. Analysis of prion factors in yeast. Methods Enzym. 351: 499-538

Chesebro, B. 1997. Human TSE disease--viral or protein only? Nature Med. 3: 491-492.

Coustou, V., Deleu, C., Saupe, S., and Begueret, J. 1997. The protein product of the het-s heterokaryon incompatibility gene of the fungus *Podospora anserina* behaves as a prion analog. Proc. Natl. Acad. Sci. USA. 94: 9773-9778.

Cox, B.S. 1965. Ψ, a cytoplasmic suppressor of super-suppressor in yeast. Heredity. 20: 505-521.

Cox, B.S. 1994. Prion-like factors in yeast. Curr. Biol. 4: 744-748.

Cox, B.S., Tuite, M.F., and McLaughlin, C.S. 1988. The Ψ factor of yeast: a problem in inheritance. Yeast. 4: 159-178.

Crist, C.G., Nakayashiki, T., Kurahashi, H., and Nakamura, Y. 2003. [PHI+], a novel Sup35-prion variant propagated with non-Gln/Asn oligopeptide repeats in the absence of the chaperone protein Hsp104. Genes Cells. 8: 603-618.

DebBurman, S.K., Raymond, G.J., Caughey, B., and Lindquist, S. 1997. Chaperone supervised conversion of prion protein to its protease-resistant form. Proc. Natl. Acad. Sci. USA. 94: 13938-13943.

Derkatch, I.L., Chernoff, Y.O., Kushnirov, V.V., Inge-Vechtomov, S.G., and Liebman, S.W. 1996. Genesis and variability of [PSI] prion factors in *Saccharomyces cerevisiae*. Genetics. 144: 1375-1386.

Derkatch, I.L., Bradley, M., Zhou, P., Chernoff, Y.O., and Liebman, S.W. 1997. Genetic and environmental factors affecting the *de novo* appearance of the [PSI+] prion in *Saccharomyces cerevisiae*. Genetics. 147: 507-519.

Derkatch, I.L., Bradley, M., and Liebman, S.W. 1998. Overexpression of the *SUP45* gene encoding a Sup35p-binding protein inhibits the induction of the *de novo* appearance of the [PSI+] prion. Proc. Natl. Acad. Sci. USA. 95: 2400-2405.

Derkatch, I.L., Bradley, M.E., Masse, S.V., Zadorsky, S.P., Polozkov, G.V., Inge-Vechtomov, S.G., and Liebman, S.W. 2000. Dependence and independence of [PSI+] and [PIN+]: a two-prion system in yeast? EMBO J. 19: 1942-1952.

Derkatch, I.L., Bradley, M.E., Hong, J.Y., and Liebman, S.W. 2001. Prions affect the appearance of other prions: the story of [PIN+]. Cell. 106: 171-182.

Dickinson, A.G., and Meikle, V.M. 1971. Host-genotype and agent effects in scrapie incubation: change in allelic interaction with different strains of agent. Mol. Gen. Genet. 112: 73-79.

Doel, S.M., McCready, S.J., Nierras, C.R., and Cox, B.S. 1994. The dominant PNM2-mutation which eliminates the Ψ factor of *Saccharomyces cerevisiae* is the result of a mis-sense mutation in the *SUP35* gene. Genetics. 137: 659-670.

Eaglestone, S.S., Ruddock, L.W., Cox, B.S., and Tuite, M.F. 2000. Guanidine hydrochloride blocks a critical step in the propagation of the prion-like determinant [PSI+] of *Saccharomyces cerevisiae*. Proc. Natl. Acad. Sci. USA. 97: 240-244.

Ebihara, K., and Nakamura, Y. 1999. C-terminal interaction of translational release factors eRF1 and eRF3 of fission yeast: G-domain uncoupled binding and the role of conserved amino acids. RNA. 5: 739-750.

Edmonds. B.T. 1993. ABP50: an actin-binding elongation factor 1α from *Dictyostelium discoideum*. J. Cell. Biochem. 52: 134-139.

Edskes, H.K., and Wickner, R.B. 2000. A protein required for prion generation: [URE3] induction requires the Ras-regulated Mks1 protein. Proc. Natl. Acad. Sci. USA. 97: 6625-6629.

Edskes, H.K., Gray, V.T., and Wickner, R.B. 1999a. The URE3] prion is an aggregated form of Ure2p that can be cured by overexpression of Ure2p fragments. Proc. Natl. Acad. Sci. USA. 96: 498-1503.

Edskes, H.K., Hanover, J.A., and Wickner, R.B. 1999b. Mks1p is a regulator of nitrogen catabolism upstream of Ure2p in *Saccharomyces cerevisiae*. Genetics. 153: 585-594.

Fernandez-Bellot, E., Guillemet, E., Ness, F., Baudin-Baillieu, A., Ripaud, L., Tuite, M., and Cullin, C. 2002. The [URE3] phenotype: evidence for a soluble prion in yeast. EMBO Rep. 3: 76-81.

Ferreira, P.C., Ness, R., Edwards, S.R., Cox, B.S., and Tuite, M.F. 2001. The elimination of the yeast [PSI+] prion by guanidine hydrochloride is the result of Hsp104 inactivation. Mol. Microbiol. 40: 1357-1369.

Glover, J.R. and Lindquist, S. 1998. Hsp104, Hsp70 and Hsp40: a novel chaperone system that rescues previously aggregated proteins. Cell. 94: 1-20.

Glover, J.R., Kowal, A.S., Schirmer, E.C., Patino, M.M., Liu, M.M., and Lindquist, S. 1997. Self-seeded fibers formed by Sup35, the protein determinant of [PSI+], a heritable prion-like factor of *Saccharomyces cerevisiae*. Cell. 89: 811-819.

Goloubinoff, P., Mogk, A., Zvi, A.P., Tomoyasu, T., and Bukau, B. 2000. Sequential mechanism of solubilization and refolding of stable protein aggregates by a bichaperone network. Proc. Natl. Acad. Sci. USA. 96: 13732-13737.

Gray, E.G. 1986. Spongiform encephalopathy: a neurocytologist's viewpoint with a note on Alzheimer's disease. Neuropathol. Appl. Neurobiol. 12: 149-172.

Guiroy, D.C., Shankar, S.K., Gibbs, C.J., Messenheimer, J.A., Das, S., and Gajdusek, D.C. 1989. Neuronal degeneration and neurofilament accumulation in the trigeminal ganglia in Creutzfeldt-Jakob disease. Ann. Neurol. 25: 102-106.

Harrison, P.M., Bamborough, P., Daggett, V., Prusiner, S.B., and Cohen, F.E. 1997. The prion folding problem. Curr. Opin. Struct. Biol. 7: 53-59.

Hochstrasser, M., Johnson, P.R., Arendt, C.S., Amerik, A.Y., Swaminathan, S., Swanson, R., Li, S.J., Laney, J., Pals-Rylaarsdam, R., Nowak, J., and Connerly, P.L. 1999. The *Saccharomyces cerevisiae* ubiquitin-proteasome system. Philos. Trans. R. Soc. Lond. B. Biol. Sci. 354: 1513-1522.

Holtzman, D.A., Yang, S., and Drubin, D.G. 1993. Synthetic-lethal interactions identify two novel genes, *SLA1* and *SLA2*, that control membrane cytoskeleton assembly in *Saccharomyces cerevisiae*. J. Cell Biol. 122: 635-644.

Howard, J.P., Hutton, J.L., Olson, J.M., and Payne, G.S. 2002. Sla1p serves as the targeting signal recognition factor for NPFX (1,2)D-mediated endocytosis. J. Cell Biol. 157: 315-326.

Jones, G.W., and Masison, D.C. 2003. *Saccharomyces cerevisiae* Hsp70 mutations affect [PSI+] prion propagation and cell growth differently and implicate Hsp40 and tetratricopeptide repeat cochaperones in impairment of [PSI+].Genetics. 163: 495-506.

Jung, G., Jones, G., Wegrzyn, R.D., and Masison, D.C. 2000. A role for cytosolic Hsp70 in yeast [PSI+] prion propagation and [PSI+] as a cellular stress. Genetics. 156: 559-570.

Jung, G, and Masison, D.C. 2001. Guanidine hydrochloride inhibits Hsp104 activity *in vivo:* a possible explanation for its effect in curing yeast prions. Curr. Microbiol. 43:7-10.

Jung, G., Jones, G., and Masison, D.C. 2002. Amino acid residue 184 of yeast Hsp104 chaperone is critical for prion-curing by guanidine, prion propagation, and thermotolerance. Proc. Natl. Acad. Sci. USA. 99: 936-941.

Kalchman, M.A., Koide, H.B., McCutcheon, K., Graham, R.K., Nichol, K., Nishiyama, K., Kazemi-Esfarjani, P., Lynn, F.C., Wellington, C., Metzler, M., Goldberg, Y.P., Kanazawa, I., Gietz, R.D., and Hayden, M.R. 1997. HIP1, a human homologue of *S. cerevisiae* Sla2p, interacts with membrane-associated huntingtin in the brain. Nat. Genet. 1: 44-53.

Kazemi-Esfarjani, P., and Benzer, S. 2000. Genetic suppression of polyglutamine toxicity in Drosophila. Science. 287: 1837-1840.

King, C.Y., Tittmann, P., Gross, H., Gebert, R., Aebi, M., and Wuthrich, K. 1997. Prion-inducing domain 2-114 of yeast Sup35 protein transforms *in vitro* into amyloid-like filaments. Proc. Natl. Acad. Sci. USA. 94: 6618-6622.

Kopito, R.R. 2000. Aggresomes, inclusion bodies and protein aggregation.Trends Cell Biol. 10: 524-530.

Krobitsch, S., and Lindquist, S. 2000. Aggregation of huntingtin in yeast varies with the length of the polyglutamine expansion and the expression of chaperone proteins. Proc. Natl. Acad. Sci. USA. 97: 1589-1594.

Kryndushkin, D.S., Smirnov, V.N., Ter-Avanesyan, M.D., and Kushnirov, V.V. 2002. Increased expression of Hsp40 chaperones, transcriptional factors, and ribosomal protein Rpp0 can cure yeast prions. J. Biol. Chem. 277: 23702-23708.

Kushnirov, V.V., Ter-Avanesyan, M.D., Telckov, M.V., Surguchov, A.P., Smirnov, V.N., and Inge-Vechtomov, S.G. 1988. Nucleotide sequence of the *SUP2* (*SUP35*) gene of *Saccharomyces cerevisiae*. Gene. 66: 45-54.

Kushnirov, V.V., Ter-Avanesyan, M.D., Didichenko, S.A., Smirnov, V.N., Chernoff, Y.O., Derkach, I.L., Novikova, O.N., Inge-Vechtomov, S.G., Neistat, M.A., and Tolstorukov, I.I. 1990. Divergence and conservation of *SUP2* (*SUP35*) gene of yeast *Pichia pinus* and *Saccharomyces cerevisiae*. Yeast. 6: 461-472.

Kushnirov, V.V., Ter-Avanesian, M.D., and Smirnov, V.N. 1995. Structure and functional similarity of yeast Sup35p and Ure2p proteins to mammalian prions. Mol. Biol. (Mosk.). 29: 750-755.

Kushnirov, V.V., Kochneva-Pervukhova, N.V., Chechenova, M.B., Frolova, N.S., and Ter-Avanesyan, M.D. 2000a. Prion properties of the Sup35 protein of yeast *Pichia methanolica*. EMBO J. 19: 324-331.

Kushnirov, V.V., Kryndushkin, D.S., Boguta, M., Smirnov, V.N., and Ter-Avanesyan, M.D. 2000b. Chaperones that cure yeast artificial [PSI+] and their prion-specific effects. Curr. Biol. 10: 1443-1446.

Lacroute, F. 1971. Non-Mendelian mutation allowing ureidosuccinic acid uptake in yeast. J. Bacteriol. 106: 519-522.

Lansbury, P.T., and Caughey, B. 1995. The chemistry of scrapie reaction: the "ice 9" metaphore. Chem. Biol. 2: 1-5.

Leroy, E., Boyer, R., Auburger, G., Leube, B., Ulm, G., Mezey, E., Harta, G., Brownstein, M.J., Jonnalagada, S., Chernova, T., Dehejia, A., Lavedan, C., Gasser, T., Steinbach, P.J., Wilkinson, K.D., and Polymeropoulos, M.H. 1998. The ubiquitin pathway in Parkinson's disease. Nature. 395: 451-452.

Liang, P., and McRae, T.H. 1997. Molecular chaperones and the cytoskeleton. J. Cell Sci. 110: 1431-1440.

Liberski, P.P. 1994. Transmissible cerebral amyloidoses as a model for Alzheimer's disease. An ultrastructural perspective. Mol. Neurobiol. 8: 67-77.

Lindquist, S., Patino, M.M., Chernoff, Y.O., Kowal, A.S., Singer, M.A., Liebman, S.W., Lee, K.H., and Blake, T. 1995. The role of Hsp104 in stress tolerance and [PSI⁺] propagation in Saccharomyces cerevisiae. Cold Spring Harb. Symp. Quant. Biol. 60: 451-460.

Liu, J.-J., Sondheimer, N., and Lindquist, S.L. 2002. Changes in the middle region of Sup35 profoundly alter the nature of epigenetic inheritance for the yeast prion [PSI⁺]. Proc. Natl. Acad. Sci. USA. 99: 16446-16453.

Maddelein, M.L., and Wickner, R.B. 1999. Two prion-inducing regions of Ure2p are nonoverlapping. Mol. Cell. Biol. 19: 4516-4524.

Marcotte, E.M., Pellegrini, M., Thompson, M.J., Yeates, T.O., and Eisenberg, D. 1999. A combined algorithm for genome-wide prediction of protein function. Nature. 402: 83-86.

Masison, D.C., and Wickner, R.B. 1995. Prion-inducing domain of yeast Ure2p and protease-resistance of Ure2p in prion-containing cells. Science. 270: 93-95.

Meriin, A.B., Zhang, X., He, X., Newnam, G.P., Chernoff, Y.O., and Sherman, M.Y. 2002. Huntington toxicity in yeast model depends on polyglutamine aggregation mediated by a prion-like protein Rnq1. J. Cell Biol. 157: 997-1004.

Meriin, A.B., Zhang, X., Miliaras, N.B., Kazantsev, A., Chernoff, Y.O., McCaffery, J.M., Wendland, B., and Sherman, M.Y. 2003. Aggregation of expanded polyglutamine domain in yeast leads to defects in endocytosis. Mol Cell Biol. 23: 7554-7565.

Michelitsch, M.D., and Weissman, J.S. 2000. A census of glutamine/asparagine-rich regions: implications for their conserved function and the prediction of novel prions. Proc. Natl. Acad. Sci. USA. 97: 11910-11915.

Moriyama, H., Edskes, H.K., and Wickner, R.B. 2000. [URE3] prion propagation in *Saccharomyces cerevisiae*: requirement for chaperone Hsp104 and curing by overexpressed chaperone Ydj1p. Mol. Cell Biol. 20: 8916-8922.

Muchowski, P.J., Schaffar, G., Sittler, A., Wanker, E.E., Hayer-Hartl, M.K., and Hartl, F.U. 2000. Hsp70 and Hsp40 chaperones can inhibit self-assembly of polyglutamine proteins into amyloid-like fibrils. Proc. Natl. Acad. Sci. USA. 97; 7841-7846.

Nakayashiki, T., Ebihara, K., Bannai, H., and Nakamura, Y. 2001. Yeast [PSI⁺] "prions" that are cross-transmissible and susceptible beyond a species barrier through a quasi-prion state. Mol. Cell. 7: 1121-1130.

Nelson, R.J., Ziegelhoffer, T., Nicolet, C., Werner-Washburne, M., and Craig, E.A. 1992. The translation machinery and 70 kd heat shock protein cooperate in protein synthesis. Cell. 71: 97-105.

Ness, F., Ferreira, P., Cox, B.S., and Tuite, M.F. 2002. Guanidine hydrochloride inhibits the generation of prion "seeds" but not prion protein aggregation in yeast. Mol. Cell. Biol. 22: 5593-5605.

Newnam, G.P., Wegrzyn, R.D., Lindquist, S.L., and Chernoff, Y.O. 1999. Antagonistic interactions between yeast chaperones Hsp104 and Hsp70 in prion curing. Mol. Cell. Biol. 19: 1325-1333.

Osherovich, L.Z., and Weissman, J.S. 2001. Multiple Gln/Asn-rich prion domains confer susceptibility to induction of the yeast [PSI⁺] prion. Cell, 106: 183-194.

Ohba, M. 1994. A 70-kDa heat shock cognate protein suppresses the defects caused by a proteasome mutation in *Saccharomyces cerevisiae*. FEBS Lett. 351: 263-266.

Ohba, M. 1997. Modulation of intracellular protein degradation by *SSB1-SIS1* chaperon system in yeast *Saccharomyces cerevisiae*. FEBS Lett. 409: 307-311.

Parham, S.N., Resende, C.G., and Tuite, M.F. 2001. Oligopeptide repeats in the yeast protein Sup35p stabilize intermolecular prion interactions. EMBO J. 20: 2111-2119.

Parsell, D.A., Sanchez, Y., Stitzel, J.D., and Lindquist, S. 1991. Hsp104 is a highly conserved protein with two essential nucleotide-binding sites. Nature. 353: 270-273.

Parsell, D.A., Kowal, A.S., Singer, M.A., and Lindquist. S. 1994. Protein disaggregation mediated by heat-shock protein Hsp104. Nature. 372: 475-478.

Patino, M.M., Liu, J.J., Glover, J.R., and Lindquist, S. 1996. Support for the prion hypothesis for inheritance of a phenotypic trait in yeast. Science. 273: 622-626.

Paushkin, S.V., Kushnirov, V.V., Smirnov, V.N., and Ter-Avanesyan, M.D. 1996. Propagation of the yeast prion-like [psi⁺] determinant is mediated by oligomerization of the SUP35-encoded polypeptide chain release factor. EMBO J. 15: 3127-3134.

Paushkin, S.V., Kushnirov, V.V., Smirnov, V.N., and Ter-Avanesyan, M.D. 1997. Interaction between yeast Sup45p (eRF1) and Sup35p (eRF3) polypeptide chain release factors: implications for prion-dependent regulation. Mol. Cell. Biol. 17: 2798-2805.

Pfund, C., Lopez-Hoyo, N., Ziegelhoffer, T., Schilike, B.A., Lopez-Buesa, P., Walter, W.A., Wiedmann, M., and Craig, E.A. 1998. The molecular chaperone Ssb from *Saccharomyces cerevisiae* is a component off the ribosome-nascent chain complex. EMBO J. 17: 3981-3989.

Prusiner, S.B. 1982. Novel proteinaceous infectious particles cause scrapie. Science. 216: 136-44.

Prusiner, S.B. 1994. Molecular biology and genetics of prion diseases. Phil. Trans. R. Soc. London Ser. B: Biol. Sci. 343: 447-463.

Prusiner, S.B. 1997. Prion diseases and the BSE crisis. Science. 278: 245-251.

Ross, C.A. 2002. Polyglutamine pathogenesis: emergence of unifying mechanisms for Huntington's disease and related disorders. Neuron. 35: 819-822.

Sanchez, Y., and Lindquist, S.L. 1990. Hsp104 required for induced thermotolerance. Science. 248: 1112-1115.

Sanchez, Y., Taulien, J., Borkovich, K.A., and Lindquist, S. 1992. Hsp104 is required for tolerance to many forms of stress. EMBO J. 11: 2357-2364.

Santoso, A., Chien, P., Osherovich, L.Z., and Weissman, J.S. 2000. Molecular basis of a yeast prion species barrier. Cell. 100: 277-288.

Satyal, S.H., Schmidt, E., Kitagawa, K., Sondheimer, N., Lindquist, S., Kramer, J.M., and Morimoto, R.I. 2000. Polyglutamine aggregates alter protein folding homeostasis in *Caenorhabditis elegans*. Proc. Natl. Acad. Sci. USA. 23: 5750-5755.

Schirmer, E.C., Glover, J.R., Singer, M.A., and Lindquist, S. 1996. HSP100/Clp proteins: a common mechanism explains diverse functions. Trends. Biochem. 21: 289-296.

Schirmer, E.C., Ware, D.M., Queitsch, C., Kowal, A.S., and Lindquist, S.L. 2001. Subunit interactions influence the biochemical and biological properties of Hsp104. Proc. Natl. Acad. Sci. USA. 98: 914-919.

Schlumpberger, M., Wille, H., Baldwin, M.A., Butler, D.A., Herskowitz, I., and Prusiner, S.B. 2000. The prion domain of yeast Ure2p induces autocatalytic formation of amyloid fibers by a recombinant fusion protein. Protein Sci. 9: 440-451.

Schlumpberger, M., Prusiner, S.B., and Herskowitz, I. 2001. Induction of distinct [URE3] yeast prion strains. Mol Cell Biol. 21: 7035-7046.

Schwimmer, C., and Masison, D.C. 2002. Antagonistic interactions between yeast [PSI+] and [URE3] prions and curing of [URE3] by Hsp70 protein chaperone Ssa1p but not by Ssa2p. Mol. Cell Biol. 22: 3590-3598.

Serio, T.R., and Lindquist, S.L. 2000. Protein-only inheritance in yeast: something to get [PSI$^+$]-ched about. Trends Cell Biol. 10: 98-105.

Serio, T.R., Cashikar, A.G., Kowal, A.S., Sawicki, G.J., Moslehi, J.J., Serpell, L., Arnsdorf, M.F., and Lindquist, S.L. 2000. Nucleated conformational conversion and the replication of conformational information by a prion determinant. Science. 289: 1317-1321.

Shaw, J.D., Cummings, K.B., Huyer, G., Michaelis, S., and Wendland, B. 2001. Yeast as a model system for studying endocytosis. Exp. Cell. Res. 271: 1-9.

Shimura, H., Hattori, N., Kubo, S., Mizuno, Y., Asakawa, S., Minoshima, S., Shimizu, N., Iwai, K., Chiba, T., Tanaka, K., and Suzuki, T. 2000. Familial Parkinson disease gene product, parkin, is a ubiquitin-protein ligase. Nat. Genet. 25: 302-305.

Sondheimer, N., and Lindquist, S. 2000. Rnq1: an epigenetic modifier of protein function in yeast. Mol. Cell. 5: 163-172.

Sondheimer, N., Lopez, N., Craig, E.A., and Lindquist, S. 2001. The role of Sis1 in the maintenance of the [RNQ+] prion. EMBO J. 20: 2435-2442.

Stansfield, I., Jones, K.M., Kushnirov, V.V., Dagkesamanskaya, A.R., Poznyakovski, A.I., Paushkin, S.V., Nierras, C.R., Cox, B.S., Ter-Avanesyan, M.D., and Tuite, M.F. 1995. The products of the *SUP45* (eRF1) and *SUP35* genes interact to mediate translation termination in *Saccharomyces cerevisiae*. EMBO J. 14: 4365-4373.

Suda, M., Fukui, M., Sogabe, Y., Sato, K., Morimatsu, A., Arai, R., Motegi, F., Miyakawa, T., Mabuchi, I., and Hirata, D. 1999. Overproduction of elongation factor 1alpha, an essential translational component, causes aberrant cell morphology by affecting the control of growth polarity in fission yeast. Genes Cells. 4: 517-527.

Talloczy, Z., Mazar, R., Georgopoulos, D.E., Ramos, F., and Leibowitz, M.J. 2000. The [KIL-d] element specifically regulates viral gene expression in yeast. Genetics. 155: 601-609.

Taylor, K.L., Cheng, N., Williams, R.W., Steven, A.C., and Wickner, R.B. 1999. Prion domain initiation of amyloid formation in vitro from native Ure2p. Science. 283: 1339-1343.

Ter-Avanesyan, M.D., Dagkesamanskaya, A.R., Kushnirov, V.V., and Smirnov, V.N. 1994. The *SUP35* omnipotent suppressor gene is involved in the maintenance of the non-Mendelian determinant [psi+] in the yeast *Saccharomyces cerevisiae*. Genetics. 137: 671-676.

Thual, C., Komar, A.A., Bousset, L., Fernandez-Bellot, E., and Cullin, C. 1999. Structural characterization of *Saccharomyces cerevisiae* prion-like protein Ure2. J. Biol. Chem. 274: 13666-13674.

Tuite, M.F., Mundy, C.R., and Cox, B.S. 1981. Agents that cause a high frequency of genetic change from [psi⁺] to [psi⁻] in *Saccharomyces cerevisiae*. Genetics. 98: 691-711.

Valouev, I.A., Kushnirov, V.V., and Ter-Avanesyan, M.D. 2002. Yeast polypeptide chain release factors eRF1 and eRF3 are involved in cytoskeleton organization and cell cycle regulation. Cell Motil. Cytoskeleton. 52: 161-173.

Van Den Hazel, H.B., Kielland-Brandt, M.C., Winther, J.R. 1996. Review: biosynthesis and function of yeast vacuolar proteases. Yeast. 12: 1-16.

Viswanathan, A., You, H.J., and Doetsch, P.W. 1999. Phenotypic change caused by transcriptional bypass of uracil in nondividing cells. Science. 284: 159-162.

Volkov, K.V., Aksenova, A.Y., Soom, M.J., Osipov, K.V., Svitin, A.V., Kurischko, C., Shkundina, I.S., Ter-Avanesyan, M.D., Inge-Vechtomov, S.G., and Mironova, L.N. 2002. Novel non-Mendelian determinant involved in the control of translation accuracy in *Saccharomyces cerevisiae*. Genetics. 160: 25-36.

Waelter, S., Boeddrich, A., Lurz, R., Scherzinger, E., Lueder, G., Lehrach, H., Wanker, E.E. 2001. Accumulation of mutant huntingtin fragments in aggresome-like inclusion bodies as a result of insufficient protein degradation. Mol. Biol. Cell. 12: 1393-1407.

Warren, D.T., Andrews, P.D., Gourlay, C.W., and Ayscough, K.R. 2002. Sla1p couples the yeast endocytic machinery to proteins regulating actin dynamics. J. Cell Sci. 115: 1703-1715.

Warrick, J.M., Chan, H.Y., Gray-Board, G.L., Chai, Y., Paulson, H.L., and Bonini, N.M. 1999. Suppression of polyglutamine-mediated neurodegeneration in Drosophila by the molecular chaperone HSP70. Nat. Genet. 23: 425-428.

Wegrzyn, R.D., Bapat, K., Newnam, G.R., Zink, A.D., and Chernoff, Y.O. 2001. Mechanism of prion loss after Hsp104 inactivation in yeast. Mol. Cell. Biol. 21: 4656-4669.

Werner-Washburne, M., Stone, D.E., and Craig, E.A. 1987. Complex interactions among members of an essential subfamily of *hsp70* genes in *Saccharomyces cerevisiae*. Mol. Cell. Biol. 7: 2568-2577.

Wickner, R.B. 1994. [URE3] as an altered Ure2 protein: Evidence for a prion analog in *Saccharomyces cerevisiae*. Science. 264: 566-569.

Wickner, R.B., Taylor, K.L., Edskes, H.K., Maddelein, M.L., Moriyama, H., and Roberts, B.T. 2001. Yeast prions act as genes composed of self-propagating protein amyloids. Adv. Protein Chem. 57: 313-334.

Wilkinson, K.D. 1997. Regulation of ubiquitin-dependent processes by deubiquitinating enzymes. FASEB J. 11: 1245-1256.

Wilson, P.G., and Culbertson, M.R. 1988. Suf 12 suppressor protein of yeast: a fusion protein related to the EF-1α family of elongation factors. J. Mol. Biol. 199: 559-573.

Yang, F., Dema, M., Warren, V., Dharmawardhane, S., and Condulis, J. 1990. Identification of an actin-binding protein from Dictyostelium as elongation factor 1α. Nature. 347: 494-496.

Young, E.T., Sloan, J.S., and Van Riper, K. 2000. Trinucleotide repeats are clustered in regulatory genes in *Saccharomyces cerevisiae*. Genetics. 154: 1053-1068.

Zadorskii, S.P., Sopova, I.V., and Inge-Vechtomov, S.G. 2000. Prionization of the *Pichia methanolica SUP35* gene product in the yeast *Saccharomyces cerevisiae*. Genetika. 36: 1322-1329.

Zhouravleva, G., Frolova, L., Le Goff, X., Le Guellec, R., Inge-Vechtomov, S., Kisselev, L., and Philippe, M. 1995. Termination of translation in eukaryotes is governed by two interacting polypeptide chain release factors, eRF1 and eRF3. EMBO J. 14: 4065-4072.

Zhouravleva, G.A., Alenin, V.V., Inge-Vechtomov, S.G., and Chernoff, Y.O. 2002. To stick or not to stick: Prion domains from yeast to mammals. Recent Res. Devel. Mol. Cell. Biol. 3: 185-218.

Index